Articulating the World

Articulating the World

Conceptual Understanding and the Scientific Image

JOSEPH ROUSE

The University of Chicago Press   Chicago and London

JOSEPH T. ROUSE is professor of philosophy at Wesleyan University. He is the author of *Knowledge and Power: Toward a Political Philosophy of Science* and *Engaging Science: How to Understand Its Practices Philosophically*.

The University of Chicago Press, Chicago 60637
The University of Chicago Press, Ltd., London

Printed in the United States of America

24 23 22 21 20 19 18 17 16 15   1 2 3 4 5

ISBN-13: 978-0-226-29367-7 (cloth)
ISBN-13: 978-0-226-29384-4 (paper)
ISBN-13: 978-0-226-29370-7 (e-book)
DOI: 10.7208/chicago/9780226293707.001.0001

Library of Congress Cataloging-in-Publication Data
Rouse, Joseph, 1952– author
Articulating the world : conceptual understanding and the scientific image / Joseph Rouse.
pages ; cm
Includes bibliographical references and index.
ISBN 978-0-226-29367-7 (cloth : alk. paper)—ISBN 978-0-226-29384-4 (pbk. : alk. paper)—ISBN 978-0-226-29370-7 (ebook) 1. Comprehension (Theory of knowledge) 2. Naturalism. 3. Concepts. 4. Science—Philosophy. I. Title.
BD181.5.R68 2015
501'.9—dc23
2015001079

∞ This paper meets the requirements of ANSI/NISO Z39.48–1992 (Permanence of Paper).

Contents

Introduction

ONE

Naturalism and the Scientific Image

I—Naturalism as a Historical Project

This book aims to advance a naturalistic self-understanding. Naturalism conjoins several core commitments. First, its advocates refuse any appeal to or acceptance of what is supernatural or otherwise transcendent to the natural world. The relevant boundary between nature and what would be supernatural or otherwise transcendent is admittedly contested, and conceptions of that boundary have shifted historically. The significance of conflicts over what is or is not "natural" nevertheless arises in substantial part from the aspiration to a naturalistic understanding. Conceptions of nature and aspirations to a naturalistic self-understanding may be mutually intertwined. Contemporary naturalists also undertake a second more specific commitment to a scientific understanding of nature. At a minimum, naturalists regard scientific understanding as relevant to all significant aspects of human life and only countenance ways of thinking and forms of life that are *consistent* with that understanding. More stringent versions of naturalism take scientific understanding to be *sufficient* for our intellectual and theoretical projects and perhaps even for practical guidance in other aspects of life. A third commitment is a corollary to recognition of the relevance and authority of scientific understanding: naturalists repudiate any conception of "first philosophy" as prior to or authoritative over scientific understanding (Quine 1981, 67).

The book develops these core commitments in ways that many fellow naturalists will find unfamiliar and perhaps even alien. I therefore need to be clear from the outset about why I still identify these proposals as a naturalistic program. Naturalism has a long and distinguished history that predates its contemporary versions. That history encompasses the earliest human efforts to understand the world and our place within it without invoking gods, mysteries, or other incomprehensible or otherworldly beings, powers, or authority. The emergence and expansion of the modern natural sciences encouraged the identification of naturalism with a commitment to the autonomy and authority of scientific understanding. Yet the constructive development of a naturalistic self-understanding extends beyond the efforts of those thinkers and inquirers who explicitly embraced a naturalistic project. Adamant critics of naturalism have developed or advanced many important aspects of what we can now recognize as a naturalistic self-understanding. Scientific achievements guided by theologically framed natural philosophies were prominent among those contributions, but philosophical objections to a naturalistic standpoint have also led to improvements in its prospects.

In retrospect, there should be no irony in the recognition that ardent critics of naturalism have constructively advanced the cause. Articulating a thoroughly naturalistic self-understanding is difficult. Throughout the history of naturalistic thought, and in some respects even today, committing to a naturalistic self-understanding required some philosophical myopia. Apart from having to cope with significant gaps in understanding the natural world, naturalists have often embraced what look in retrospect to be oversimplified conceptions of what a defensible naturalism would require. Some proponents were overly optimistic about the capacities of austere scientific and philosophical resources. Others overlooked residual theological or supernatural commitments in their own efforts. Many have not fully recognized or understood the complexity of the phenomena a naturalist must account for or the sources of incoherence within their projects. How else could they endorse and defend commitments that would otherwise outrun the limits of recognizable feasibility? It should be no surprise that the challenges confronting a more adequate philosophical naturalism have often been most carefully and insightfully understood by those who therefore eschewed any commitment to naturalism. As Charles Taylor noted, "In philosophy at least, a gain in clarity is worth a thinning of the ranks" (1985, 21).

Recognizing the dialectical complexity of philosophical naturalism throughout its history has important consequences. Naturalism as a philosophical and scientific project cannot simply be identified with any of its various formulations, including currently prominent versions. Some of the most important achievements within the naturalistic tradition have reformulated which commitments a genuinely naturalistic project must undertake. Many of these reformulations had philosophical roots. Hume's criticisms of causal necessity and of derivations of "ought" from "is," Kant's "Copernican Revolution," Frege's and Husserl's arguments against psychologism, Quine's criticism of the analytic/synthetic distinction, and Wittgenstein's reflections on rule following, among others, have left their mark upon subsequent formulations of naturalism.[1]

Other revisions in then-predominant conceptions of naturalism call attention to implicit tensions between naturalistic philosophy and the empirical sciences. The establishment and pursuit of new scientific inquiries have been crucial to the advance of naturalism. Indeed, naturalism is nowadays often simply identified with a scientific or even scientistic conception of the world. Yet the potential tensions between philosophical naturalism and the empirical sciences are apparent from the many occasions when scientific developments have stranded scientifically based philosophical programs. Philosophical naturalisms have often confronted disciplinary, theoretical, methodological, or empirical innovations and discoveries in the sciences that challenged their version of naturalistic understanding. Examples of broadly naturalistic scruples undermined by scientific developments include seventeenth-century mechanistic hostility to "occult" gravitational action-at-a-distance, causal determinisms grounded in classical physics, Quine's commitment to behaviorism, or the rejection of biological teleology. Moreover, the proliferation of relatively autonomous scientific disciplines and research programs leaves open the question of which approaches to which sciences would most constructively advance a naturalistic point of view. Fundamental physics, evolutionary biology, neuroscience, cognitive psychology, and the sociology of knowledge are prominent among contemporary

1. I would argue that Hegel's criticisms of Kant's transcendental idealism, Nietzsche's relentless exposure of residual theological assumptions within putatively naturalistic or "free-thinking" projects, Heidegger's conception of Dasein as being-in-the-world, and Merleau-Ponty's reflections upon the bodily basis of intentionality should also serve as important contributions to the articulation of a more adequately naturalistic philosophical standpoint, but their work is only beginning to be assimilated by philosophers who aspire to a naturalistic understanding.

contenders, but the relations among these scientific orientations and their respective philosophical significance for naturalists remain contested.

I regard naturalism as a historically situated philosophical project.[2] We find ourselves in the midst of ongoing conflicts over what naturalism's commitments are and why they matter, along with challenges to those commitments. I do not here defend a naturalistic self-understanding against those who regard it as unattractive. I endorse a broadly naturalistic stance, but my reasons for doing so are familiar, and I have nothing especially original to say on that topic. I am instead concerned to respond to the possibility that a consistent and thoroughgoing naturalistic self-understanding is unattainable. This book proposes revised conceptions of ourselves and of the sciences that are directly responsive to conflicts over the viability of a naturalistic stance. It reformulates the dominant contemporary philosophical conceptions of naturalism, both by reworking received philosophical approaches to science, intentionality, and conceptual understanding and by drawing upon recent scientific work that has mostly not yet been assimilated philosophically. I endorse the resulting conception, but I do not propose that it would or should settle these issues once and for all. The questions of what philosophical naturalism is, what must be done to sustain a viable naturalistic orientation, and whether and why to be a naturalist will undoubtedly remain at issue within the tradition. My aim is instead to refine and clarify these issues to avoid recognized or recognizable problems and to propose and defend new directions for further philosophical and scientific work in response. These limited aspirations do not merely result from modesty about what I did or could accomplish, though modesty is undoubtedly appropriate. These aspirations are instead shaped by the conceptions of science and philosophy developed in the course of the book, which emphasize that conceptual understanding is always contested and future directed in ways oriented by what is at issue and at stake in those conflicts.

The book is motivated by a specific conception of the current situation in the philosophical understanding of naturalism. The most pressing challenge for naturalism today is to show how to account for our own

2. Some readers might infer from this claim that the issues raised by naturalism are not real or abiding concerns. This inference would be mistaken. That an inquiry bears the marks of its history is no objection to the seriousness of the issues it raises, even if there have been and will be historical shifts in the conception of what those issues are. Moreover, the appeal to a philosophical standpoint outside of our natural history as inquirers, as the standard for which philosophical issues are important, is not one that naturalists should accept. Thanks to Willem deVries for calling attention to this possible objection.

capacities for scientific understanding as a natural phenomenon that could be understood scientifically. Naturalist views that cannot meet this challenge would be self-defeating. The principal claim of the book is that meeting this challenge requires substantial, complementary revisions to familiar philosophical accounts of both of its components: how to situate our conceptual capacities within a scientific understanding of the world and what a scientific conception of the world amounts to. The two parts of the book develop a broad overview of these revisions and their rationales. These are relatively new approaches to the issues, and much work remains to be done on both sides. In the first part of the book, I reconsider how to think philosophically and scientifically about conceptual understanding. In place of more familiar appeals to a functional teleology of cognitive or linguistic representations, I emphasize the normativity of discursive practice within an evolving developmental niche and take both language and scientific practices to exemplify the evolutionary process of niche construction. In the second part of the book, I reconsider the sense of scientific understanding embodied in naturalists' core commitment to situating philosophical work within a scientific conception of the world. The ongoing practice of scientific research encompasses the relevant form of scientific understanding; efforts to extract an established body of scientific knowledge from that practice are among the philosophical impositions upon science that naturalists should reject. The two parts of the argument are presented sequentially, but they should be understood as mutually reinforcing. The first part situates conceptual understanding within a scientific conception of nature. The second part explicates what it is to have a scientific conception of nature in terms of that account of conceptual understanding. This preliminary chapter prepares the ground by working out the conception of our current philosophical situation as aspiring naturalists, which motivates the remainder of the book.

II—Sellars and the Prospects for Philosophical Naturalism

Wilfrid Sellars (2007, ch. 14) provocatively framed contemporary philosophical discussions of naturalism by recognizing tensions between two alternative conceptions of human beings and our place in the world. The philosophical tradition has inherited what Sellars calls the "manifest image" of ourselves as *persons* whose involvement in the world incorporates sentient experience, conceptual understanding, and rationally reflective agency. A different self-conception emerged from the natural sciences, even though the sciences arose as exercises of our "manifest" capacities. This emergent "scientific image" takes our life activities and achievements to be comprehensible without residue in theoretical terms drawn from physics, chemistry, biology, and perhaps psychology and the social sciences. The manifest and the scientific images each purport to completeness and autonomy. The manifest image takes the world as the setting for our experience, understanding, and action, incorporating the scientific image in "manifest" terms as a rationally explicable achievement of human understanding. From within the manifest image, scientific understanding is accountable to sense experience and is only meaningful to and authoritative for us through a shared commitment to think and act in empirically accountable terms. From the other direction, the scientific image proposes that experience, thought, and action are explicable in theoretical terms drawn from the relevant scientific disciplines. For Sellars, both conceptions express insights we ought to endorse. Each is nevertheless comprehensive in ways that may leave no space for the other's insights within its own account of our place in the world. Sellars thus identified a preeminent contemporary philosophical task as doing justice to the comprehensiveness and apparent autonomy of both images in a stereoscopically unified vision of ourselves in the world. Sellars also insisted that this stereoscopic conception should be naturalistic in a strong sense. An adequate fusion of the images should give priority to the scientific image, situating our self-conception as sentient, sapient, rational agents within the horizons of a scientific conception of ourselves as natural beings.

Despite the prominence of W. V. O. Quine as an advocate of naturalism, Sellars's philosophical vision predominantly sets the terms in which naturalism is nowadays conceived and discussed. First, Sellars recognized that naturalism cannot simply culminate in the replacement of philosophy by some empirical scientific discipline, as Quine proposed that scientific psychology might replace epistemology. Philosophical questions go beyond the interests and the locus of the various scientific disciplines. Sellars's expression of the distinctively philosophical task in relation to the sciences is well known: "to understand how things in the broadest possible sense of the term hang together in the broadest possible sense of the term" (2007, 369). That task receives a more determinate form, however, in Sellars's aspiration to account for the legitimate insights of both the manifest and the scientific images. The manifest image locates us within the "space of reasons" in which normative authority is constituted, including the normative authority of science itself. In understanding and expressing normative authority, however, philosophy does not further describe or explain things but instead articulates and contributes to a shared project. A naturalist will of course conceive that philosophical project in ways that rely upon scientific understanding, and that do not claim independent authority over scientific inquiry, but will not be able to dispense with philosophical work.

Sellars's conception of a naturalistic philosophy was also influential in two further respects. For Quine, the primary task of scientific theory is descriptive. The psychological theories that would replace epistemology aim to describe how we actually construct systematic and far-reaching scientific theories from a meager base of evidence. Sellars offered a more expansive conception of scientific aspirations. Science aims to explain what happens in the world. The priority that Sellars accords to the scientific image derives from its greater explanatory power: science enables us to understand and explain as well as describe the phenomena within its purview. This difference in turn accounts for the more expansive intellectual resources that Sellars accords to the scientific image. Whereas Quine would restrict science and philosophy to the most austere theoretical vocabulary sufficient to characterize actual events and dispositions, Sellars insists that the modal locutions of scientific laws are indispensable. The philosophical rehabilitation of modal language and inference from empiricist critics is a much longer story than I need to tell here.[3] One clear outcome of that rehabilitation, however, has been to lead most naturalists toward Sellars's conception of the scientific image as a framework of explanatory laws rather than Quine's vision of efficiently systematized resources for theoretical description.[4]

Sellars not only set the terms in which most naturalists understand the scientific image and its philosophical authority, however. His work also guided several prominent challenges to naturalism. In his provocatively titled book *The Scientific Image*, for example, Bas van Fraassen (1980) proposed an epistemological challenge to Sellars's naturalism on two fronts. He first argued that the explanatory power of the scientific image does not confer upon it a philosophical priority over the manifest image of ourselves as rational knowers and agents. Explanation is only a

3. For a historical discussion of modal logic and modal concepts during the relevant parts of the twentieth century, see Shieh (forthcoming). For a discussion of the philosophical significance of this history from a distinctive point of view, see Brandom (2008, ch. 4).

4. The issue of whether there are laws (or "strict laws") outside the physical sciences still divides many naturalists who agree that scientific understanding has a modal dimension. In chapter 8, I argue that a conception of laws of nature that recognizes their role in scientific practice shows that laws are pervasive even in the life and human sciences. This conception of laws itself draws upon the central Sellarsian theme of attending to the role that various concepts or locutions play in reasoning.

pragmatic virtue responsive to contextually specific questions and concerns and cannot sustain the ontological priority naturalists ascribe to the scientific image. Second, van Fraassen argued that a scientific conception of the world should remain tethered to its rational accountability to human observation, even though the conceptual content of scientific theories legitimately outruns the limits of human observation. As rational agents with limited sensory access to the world, we should only believe what our best scientific theories tell us about what we can observe. Accepting van Fraassen's constructive empiricism would thereby restore philosophical priority to the manifest image as the source of rational epistemic norms to which the scientific image must answer. Empiricist epistemology would set limits to scientific understanding.

Several of Sellars's former colleagues at the University of Pittsburgh defend the philosophical autonomy of the manifest image by a different route. Unlike van Fraassen, John McDowell, Robert Brandom, and John Haugeland would not legislate rational constraints upon the scope of scientific beliefs. Indeed, their work shifts the primary philosophical concerns with science away from epistemology. These "left-Sellarsians"[5] instead seek to comprehend the normativity of conceptually articulated understanding. Each situates conceptual normativity within the manifest image of ourselves as reflective rational agents. Each places scientific understanding among our most important achievements as concept users but regards its explanatory resources as insufficient to understand the normative authority that constitutes an intelligible "space of reasons." To be sure, our self-conception as rational agents who answer to norms must be consistent with our self-conception as scientifically explicable natural beings. They see nothing mysterious, ineffable, or metaphysically transcendent about conceptual normativity and to that extent espouse a minimalist naturalism. Yet that consistency is a rational demand we should impose upon ourselves from within the space of reasons.

5. The distinction between left- and right-Sellarsians tracks two loosely defined groups of philosophers, each strongly influenced by the work of Wilfrid Sellars. Right-Sellarsians (exemplified by Ruth Millikan, Daniel Dennett, Paul Churchland, William Lycan, or Jay Rosenberg) draw especially upon Sellars's commitment to scientific realism, his thoroughgoing naturalism, his insistence upon accommodating a more sophisticated empiricism and a prominent role for conceptual rationality within a broadly reductionist conception of the scientific image, and in some cases, his retention of a role for representational "picturing." Left-Sellarsians (exemplified by Richard Rorty, Robert Brandom, John McDowell, or John Haugeland) emphasize his rejection of the empiricist Myth of the Given, the irreducibility of the logical space of reasons to causal or law-governed relations, his emphasis upon inferential roles as determinative of conceptual content, and the role of social practice in interpreting and justifying conceptual content while downplaying or rejecting his naturalism, scientific realism, and pictorial representationalism.

The opening of a conceptually articulated space of reasons, although identifiable in retrospect as an event within a natural history of the human species, cannot be properly understood in terms of its natural history or of the laws, causes, or symmetries that govern it. Moreover, from within the space of reasons we can then recognize scientific understanding as only one among many domains of conceptually articulated human activities or achievements, each normatively accountable in its own characteristic ways.

Other critics prominently challenge philosophical naturalism in a different, more limited way by arguing that consciousness, sensory experience, moral obligation, aesthetic judgment, religious transcendence, or some other aspect of the world resists assimilation within a scientific understanding of nature. Such criticisms nevertheless usually presuppose familiar conceptions of scientific understanding in order to argue for their limits. Van Fraassen and the left-Sellarsians thus challenge naturalism in a deeper way. They do not merely discern residual pockets of resistance to an otherwise inclusive scientific conception of the world. They instead conclude that a comprehensively natural-scientific conception of the world would render incomprehensible the authority of the scientific image itself. Van Fraassen would revise a Sellarsian conception of the scientific image to acknowledge limits upon reasonable belief. McDowell, Brandom, and Haugeland challenge the philosophical priority of the scientific image more comprehensively. They argue that a "baldly naturalistic" (McDowell 1994) conception of ourselves as part of nature as scientifically understood not only overreaches its empirical justification. A radically comprehensive naturalism would undermine its own intelligibility as a conception of the world. The scientific image and the understanding that it promises depend upon our capacities for conceptual understanding and its rational accountability. These very capacities for conceptual thought cannot be fully assimilated within the terms of a scientific understanding of nature.[6] In chapter 5, I argue that the left-Sellarsians are right to focus philosophical attention upon conceptual capacities more generally rather than empirical justification. For now, my point is only that among contemporary philosophical critics of naturalism, they provide the most fundamental and far-reaching

6. Brandom, McDowell, and Haugeland each rejects Sellars's own proposed "fusion" of the manifest and scientific images as dependent upon an untenable distinction between describing what is the case and "rehearsing a [shared] intention" (Sellars 2007, 408). I agree but will not argue for that claim here.

challenge to the philosophical priority naturalists accord to scientific understanding.

I nevertheless take Brandom, McDowell, and Haugeland to advance the naturalist cause constructively despite their rejection of stringent forms of philosophical naturalism. By showing where currently influential versions of naturalism fall short, they highlight the requirements for a more adequate naturalistic self-understanding. Their critical arguments also focus attention upon an indispensable but challenging desideratum for any viable philosophical naturalism. If Sellars is right about the comprehensiveness of scientific understanding, then a crucial philosophical task for naturalists is to comprehend how the capacity to understand the world scientifically fits within the purview of that scientific conception. In pursuing that task, we cannot take for granted either the scientific terms in which nature and ourselves should be understood or any particular account of what it is to understand the world scientifically. Not only do the sciences continue to refine and develop our understanding of the world, but empirical and philosophical studies of the sciences in turn are refining and developing our conception of scientific understanding.

McDowell (1994) directly disputes this Sellarsian conception of our philosophical situation and the tasks it poses. A central theme of his lectures is that no constructive philosophical or scientific work is needed to grasp how conceptual capacities, including capacities for scientific understanding, are compatible with a scientific understanding of nature. McDowell first rejects in advance the possibility that the rational spontaneity of human conceptual capacities could ever become scientifically intelligible within a "disenchanted" conception of nature as governed by law. The effort to incorporate reason within nature must yield either a "bald naturalism" that repudiates conceptual normativity altogether or a philosophical revisionism that constructs an inadequate simulacrum of the conceptual domain in "disenchanted" terms. He then argues that no such efforts are called for. We are already entitled to a conception of "second nature" through which human animals are brought into language and cultural tradition as part of their normal development: "Second nature could not float free of potentialities that belong to a normal human organism. This gives human reason enough of a foothold in the realm of [natural] law to satisfy any proper respect for modern natural science" (McDowell 1994, 84). Only misguided philosophical anxieties could drive further inquiry into *how* rational spontaneity is grounded in human biological potentialities or suggest that such inquiry might constructively inform our understanding of science, nature, or reason.

McDowell thereby closes off further philosophical reflection or scientific inquiry into the relations between a scientific understanding of nature as the realm of law and the scope and character of human conceptual capacities. Despite his insistence upon a "standing obligation to reflect about the credentials of the putatively rational linkages that govern [active empirical thinking]" (1994, 12), McDowell repudiates that obligation at the point where first and second nature meet. He takes for granted both our received philosophical accounts of scientific understanding as the disenchanted realm of law and our received biological accounts of human organisms, which together inform his insistence that a more thoroughly naturalistic account of conceptual understanding cannot be satisfactory.

This book instead takes up the obligation for critical reflection upon human conceptual capacities at the very point where McDowell urges philosophical and scientific forbearance. My arguments in the book aim to advance a broadly Sellarsian philosophical naturalism by rethinking both the scientific and the manifest images in light of possibilities for their philosophical and scientific reconciliation. I aim to show how we might better situate our self-understanding as persons responsive to normative considerations within a broadly scientific understanding of nature that incorporates our conceptual capacities as natural phenomena. McDowell is right that our received conceptions of nature and science foreclose a more thoroughly naturalistic incorporation of scientific understanding within nature as scientifically understood. I take that conceptual impasse to call for renewed philosophical reflection and scientific inquiry rather than acquiescence. Such reflection and inquiry should also aspire to advance our self-understanding constructively and not merely to relieve recurrent philosophical anxieties about our conceptual footing in the world.

III—Reconceiving the Fusion of the Manifest and Scientific Images

A broadly Sellarsian philosophical project would overcome the apparent incompatibility of the manifest and scientific images by fusing them into a more coherent conception of ourselves and our capacities, which nevertheless acknowledges and accommodates the insights of both. My approach to that project is avowedly naturalistic in the sense that the resulting fusion incorporates our developed and developing capacities for scientific understanding within the natural world as scientifically understood. Achieving a more adequately naturalistic self-understanding nevertheless requires some reformulation of the terms in which that task has previously been conceived. My reformulation is primarily responsive to new philosophical and empirical work arising from three directions. These developments bear upon one another in revealing and complementary ways, even though they are rarely considered together.

First, Haugeland, McDowell, and Brandom have further developed the "manifest" conception of ourselves as agents who perceive, understand, and act within the world as responsive to conceptually articulated norms. Their work thereby complicates as well as enriches the task of achieving a naturalistic fusion of the scientific and manifest images.[7] Each of them takes his account of conceptual capacities to block any stringent or (in McDowell's 1994 phrase) "bald" naturalism. They endorse a minimalist naturalism, arguing that nothing in their views is *inconsistent* with what we learn from the natural sciences. Conceptual normativity nevertheless remains autonomous in their view, without need or expectation of further scientific explication. This opposition to a more thoroughgoing philosophical naturalism presumes familiar conceptions of scientific understanding, however, and also does not consider some new theoretical and empirical resources for a scientific account of our conceptual capacities. The other two developments guiding this book suggest that these presumptions are misguided.

A second body of work that centrally informs my project comes from recent philosophy of science and interdisciplinary science studies. This work offers compelling reasons to reconceive familiar accounts of scientific understanding exemplified by the Sellarsian scientific image. A central concern of philosophical naturalism has been to let the sciences speak for themselves, freed from the prejudices and constraints of inherited philosophical or other preconceptions. Naturalists' widespread rejection of classical empiricist epistemology strikingly exemplifies this commitment. Empiricists are skeptical of concepts or claims whose content and justification are inferentially distant from what is observable

7. As I argue in chapter 2, their work in this respect is constructively supplemented by other recent accounts of the role of practical bodily skills in perception and action, building upon Heidegger's and Merleau-Ponty's phenomenology. Most of the phenomenological discussions of "skilled coping" with our surroundings have concluded that the importance of unreflective bodily skill in disclosing the world to us challenges the priority that the left-Sellarsian tradition ascribes to conceptual understanding (Dreyfus 1979, 1991, 2005; Kelly 1998, 2001, 2000; Carman 2008; Schear 2013). I argue below that these oppositions are misplaced, and thus that work by Dreyfus, Kelly, Carman, Thompson (2007), or Gallagher (2005) constructively engages with the insights of the left-Sellarsians. Some philosophers, notably Nöe (2004, 2009), Lance (2000), and Haugeland (1998, ch. 9), already proceed in ways that build upon that continuity.

with human sensory capacities. Naturalists instead highlight the robust successes of theoretical science that appeal to unobservable entities or processes and argue that we should instead be skeptical of these empiricist scruples. If the best scientific research successfully draws upon a richer set of conceptual and methodological resources than empiricist epistemology would countenance, so much the worse for empiricist epistemology.

Van Fraassen's challenge to naturalists arose from this clash between scientific practice and empiricist presuppositions; he sought to restore the governance of empiricist epistemological norms over the scientific image. I draw an opposing moral from recent work in philosophy of science and science studies: naturalist criticism of empiricist epistemology has not yet gone far enough. The predominant philosophical conception of the scientific image still reflects a long-standing philosophical preoccupation with epistemology, which is in tension with the practices and achievements of the sciences. Recent philosophical, social, and historical studies of science shift attention away from scientific knowledge as a detachable product of inquiry, focusing instead upon the ongoing articulation and development of scientific practices. An important aspect of this shift is temporal. The temporal orientation of epistemology is largely retrospective: To what extent is an already established body of scientific knowledge claims reliable or justified?[8] Yet that retrospective orientation is at odds with the practical orientation of scientific research. Scientists are also concerned with questions of justification and reliability, but from a different direction. Their work is governed by the prospective orientation of a research program, and they want to know whether and how past practice can successfully guide further exploration and disclosure of the world that will likely revise that guiding understanding. Philosophical and empirical studies of the sciences thereby encourage reconceiving the scientific image as incorporating a situated practical capacity to extend and refine current understanding of ourselves-in-the-world rather than consisting in a systematic representation.

8. Epistemologists recognize that knowledge continues to grow and are concerned to recognize and promote that openness to further development. Some epistemologists (empiricists are a prominent example) also recognize limits to scientific knowledge, which therefore constrain future inquiry. Yet even in looking ahead, epistemology characteristically does so in the future perfect tense—that is, from the projected standpoint of one looking back upon the prior achievement and justification of knowledge claims. The standpoint of research has a different temporal orientation in which key concepts, methods, and claims are at issue in ongoing inquiry and provide a more or less determinate direction to inquiry despite, or even because of, their partial indeterminacy and open-endedness.

The third primary resource for my reformulation of naturalism comes more directly from the sciences themselves. The emergence of the mid-twentieth-century evolutionary synthesis provided powerful new conceptual resources for philosophical naturalists. Evolutionary theory offered promising possibilities for understanding the normativity of knowledge and conceptual content in terms of genetic processes that secure biological adaptation to an organism's environment. A neo-Darwinian conception of ourselves has thereby become central to the scientific image as we know it, both for naturalists such as Ruth Millikan (1984) or Daniel Dennett (1987, 1995) whose philosophical vision was explicitly evolutionary and for others for whom evolution merely provided a broader horizon for their accounts of intentionality and knowledge. New theoretical developments within evolutionary theory (e.g., developmental evolution, developmental systems theory, ecological-developmental biology, and niche construction theory), along with new empirical work on animal behavior, human evolution, and language, now challenge familiar ways of thinking about cognition and knowledge in evolutionary terms and suggest alternative approaches for situating human understanding within our evolutionary trajectory. Philosophical naturalism commits us to maintaining an ongoing engagement with scientific work in this way without settling for familiar and congenial conceptual horizons that the sciences continue to surpass.

My interests in these three projects arose independently. Left-Sellarsian accounts of conceptual normativity, philosophical and empirical work on scientific practice, and the extended evolutionary synthesis (Müller and Pigliucci 2010) each stands on its own as a well-developed line of inquiry. All three bodies of work are nevertheless mutually supportive in ways that strengthen the case for reformulating a naturalistic fusion of the manifest and scientific images. Their constructive contributions to understanding conceptual normativity and scientific practice encourage thinking differently about ourselves and our capacities for understanding the world scientifically. They also help us recognize and overcome residual theological or "supernatural" commitments that still shape avowedly naturalistic projects in philosophy, science, and science studies. In this respect, the convergence of these ways of thinking about ourselves and the sciences from within a scientific understanding of nature promises a more thoroughly naturalistic conception of ourselves and the world.

This approach to a naturalistic fusion of the manifest and scientific images draws together several mutually reinforcing themes. One theme is the need to reorient the place of scientific understanding within the manifest image of ourselves as persons responsive and accountable to norms. A familiar conception of science emphasizes its role in justifying belief; we are accustomed to thinking of ourselves as believers who formulate and accept representations of how things are. The meaning and justification of those beliefs would then be the primary target for philosophical explication and assessment. Sellars, Brandom, McDowell, Haugeland, and others within this tradition suggest a different conception of ourselves, which also changes the central tasks for science and philosophy. We are concept users who engage others and our partially shared surroundings in discursive practice. The primary phenomenon to understand naturalistically is not the content, justification, and truth of beliefs but instead the opening and sustaining of a "space of reasons" in which there could be conceptually articulated meaning and justification at all, including meaningful disagreement and conceptual difference. This "space of reasons" is an ongoing pattern of interaction among ourselves and with our partially shared surroundings. As Ian Hacking once noted, "Whether a proposition is as it were up for grabs, as a candidate for being true-or-false, depends on whether we have ways to reason about it" (2002, 160). The space of reasons encompasses not only the claims that we take to be true or false but also the conceptual field and patterns of reasoning within which those claims become intelligible possibilities whose epistemic status can be assessed. Any determination of the content, justification, or truth of beliefs emerges from that larger process of ongoing interaction. Whether conceived as second nature (McDowell 1994), discursive practice (Brandom 1994), constituted domains (Haugeland 1998), or a functional linguistic pluralism (Price 2011), the space of reasons cannot be reduced to the various contents expressed or expressible within it. The familiar epistemological conception of us as believers, who might ideally share a common representation of the world in the scientific image, thus conflates particular moves within discursive practice or the space of reasons with the space or practice itself.

A product-oriented conception of scientific understanding might appeal to the concept of a language (including a language of thought if there were such a thing) to express this difference between the space of reasons and the claims that can be assessed within it. We could then distinguish beliefs within a language from the language itself as a larger space of possibilities within which those beliefs can be intelligibly expressed and assessed.[9] As we shall see, however, this appeal to the determinate

9. An influential example of such an approach received classic formulation in Carnap's (1950) "Empiricism, Semantics and Ontology." Carnap distinguished internal questions, which can be

structure of a language in place of the dynamic configuration of a space of reasons is not adequate for multiple reasons. Utterances or marks only become linguistically interrelated through their place within discursive practice, which extends beyond language to incorporate perception and action. Moreover, appeals to language as the horizon within which beliefs acquire content inappropriately reify a structure abstracted from the dynamics of ongoing discursive interaction. Above all, such a conception mistakenly separates language, taken as a space or structure of possible representations of the world, from the world to which it is semantically accountable.[10]

Recent work in evolutionary biology and the philosophy of biology resonates with this shift of attention away from beliefs as mental or linguistic representations toward a conception of discursive practice or the space of reasons. Earlier philosophical work on the evolution of cognitive capacities tended first to focus upon "intelligence" as a general cognitive capacity and more recently upon the functional and adaptive role of mental representations within the behavioral economy of an organism's way of life. Whether taking language and conceptual understanding as continuous with the cognitive capacities of many nonhuman animals or as marked by a sharp break due to the symbolic or recursive character of language, philosophers have typically construed the cognitive capacities of animals (including human animals) in terms of self-contained abilities for perceiving, representing, and responding to the world "external" to the organism.[11]

This entrenched way of thinking about cognition as self-contained becomes increasingly problematic biologically, with closer attention to the developmental, physiological, and evolutionary entanglement of organisms with their environments. The resulting reconceptions challenge

expressed with determinate truth values and assessed within a language, from external questions about the existence of entities independent of any linguistic framework and distinguished both kinds of question from the pragmatic issue of which language to adopt.

10. Price (2011) rightly emphasizes a functional pluralism within language, which cannot be accounted for in terms of a general account of representation, even though one does need to account for the role of a common assertoric form that can be used in functionally diverse ways. Yet I am also emphasizing a further shift in this direction, from thinking about language as a structure with diverse uses, to discursive practice, which gives philosophical centrality to these patterns of situated use.

11. Sometimes, as Godfrey-Smith (2002) notes, the representational structures and processes that supposedly constitute cognition have been taken as embedded in patterns of neurological "wiring-and-connection" (Sterelny 2003, 4); for other theorists, they were instead global explanatory attributions needed to make sense of organisms' overall patterns of responsiveness to their surroundings, for which Dennett's (1987) "intentional stance" is an exemplary case. For my purposes, the recent challenges to representationalist accounts of cognition need not differentiate between these two opposing versions of cognitive representationalism.

traditional cognitive internalism from two different directions.[12] Understanding the close intertwining of organisms' sensory systems with their repertoires for behavioral and physiological responsiveness shows how organisms are closely coupled with their environments. An organism's biological environment does not consist of objectively independent features of its physical surroundings. Biological environments are bounded and configured as the settings to which organisms' ongoing way of life is responsive. An organism's environment consists of those features or aspects of its surroundings that matter to its development, physiology, reproduction, and consequent evolution across generations. The organism and its way of life can in turn only be explicated as part of a larger biological pattern that encompasses its environment.

Understanding this close coupling of organism and environment shows how some organisms can develop robust capacities for tracking and flexibly responding to multiple environmental features, which can account for very sophisticated responsiveness to variable environmental conditions without postulating representational intermediaries (Sterelny 2003). Such perceptual and practical capacities are adaptively *directed* toward and *responsive* to a selective configuration of the organism's physical environment.[13] These capacities nevertheless contribute to organisms' behavioral and physiological economy in ways that do not differentiate how the organism *takes* its surroundings to be (which might be *mis*taken) from what we can provisionally call an extensional determination of those aspects of its physical and behavioral surroundings to which its way of life is responsive.[14] Organisms are selectively oriented toward aspects

12. A substantial and growing body of philosophical and cognitive-scientific work on "extended" or "enactive" cognition (Clark 2003, 2008; Nöe 2004, 2009; Thompson 2007; Chemero 2009; Rowlands 2010; Shapiro 2011, among others) complements and reinforces many of the themes in my argument. I have not attempted to develop those connections because they are not needed to explicate the primary revisions of philosophical naturalism advanced in the book, even though their work supports or further develop many of my more specific themes.

13. Organisms' perceptual/practical capacities are "selective" in a dual sense. They are directed toward only some features or aspects of their physical surroundings and do not register or respond to others. These relevant aspects of their physical surroundings in turn constitute the organism's "selective environment" (Brandon 1990; Brandon and Antonovics 1996)—that is, the environmental configuration that is selectively relevant to the organism's adaptive fitness, both in being relevant to the evolutionary prospects of the reproductive populations to which they belong and as having themselves arisen in response to past selective regimes. The selective environment of an organism or its lineage incorporates those aspects of its environment that differentially affect its physiological functioning and population size, its normal developmental patterns, and its comparative fitness in relation to other organisms and lineages.

14. The distinction of extension from intension differentiates the object of an intentional directedness from its manner of presentation (how the object is "taken" to be in that intentional comportment toward it). This provisional formulation suggests that organisms' constitutive interaction with their environments "picks out" which aspects of their surroundings belong to their environment

of their surroundings without also taking them *in* or *as* some determinate conception.[15] The recognition that organism/environment coupling is selective but not conceptually articulated might seem initially to sever the connection between nonhuman organisms' perceptual/practical involvement with their surroundings and our own conceptually articulated intentional directedness. Proposing a sharp divide between human and nonhuman cognition might then seem to conflict with a naturalistic emphasis upon understanding us as animals in evolutionary continuity with our primate kin.

The development of niche construction theory (Odling-Smee, Laland, and Feldman 2003) and its application to understanding the evolution of language and symbolic-conceptual understanding (e.g., Deacon 1997; Dor and Jablonka 2000, 2001, 2004, 2010; and Bickerton 2009, 2014) restores this connection in a way that reinforces the left-Sellarsian turn from mental representation to public discursive practice. Niche construction is the transformation of the developmental, selective environment of an organism and its lineage by ongoing, cumulative interactions of other organisms with that environment. The biological environment of an organism's lineage thus is not simply given but is instead dynamically shaped by ongoing interaction with the organisms in that lineage. Such transformations are not limited to enduring physical effects on the abiotic environment but also include persistent forms of behavioral niche construction. Behavioral niche construction requires only that the presence of behavioral patterns, and their selective significance for individual organisms' evolutionary fitness, be reliably reproduced in subsequent generations.[16] The emergence of

without thereby having a "sense" or manner of presentation that might incorrectly characterize the very aspect of the world that it picks out. The claim may seem odd, because organisms do respond differently to different aspects of the world: eating some, fleeing others, using still others as concealment. Despite its initial, provisional usefulness, what is ultimately misleading about describing organisms' way of life as determining the organisms' selective environment "extensionally" is that in semantic contexts, extensions are understood to consist in sets of *objects* with multiple determinations. What the organisms' way of living "picks out" is not an object, however, but what Gibson (1979) calls an "affordance," defined only in relation to what it "affords" the organism. See chapters 2–4 for further discussion and clarification.

15. One could put the point another way by saying that its normal way of life, as responsive to and dependent upon interconnected features of its surroundings, *is* a holistic pattern of "taking as." I prefer instead to distinguish organisms' "one-dimensional" selective directedness toward what thereby becomes part of their environment from a further two-dimensional articulation and tracking of different ways of conceiving aspects of its environment *within* the larger pattern of its selective interaction with that environment. This distinction is developed in greater detail throughout part 1.

16. The level of "reliability" of reproduction need not match the evolved replicative fidelity of cellular transcription, translation, and expression of DNA sequences, which is itself a dynamic and only partially reliable process. So long as there is sufficient stability to affect the cumulative selective

communicative-cooperative practices that evolve into language is a pre-eminent example of niche construction. Language is a persisting public phenomenon that coevolves with human beings. Human beings normally develop in an environment in which spoken language is both pervasive and salient, while languages only exist in gradually changing forms that can be learned and thereby reproduced. Human abilities to acquire and take up the skills and discriminations that enable the ongoing reproduction of that phenomenon are integral to our overall practical/perceptual responsiveness to our environment, which has thereby become a discursively articulated environment. The evolutionary emergence of this capacity and its ontogenetic reconstruction in each generation rely on the same close coupling with our discursively articulated environment that characterizes other organisms' capacities for perceptual and practical responsiveness to their selective environments. There is nothing mysterious or even discontinuous about the gradual development of the linguistic capacities and performances that enable conceptual understanding. Yet the only partial autonomy of discursive practice from systematic interconnectedness with other aspects of our perceptual-practical immersion in an environment allows for the emergence of symbolic displacement and conceptual understanding.[17]

Conceptual understanding thus emerges biologically as a highly flexible, self-reproducing and self-differentiating responsiveness to cumulatively constructed aspects of our selective environment. Discursive niche construction is not limited to our abilities to perceive and produce linguistic expressions. Other symbolically significant expressive capacities (e.g., pictorial, musical, corporeal, equipmental, and more) are also integral forms of human niche construction.[18] More important, however, is that the resulting capacities for symbolic displacement also incorporate practical-perceptual immersion in an environment. Our perceptual and practical capacities are not themselves different in kind from those of other organisms, but they are transformed by their uptake within discursive practice. McDowell (1994) characterizes the possible discursive significance of everything we do as "the unboundedness of

pressures on the organism's developmental patterns, niche construction can have a significant evolutionary effect, often on more rapid time scales than is possible for evolution that is directly driven by genetic mutations and duplications or regulatory shifts in gene expression.

17. Chapters 3–5 work out how to think about linguistic understanding and conceptual normativity as examples of niche construction and how that matters to a naturalistic conception of our linguistic and conceptual capacities.

18. My argument below does not depend upon whether our various expressive repertoires evolved together or if one of them arose earlier in ways that enabled others.

the conceptual." Our discursively articulated practical/perceptual involvements are pervasive throughout and integral to the world in which we develop as and into adult human beings. Their cumulative effects dramatically transformed us as a species and indirectly affected many others, including some thereby driven to extinction. The verbally articulated differences that are so central to our developmental, selective environment are nevertheless almost entirely opaque to our various "companion species" (Haraway 2008) and, to that extent, not part of their biological environments. Our inherited responsiveness and massive ongoing contribution to this peculiar cumulative history of niche construction, and not any general cognitive capacities, are what primarily differentiate us as concept users from any other known organism.[19]

The emergence of scientific inquiry within the recent history of the human species has contributed extensively and intensively to our ongoing niche construction. Philosophical attention to these contributions was long focused primarily upon the production and justification of scientific knowledge. New philosophical studies of scientific practice, augmented by the rapid growth of empirical research on scientific practice in multiple disciplines, have now emerged alongside traditional epistemology and epistemological philosophy of science. Studies of scientific practice promise constructive mutual engagement with both the left-Sellarsian philosophical tradition and the emerging understanding of discursive practice as a form of evolutionary niche construction.

Studies of scientific practice share with the left-Sellarsians a primary focus upon the articulation of conceptual understanding rather than the justification of knowledge claims. Quine's famous image of science as a "totality of knowledge or beliefs [that] is a man-made fabric which impinges on experience only along the edges" (1953, 42) exemplifies the widespread construal of scientific knowledge as composed of systematically interrelated statements. Recent work on scientific practice revises and expands this familiar embodiment of the scientific image. Attention to scientific practice challenges familiar conceptions of theories as systematic sets of statements. Scientific understanding is instead mediated

19. We cannot easily isolate the role of discursive niche construction from other physiological and cognitive changes that were involved in enabling and sustaining it. The long history of coevolution of language and *Homo sapiens* has enhanced and reconfigured our perceptual and cognitive capacities, from the structure of the human brain and its interconnectedness with various sensory and motor capacities (such as our refined capacities for voluntary vocal articulation and auditory discrimination and diminished olfactory sensitivity), to our highly neotenous bodily and neurological development with its associated forms of extended child-rearing, and the relative stability of social groups from the familial to the linguistic.

by mathematical, visual, physical, verbal, and other kinds of models (Morgan and Morrison 1999) and by the coordination of such models with laboratory phenomena that display the conceptual relations that the theories express. The experimental systems used in research also serve as models of a scientific domain in ways that affect which conceptual relations can show up clearly in that context.[20] The models employed are often mutually inconsistent and overlap in their domain of applicability while also leaving gaps where no models adequately articulate theory (Giere 1988; Cartwright 1999; Wilson 2006). Theoretical understanding encompasses not merely a grasp of truth claims but also abilities to use and extend the standard models and to recognize which models are appropriate for which situations and purposes. Studies of the role of discipline formation and conceptual exchange across disciplinary boundaries in shaping the conceptualization of scientific domains (e.g., Bono 1990; Bechtel 1993; Galison 1997; Lenoir 1997; Rheinberger 1997) have gone further beyond the more limited scope of epistemological conceptions of scientific understanding. These extensive patterns of discursive exchange also embed scientific practices in larger patterns of cultural practice and understanding, which further contribute to the content and significance of scientific understanding.[21]

In more traditional accounts of the scientific image, laws of nature often distinguish the domain of nature from the forms of social and cultural life that have emerged within it. Kant's distinction between phenomena governed by natural laws and thoughts and actions governed by a rational conception of laws that we give to ourselves is an influential precursor in this respect to Sellars's distinction of the scientific and manifest images. In chapter 8, I argue that from the standpoint of scientific practice, "laws of nature" are best understood as scientific laws expressing commitments undertaken and deployed in scientific reasoning in various contexts of inquiry (Lange 2000a, 2007).[22] Such reasoning is

20. Bolker (1995) offers the telling example of the standard model organisms for developmental biology, whose common features of rapid, highly canalized development, short generation times, small adult size, and single-stage developmental process effectively isolate "development" from environmental interaction and block the factors that govern developmental canalization from experimental scrutiny within this scientific domain.

21. A growing literature in the anthropology of science and in cyborg anthropology has been especially attentive to how scientific work is situated within broader cultural patterns of conceptualization and significance. Prominent anthologies in this field include Downey and Dumit (1997); Layne (1998); Goodman, Heath, and Lindee (2003); and Franklin and Lock (2003). Relevant monographs include Haraway (1997), Rabinow (1996), Traweek (1988), Dumit (2004), Helmreich (1998, 2009), and Martin (1994, 2007), among many others.

22. Lange's understanding of laws does not reject the connection of laws with necessity but instead treats nomological necessity as a holistic stability of the truth of laws under various

integral to our ongoing interaction with our environment, however, in ways that further articulate it discursively. Understanding laws within scientific practice can thereby help conjoin the manifest and scientific images rather than to divide them.

The proposal that different laws of nature are salient and authoritative within different domains of scientific practice exemplifies a more widespread recognition of the disunity of scientific practice and understanding (Dupre 1993; Galison and Stump 1996; Cartwright 1999; Lange 2000a; Hacking 1992; Wilson 2006; Giere 2006; Bechtel 1993). Here we encounter a fracture in recent philosophical discussions of naturalism that will play an important role in the book. In many areas of recent philosophy (e.g., metaphysics, epistemology, philosophy of language and mind, ethics), naturalists and their critics mostly take for granted that scientific understanding aspires to the comprehensiveness suggested by Sellars's account of the scientific image. Philosophers of science, however, often regard naturalism as guiding our understanding of the sciences in a different direction than naturalism as conceived elsewhere in philosophy. Emphasizing naturalists' commitment not to impose philosophical preconceptions or constraints upon the sciences, they often take at face value "science as we know it: apportioned into disciplines, apparently arbitrarily grown up; governing different sets of properties at different levels of abstraction; pockets of great precision; large parcels of qualitative maxims resisting precise formulation; here and there, once in a while, corners that line up, but mostly ragged edges; and always the cover of law just loosely attached to the jumbled world of material things" (Cartwright 1999, 1).

Recognition of disunity among the sciences may then seem to threaten the very idea of a comprehensive "scientific image" and with it the notion of a stringent philosophical naturalism. If the sciences yield powerful insights but only within patchy, relatively disconnected domains of inquiry, then perhaps only a minimalist naturalism is called for. Naturalists could take scientific understanding to be authoritative wherever it can be achieved but would not expect its authority to extend everywhere. The supposed need to accommodate a scientific un-

counterfactual suppositions and inductive extensions. The result is to understand nomological necessity as expressing a norm of reasoning in scientific practice rather than a special kind of truth independent of the practical contexts within which the laws are used. In this approach to laws, counterfactuals, and nomological necessity, which gives priority to their role in scientific reasoning, Lange builds upon a central theme in Sellars's own thinking about conceptual understanding. See especially Sellars (1948, 1957). I discuss laws and modalities more extensively in chapter 8.

derstanding of nature within nature as scientifically understood might then also dissolve; there would be no reason to expect scientific work in its messy complexity to be among the phenomena readily accessible to natural scientific understanding. I endorse and draw upon many of the "disunifiers'" claims about scientific practice and understanding but draw a different inference. We need to reconceive rather than abandon the comprehensiveness of scientific understanding. We can recognize the kinds of disunity that science studies research has rightly identified while also recognizing an indispensable mutual openness and accountability across the various domains of scientific work. The result, introduced in chapter 6 and developed throughout part 2, is a different sense of how scientific understanding is comprehensive. Scientific understanding "as a whole" yields neither a unified theoretical representation of the world nor a disconnected collection of disciplinary practices but instead articulates and refines the space of reasons as interconnected and indefinitely extensible.[23]

This conception of scientific practice suggests that we understand the sciences themselves as part of our ongoing niche construction in ways that conjoin the material and behavioral-discursive reconstruction of our developmental environment. The establishment of reliable, reproducible experimental phenomena that manifest clear patterns in the world plays a significant role both in the articulation of specific scientific concepts and in opening whole domains of scientific work to conceptual articulation and understanding. Moreover, the development of various kinds of models—physical models, visual diagrams or images, mathematical models, and more—is integral to the articulation of conceptual understanding. Scientific conceptual understanding is never just a matter of "mental" representation but always involves changing the world around us in ways that enhance its intelligibility. These changes take many forms: rearranging things to reveal informative patterns in the laboratory and the world outside it, building and deploying new kinds of models in new ways, or extending and refining the inferential entanglements of scientific concepts and other discursive elements with other domains of discursive practice. Moreover, these material transformations of the world's intelligibility are not merely important as aids to

23. In this emphasis upon doing justice to both the disunity of the various scientific disciplines and other specialties and their mutual accountability, my argument develops a central theme from Sellars's own account of the scientific image (2007, 370–71). Thanks to Willem deVries for reminding me of this important continuity with Sellars.

initial discovery or subsequent pedagogy. Many uses of well-established scientific concepts beyond the laboratory require transforming the circumstances of application to resemble sufficiently the laboratory or other experimental settings where the relevant conceptual distinctions were developed and stabilized (Latour 1983; Rouse 1987). This role for experimental, technological, and metaphorical activity in conceptual understanding reinforces the notion that conceptual articulation is a phenomenon of niche construction through which we inherit, reproduce, and extend physical and behavioral transformations of the world that allow it to be intelligible in new ways.[24]

These conceptions revise the scientific image in ways that undercut Sellars's original metaphorical use of the term 'image' for a scientific conception of the world. On the conceptions of science and naturalism proposed in this book, the sciences do not offer a systematic representation of the world as a whole, even as a promissory note. They instead make a decisive contribution to our ongoing efforts to transform our environmental niche in ways that allow it to be conceptually intelligible, and these forms of intelligibility are integral to the ongoing natural history of our species. The sciences introduce new experimental systems, practices, and skills, their technological extension both within and beyond the research context and conceptual revisions that engage and mutually transform other discursive practices. They develop new models and more general theoretical formulations along with the mathematical and other inferential understanding needed to work with them. They raise new conceptual and practical issues that people must respond to in various ways, including closing off some ways of thinking and acting that once seemed intelligible and attractive. Together they help reconstitute a space of intelligible possibilities for understanding and articulating ourselves and the world, including possibilities for reasonable disagreement. The concepts developed and deployed in those practices are always only partially determinate, open to more extensive and intensive articulation and inferential development with respect to other aspects of our discursive involvement in the world. This

24. It may seem odd initially to think of metaphorical uses of theoretical concepts as material transformations akin to the building of experimental systems or new theoretical models. Yet Davidson (1984) and his followers (especially Rorty 1991) and Bono (1990) have each in different ways emphasized that metaphor is a phenomenon of the use of language in which a nonstandard use is an "unfamiliar noise" (Rorty 1991) or a material exchange between discourses (Bono 1990). Once we understand language itself as a form of behavioral niche construction, all linguistic uses have to be understood as material components of our developmental and selective environment.

open-endedness is not only a matter of human limitation. Part of what the sciences open and sustain is a grasp of scientific significance within a broader intellectual and cultural milieu such that some truth claims matter more than others and matter to us in different ways. This dimension of scientific significance does not simply involve the shaping of inquiry by prior interests and involvements, however, for our interests and involvements are also themselves always at issue for us. In recognizing scientific understanding as materially and conceptually shaping the world we live in, as our biological environment, we understand both our vulnerable dependence upon our worldly circumstances and our openness to partial self-transformation. Just what a human way of life is, and could become, is ultimately part of what is at issue for us in our ongoing niche construction. Understanding both our situated dependence upon a historically evolving environment and our partial openness to remaking our way of life within that world is the most important outcome of recognizing how our biological niche is also a conceptually articulated space of reasons.

IV—Advancing a Naturalistic Self-Understanding

My project of working out and defending a revised conception of the scientific image and its place within a naturalistic self-understanding is also situated within this historically constituted conceptual space. Philosophical naturalism is an evolving project whose characterization is itself at issue in ways that are accountable to its own history and its prospects for further development. The issues relevant to defending or opposing naturalism, and how their resolution matters, are shaped by a tradition within science, philosophy, and science studies as an intelligible space of reasoning and understanding. The justification for understanding philosophical naturalism and the scientific image in the ways I propose is that it responds constructively to recognized conceptual and empirical issues within that tradition, brings out significant new concerns and ways of addressing them, and does both in ways that open new possibilities for further development and refinement.

How does my reformulation of philosophical naturalism claim to answer more adequately to the issues that have emerged within naturalism as a historical project? A central consideration within the naturalistic tradition is the rejection of "first philosophy" and the consequent insistence that philosophy should take direction from our best scientific understanding of the world rather than impose philosophical preconceptions upon the sciences. That distinction nevertheless has a contested history. As one prominent example, an empiricist epistemological stance has long been associated with a scientific conception of the world, from Locke through Mach, Neurath, and Quine, to van Fraassen. Philosophers now increasingly recognize that empiricism is not coextensive with a scientific conception of the world, however, but instead might be a philosophical orientation imposed upon science. Naturalism and empiricism provide *opposing* philosophical orientations.

Sellars's original conception of the scientific image as a systematic theoretical representation is also a dispensable philosophical imposition that the sciences do not need and perhaps do not even accommodate. The sciences do seek to develop retrospective compilations and systematizations of scientific understanding in multiple contexts: textbooks, review articles, encyclopedias, handbooks, policy analysis, or more locally in efforts to find common ground for an interdisciplinary collaboration. In each case, the retrospective compilation of the current state of knowledge is partial, perspectival, and oriented toward a task at hand. Yet there may then be no need, and no scientific basis, for how to specify *the* scientific image in the form of a general, all-encompassing scientific representation for no purpose in particular. Indeed, there may be real conflict between the sciences and a philosophical conception of a unified theoretical representation of the world as a whole. The sciences often employ mutually inconsistent treatments of similar situations, which cannot be accommodated within a single, consistent theoretical representation. Moreover, the sciences incorporate a significant range of disagreement. Even where they seek the resolution of specific disagreements within or among disciplines, that resolution may open up further issues not resolvable in the same way. In this respect, a conception of scientific practice as encompassing room for legitimate disagreement within and across various scientific disciplines, theoretical orientations, or research programs seems a more appropriate stance for naturalists. Empirically, the expectation that an expression of contemporary scientific understanding would take the form of a systematically unified theoretical "image" of the world seems not to accord with how scientific work is actually done. A conception of the scientific image in these traditional terms may thus fall short of naturalistic deference to science, as does the empiricist hostility to unobservable entities that once seemed to define a "scientific conception of the world" for many philosophers. Both may instead be a vestige of long-standing philosophical commitments to epistemology as "first philosophy" (Quine 1965).

The turn to scientific practice also better accommodates the empirical contingency of scientific understanding. Naturalists' commitment to be guided by the best contemporary scientific understanding brings with it a recurrent temptation to reify its terms, concerns, disciplinary orientations, and methodological strategies and constraints. The synecdoche that would mistakenly identify a scientific conception of the world with its most recent incarnations blocks an important virtue that naturalism should promote—namely, openness to empirical discovery, conceptual innovation, and their possible challenge to familiar ways of life and ways of thinking. Arthur Fine succinctly characterized this important aspect of naturalists' commitment not to countenance philosophical impositions or constraints upon the sciences, even though he does not explicitly identify it with naturalism: "[The Natural Ontological Attitude] sees . . . science as a set of practices with a history. That history constrains our understanding of current practice and structures our evaluation of promising problems and modes of inquiry. Because the practice is varied and self-reflective, it encompasses the possibility of moving on in virtually any direction that can be rationalized in terms of current practice and past history" (1986a, 10). This book's conception of a scientific understanding of the world (the scientific image) builds such conceptual and methodological open-endedness into our understanding of the sciences.

Insistence upon the conceptual and methodological open-endedness of scientific understanding points toward another important aspect of the naturalism advocated here. Despite rejecting familiar conceptions of the scientific image as a comprehensive theoretical representation of the world, I still insist upon a residual sense in which scientific understanding remains comprehensive. In this respect, the naturalism I advocate is partially at odds with the more minimalist forms of naturalism promoted by the left-Sellarsians and for different reasons by many of the advocates of scientific disunity. The sense in which a naturalistic self-understanding remains both comprehensive and comprehensively scientific can be seen from two complementary directions. The first is that our way of life as a *biological* lineage shares a single, comprehensive biological niche for all its internal variegation. We are mutually dependent upon and vulnerable to one another and our shared environment for whether and how that way of life will continue. The second is that the *conceptual* character of our discursive practices, including scientific practices, depends upon their significance for, and openness to challenge from, what we say and do in other aspects of our lives. The "space of reasons" is and must be a unified space. This sense of the constitutive comprehensiveness of conceptual understanding generally, and scientific understanding specifically, depends upon the specific arguments developed below, especially in chapters 7 and 10. Two points are important to make now, however, both for understanding my commitment to naturalism and for foreshadowing an important line of argument in the book. First, these two ways of understanding the comprehensiveness of the scientific image make the same point from different directions. On the view developed in this book, our biological environment and the conceptually articulated space of reasons are the same natural phenomenon. Second, part of the importance of recognizing the unity and comprehensiveness of conceptual space is that it blocks the temptation to insulate our various scientific practices and other aspects of our way of life from one another. 'Naturalism' has long been a "fighting word," often motivated by challenges to various claims or practices as inconsistent with or inappropriate for our self-understanding as natural entities. More minimalist naturalisms can too readily shield various conceptual or practical domains (including scientific practices) from such criticism by allowing them too much autonomy. On the view developed in this book, it matters to discursive practices generally, and scientific understanding specifically, that their conceptual autonomy is only partial. They remain accountable to other discursive practices and to their involvement within our partially shared way of life. In this way, the Sellarsian conception of a naturalistic philosophy as aspiring to understand "how things in the broadest possible sense of the term hang together in the broadest possible sense of the term" takes on renewed importance.

This way of accounting for scientific understanding as part of nature scientifically understood also calls attention to other possibly non-naturalist vestiges within some alternative conceptions of naturalism. A comprehensive-representational conception of the scientific image may no longer need or permit reference to God. Yet the very idea of a systematic theoretical representation of the world as a whole may still express a theological understanding of the standpoint of scientific knowledge. God's understanding would represent the world from a standpoint outside of the world represented, and scientific understanding is too often conceived as aspiring to a comparably external position (what McDowell [1994] sometimes describes as a view from "sideways on"). To the extent that scientific understanding is conceived as having determinate content intralinguistically, apart from its involvement in broader patterns of material interaction with the world, for example, it may still aspire to such an external, "sideways-on" conception of the world. Such accounts might also violate naturalistic commitments by thinking of concepts as having "immaculate" content, freed from any bodily or worldly entanglements.[25] Efforts to remove conceptual contents and norms from their embeddedness within a scientifically understood natural world have a long history. These efforts include Descartes's conception of "thinking substance" as noncorporeal, Kant's account of action and belief as noumenal in answering to a conception of law rather than to laws of nature, Husserl on transcendental consciousness as a "region of being" outside of nature, Frege's and other quasi-transcendental conceptions of logic as laws of pure thought, and uses of the distinction between computer hardware and software as models for the supposed immateriality of thought. In thinking of scientific practice as part of the natural history of discursive niche construction, we can instead advance further toward a thoroughly worldly, naturalistic conception of our own capacities to understand ourselves and our surroundings: scientific understanding *articulates* the world from within rather than representing it from an imagined external standpoint.

Some accounts of scientific understanding that aspire to a naturalistic conception also lapse into a fideistic conception that marks its origins within a Christian theological tradition. Why should a commitment to scientific understanding be expressed as "belief in" a systematically expressed body of scientific doctrine in whole or part? A scientific understanding of the world commits one to working with the conceptual resources provided by scientific practices but may only require that one work within a partially shared conceptual space. In many contemporary collisions between scientific understanding and theological commitments, naturalists already concede too much to their theologically minded opponents in accepting questions about what to believe as appropriate expressions of scientific understanding. Paralleling Nietzsche's acerbic response to the residually Christian moral commitments of nineteenth-century atheistic freethinkers such as George Eliot, we might conclude that many philosophical naturalists today "are rid of the Christian God and now . . . all the more firmly . . . cling to Christian [epistemology]" (Nietzsche 1954, 515).

These suggestions for how we might revise the scientific image to conjoin it with a conception of ourselves as rational concept users are still

25. Philosophers of language often do appeal to causal interaction with the world as the basis for understanding linguistic reference. Yet it is difficult to articulate both reference and conceptual content via the same causal entanglements so as to differentiate what we are talking about from how we take it to be. The conceptual content of theoretical understanding is often thereby taken to be entirely intralinguistic, thereby implicitly expressing an "immaculate" conception of the world from the outside. See especially chapter 7.

only promissory notes. The remainder of the book pursues this strategy for how to conceive scientific understanding as itself part of nature as scientifically understood in two parts. The first part develops an account of intentionality and conceptual normativity as scientifically comprehensible phenomena. Chapter 2 sets the stage by reviewing some of the principal dividing lines in philosophical accounts of intentionality and conceptual normativity and providing initial arguments for the strategy undertaken in what follows. This philosophical strategy has two characteristic features: First, it begins with practical and perceptual interaction with the world and asks how that interaction can become conceptually articulated, rather than beginning with representational or inferential content and then asking how that content is fulfilled or disconfirmed in perception and action. Second, it understands such conceptually articulated intentional directedness as a normative status within discursive practice rather than an operative process in cognition. Chapters 3 and 4 then develop a single extended line of argument for understanding language, thought, and other conceptually articulated performances as forms of behavioral niche construction that have coevolved with human ways of life. Conceptually articulated understanding is part of our natural history as a lineage, marked by a characteristically two-dimensional responsiveness to a changing developmental, ecological, and hence selective environment. This two-dimensionality differentiates the features of our environment to which we are responding with some performance from how we thereby "take" those features to be.

The first part concludes with chapter 5, which shows how to understand the normativity of our conceptual capacities in these terms. The chapter begins by distinguishing two approaches to the objective accountability of discursive practices. In contrast to traditional efforts to establish the epistemic objectivity of articulated judgments, Davidson, Brandom, McDowell, Haugeland, and others rightly give priority to the objectivity of conceptual content and reasoning. They nevertheless mistakenly attempt to understand conceptual objectivity as accountability to objects understood as external to discursive practice. A more expansive conception of discursive practice, as organismic interaction within our discursively articulated environment, shows how conceptual normativity involves a temporally extended accountability to what is at issue and at stake in that ongoing interaction. "Issues" and "stakes" are anaphoric concepts that enable reference to people's mutual but partially incompatible directedness toward the future development of their ongoing practices and way of life. Most organisms act to maintain and reproduce their lineage through ongoing responsiveness to life-relevant features of what thereby becomes their biological environment. Conceptually articulated ways of life are two-dimensional in the deeper sense that they are oriented not only toward continually maintaining their biological lineage but also toward determining what that way of life is and will be. This sense of two-dimensionality is "deeper" in that it enables those organisms to differentiate how they take their environment to be from how it is.

The second part of the book shows how to situate scientific practice within this account of conceptual understanding, yielding a corresponding reconception of the scientific image, which embeds scientific understanding within scientific practice. First and foremost, I argue in chapter 6, what the sciences provide is not a single, integrated position "within" the Sellarsian space of reasons. The sciences instead continually reconfigure the space of reasons itself, changing how aspects of the world are intelligible to us and which aspects stand out as scientifically and culturally significant. They do so not merely by changing how we think and talk about the world theoretically, as I argue in chapter 7. Experimental practice makes important contributions to conceptual articulation in the sciences by creating phenomena that allow new aspects of the world to be intelligible. New ways of thinking and talking would make no sense apart from the experimental systems that mediate the applicability of scientific concepts and models. The sciences allow the world to show itself intelligibly in new ways in significant part by making new things happen. It is in this sense that scientific understanding articulates the world itself, rearranging it in ways that allow new conceptual possibilities to emerge. That chapter also begins to develop my reasons for retaining a sense of the comprehensiveness of scientific understanding despite the apparent "disunity" of scientific-understanding-in-practice.

Chapter 8 works out the modal character of scientific understanding. Instead of beginning with a philosophically determined conception of laws of nature, and then asking which sciences discover laws, the chapter follows Marc Lange and John Haugeland in asking what roles laws play in scientific practice and identifying as laws whatever plays those roles in a given science. Lange and Haugeland each make indispensable contributions to understanding the conjoined alethic-modal and normative dimensions of scientific understanding. Taken together, this conception of scientific laws shows why scientific concepts are developed within partially autonomous disciplinary domains. The holistic patterns of counterfactual stability that mark the "necessity" of laws within a scientific domain also establish domain-constitutive norms of scientific reasoning. The chapter then concludes by showing why we should think of such laws neither as linguistic or mathematical representations nor as invariant patterns in the world apart from us but instead as more inclusive worldly patterns that also incorporate scientific practices of pattern recognition. We can thereby understand in a more detailed way why scientific understanding is a form of niche construction that changes the world and ourselves in ways that enable its novel forms of intelligibility.

Chapters 9 and 10 together conclude the main argument of the book by showing how scientific understanding exemplifies the temporality of conceptual normativity discussed in chapter 5. Chapter 9 shows how the sciences initially open new law-governed conceptual domains, which can nevertheless be already authoritative over scientific and other discursive practices, by developing "fictional" experimental or other practical contexts that come to exemplify conceptual norms. Chapter 10 shows how scientific significance expresses a future-directed accountability to what is at issue and at stake in scientific practices and in the larger patterns of cultural niche construction to which they belong. Scientific significance accrues to both the "homonomic" conceptual development internal to a law-governed scientific domain and its "heteronomic" conceptual relations to other practices and concerns that indicate what is at stake in understanding that domain.

Taken together, these two parts of the book's argument provide what I believe to be a coherent and more philosophically and empirically promising framework for a naturalistic self-understanding. An important part of the rationale for the book's approach to recognizing conceptual understanding as a natural phenomenon and accounting for its normative authority is that this approach yields a more adequate account of conceptual understanding in scientific practice. Part of the rationale for this conception of scientific practice and understanding in turn is that it enables a more satisfactory conception of ourselves as part of nature as scientifically understood. I take this way of mutually calibrating our best scientific understanding of our own conceptual capacities with our best understanding of the practices and achievements of the sciences to be integral to the very idea of a philosophical naturalism. The concluding chapter of the book presents its constructive vision of naturalism and of the place of conceptual capacities and scientific understanding within a naturalistic self-conception. This summation is followed by a brief epilogue reflecting upon one especially important way in which this naturalistic account of ourselves and our scientific achievements and projects makes a difference to our self-understanding. A naturalistic account of language and science as forms of niche construction highlights the particularity, contingency, and vulnerability of conceptual understanding as part of the natural history of our species. It does so in a way that not merely accounts for the normative authority of scientific understanding, however, but recognizes the sciences' centrality to our way of life and our current political, cultural, and environmental situation and prospects.

PART ONE

Conceptual Understanding as Discursive Niche Construction

TWO

What Is Conceptual Understanding?

Questions about the nature of conceptual understanding and its adequacy are among the oldest and most central issues in philosophy. Their entanglement with questions about the role of reasoning and rational norms in human life makes them pervasive in both philosophy and broader conversations about the human condition. In philosophy, at least, these issues are also remarkably divisive. In just the past two decades, for example, a sequence of prestigious Locke Lecturers at Oxford (McDowell 1994; Fodor 1998; Jackson 1998; and Brandom 2008) have presented and defended very different accounts of concepts or conceptual understanding. The disconnection among their views is so substantial that other readers might wonder whether we philosophers have any idea (or at least any one idea) of what we're talking about when we talk about concepts.

My own attempt to disentangle some confusions and misunderstandings that pervade philosophical treatments of the conceptual domain begins with some recent exchanges between John McDowell and Hubert Dreyfus concerning the scope of conceptual understanding (Dreyfus 2005, 2007a, 2007b; McDowell 2007a, 2007b; Schear 2013). Despite approaching one another's work with seriousness, respect, and goodwill, McDowell and Dreyfus often talk past one another unproductively. Each starts from different presuppositions about conceptual understanding and what it means for concepts to play a role in some domain of human life. The disconnection in their conversation exhibits one of several fault lines among philosophical accounts of conceptual understanding. The exchanges among McDowell and Dreyfus are especially revealing because each gets something importantly right that must be accommodated in any more adequate account of the conceptual domain. An important concern of this book is to do justice to both McDowell's and Dreyfus's insights in the context of a broadly naturalistic understanding of scientific practices and human capacities.

I—McDowell and Dreyfus on the Scope of Conceptual Understanding

John McDowell aims to understand how rational norms and a capacity for reflective criticism engage perceptual openness to the world so as to render knowledge and action accountable to perception. His Locke Lectures (McDowell 1994) and subsequent work (McDowell 2009) argued that prevailing philosophical treatments of conceptual understanding render unintelligible how experience bears upon conceptually articulated judgments. These positions consequently fail to comprehend how judgments have conceptual content. McDowell endorses the Kantian insistence that spontaneous discursive thought and judgment are idle and empty unless "externally" constrained by sensory experience. He then argues that traditional empiricist accounts of perception as providing "nonconceptual content" block any rational bearing of such content upon thought and action, even in their sophisticated versions. McDowell thus reiterated Sellars's (1997) rejection of any epistemic role for a nonconceptual Given. Yet McDowell also argued that Donald Davidson's (1984, 2001) alternative account of conceptual understanding as entirely self-contained ("only a belief can justify a belief") leaves conceptual thought bereft of rational responsiveness to perception. Davidson understood perception as a merely causal prompting of discursive judgment in thought and talk. If Davidson were right, McDowell picturesquely proclaimed, conceptual thought could only be a "frictionless spinning in a void" (1994, 66). Moreover, McDowell takes a third prominent alternative strand in contemporary philosophy, "baldly naturalistic" efforts to account for conceptual understanding as a scientifically comprehensible natural phenomenon, to be self-defeating. Bald naturalists presuppose the achievements of the natural sciences as rationally justified, yet he thinks they are forced to describe conceptual understanding in ways that allow no role for such rational accountability.

McDowell's own response to these philosophical failings is therapeutic rather than constructive. What blocks a more adequate recognition of how conceptual understanding engages perceptual receptivity is not the lack of a good philosophical or scientific theory but the blinders imposed by mistaken philosophical assumptions. Rejecting these assumptions dissolves the problem rather than solving it. Against Evans (1982) and other advocates of nonconceptual content, McDowell argues that human perceptual experience is already conceptually articulated. Against rationalists such as Davidson, McDowell can accept the holism of conceptual normativity that they advocate while concluding that its applicability is not limited to spontaneous judgment and action. Judgments are accountable to a perceptual receptivity whose deliverances are already conceptually articulated. Finally, against "bald naturalists," McDowell concludes that nothing in his account contravenes legitimate respect for our scientific self-understanding as animals whose capacities were formed by biological evolution. We are not philosophically confined, as "bald naturalists" mistakenly believe, within a conception of "first nature" expressed by the inexorable laws of physics, chemistry, and biological evolution. We are entitled to recognize the acculturated "second nature" of our habituation into practices of discursive performance and responsiveness to our surroundings as being fully compatible with our scientific conception of first nature. First nature can be comprehensive (nothing supernatural violates its laws) without being exhaustive (not all justifiable descriptions of events or patterns in the world can be expressed in its terms). Our second-nature habituation enables us to recognize and respond to higher-level patterns in the world that would be gerrymandered if they could be regimented in scientific terms at all.[1] The morally significant patterns that express virtues, or the conceptually articulable performances that express reasoning and judgment, mark genuine resemblances among physical happenings in the world that are nevertheless not physical resemblances. Once these mistaken assumptions (that perception provides nonconceptual content or is "external" to rational norms, or that the domain of natural law is exhaustive) have been set aside, no philosophical problems remain for us to solve concerning the rationality of knowledge and action.

1. For an illuminating discussion of McDowell on the relations between characterizations of the world in terms of natural laws and the gerrymandering of higher-order classifications if they are expressible at all in terms of laws at lower levels, see Lange (2000b).

Dreyfus fundamentally objects to what he regards as McDowell's blithe presumption that rational reflection and conceptual articulation are pervasive in human life. The dominant strain in philosophy since Plato has taken human beings to be rational creatures for whom nothing we do is immune to explicit articulation, reflection, and critical assessment. We often fall short of rational ideals, but at our best, we would hold ourselves up to critical, rational scrutiny in every aspect of our lives. Dreyfus (1979) long ago saw the apotheosis of this tradition in the aspiration to simulate, model, or even supplant human judgment and experience with digital computers. If reasoning is formally explicable rule following, and the highest human capacities are manifest in our rationality, then our intelligence could be theoretically modeled and even practically instantiated by computer programs. Eventually, the growing computational power of digital machines would thereby dramatically enhance both our self-understanding and our practical capabilities. Dreyfus's philosophical career was forged by and grounded in his vocal and prophetic assessment of the failure of this empirical research program in artificial intelligence (AI) as an inevitable outcome of faulty philosophical commitments.

Dreyfus argues that a prereflective, nonconceptual bodily involvement in the world is the enabling condition for our more limited capacities for conceptual understanding and rational reflection. Rationalist philosophical projects foundered precisely when venturing into everyday perceptual involvement in and practical responsiveness to our surroundings. Moreover, these everyday skills at pattern recognition and flexible situational responsiveness can be cultivated and enhanced to produce the exceptional levels of skilled performance exemplified by grandmasters in chess or elite athletes. Chess provides an especially striking case as an apparently promising venue for formal simulation. As a formally specifiable, rule-bound game, chess play seems tailor-made for computer simulation of a distinctively human excellence. Massive brute computational power eventually prevailed, but its very manner of success showed the irrelevance of classical AI to understanding human intelligence. Human chess players circumvent rather than solve the need for massive algorithmic computation by responsiveness to high-level patterns on the board that they can learn to recognize without having, or even being able, to analyze them into explicitly articulable components.

Skillful responsiveness to situations must be learned, but the learning process supposedly illustrates how genuinely skillful engagement with the world is nonconceptual. When first exploring an unfamiliar domain or attempting a new skill, humans are reflective, and where explicit rules are available, they are attentive to those rules to guide their performance. The result, however, is halting, bumbling performance, just the opposite of the skillful competence often ascribed to rational control. Dreyfus argues that the achievement of genuine expertise in most human domains circumvents explicitly articulated rules or norms rather than internalizing them. Experts attune themselves to meaningful patterns in the world, which show up as a practical solicitation experienced and taken up at the level of the body, not as articulated representations. As Dreyfus long ago expressed his claim, "In acquiring a skill . . . [there] comes a moment when we finally can perform automatically. At this point we do not seem to be simply dropping these [previously deployed] rules into unconsciousness; rather we seem to have picked up the muscular gestalt which gives our behavior a new flexibility and smoothness. The same holds for acquiring the skill of perception" (1979, 248–49). For Dreyfus, recognizing our bodily responsiveness to circumstances in perception and action shows why any attempt to insinuate conceptually articulated representations amid everyday practical-perceptual skills or the extraordinary performances of experts would be doubly mistaken. Conceptual understanding is superfluous wherever we have become skillfully responsive to circumstances, since we can respond flexibly and appropriately without any intervening conceptualized representations. Reflective conceptual expression is also antagonistic to skilled engagement with the world; in stopping to think, we would dissolve the smooth-flowing, skilled bodily attunement to what is taking place.

Such a prereflective, unarticulated flow of skillful bodily responsiveness to the solicitations of one's situation is not limited to the extraordinary capacities of expert performers, however. Or rather, in negotiating our way perceptually and practically around everyday circumstances, all normal human beings are experts. Dreyfus follows Merleau-Ponty (1962) in emphasizing that perception is not merely receptive but instead requires skillful performance that is appropriately responsive to what is perceived. Feeling a texture requires appropriate movement of the hand over a surface, where its appropriateness is guided by skillful responsiveness to the surface itself. We tend to think of vision as mostly passive, but here, too, one must learn to direct one's vision to focus, scan, and track things. We only see what we have learned to explore and track visually. Perception is a skilled bodily activity, and like all forms of skillful coping with circumstances, Dreyfus argues, it is a prereflective bodily responsiveness to the world without any intervening conceptualization or reflective assessment. He does not deny that we can step back reflectively from our ordinary perceptual immersion in our immediate surroundings and a task at hand or guide exploration by explicitly articulated conceptualizations, as in looking for something that fits a description. He only insists that reflective and conceptually explicit observation involves a break from, but also a dependence upon, ordinary skillful practical-perceptual immersion in the world.

Dreyfus claims that the nonconceptual character of ordinary practical-perceptual coping with circumstances is directly accessible phenomenologically. Careful description of one's own ordinary perceptual experience shows the absence of conceptually articulated thoughts or reasoning or of reflective distance between how circumstances solicit bodily activity and our ordinary smooth, flexible responsiveness to such solicitation. To claim, as McDowell does, that conceptual understanding and the capacity for rational reflection are pervasive in perceptual experience would interpolate traditional rationalist philosophical prejudices into our experience. These prejudices are a philosophical construction imposed upon experience, an all-too-familiar "myth of the pervasiveness of the mental" that arises from too much reading of Plato, Descartes, and Kant and insufficient attention to how we engage the world perceptually.

McDowell responds to Dreyfus in turn that if skillful practical-perceptual coping with our surroundings really were impervious to conceptual normativity, it would be utterly incomprehensible how the experiential flow of skillful responsiveness could ever bear evidentially upon judgments or how skills could be accountable to assessment. Any reflective break from inarticulate immersion in experience would then break with it completely, miring us in a realm of discursive spontaneity with no purchase upon experience and thus without conceptual content. In his concern to avoid what McDowell agrees is a philosophical tendency to overintellectualize and overrationalize bodily engagement with the world, Dreyfus allegedly resurrects Sellars's Myth of the Given, a long-familiar but unacceptable form of philosophical storytelling about experience. McDowell also insists that his claim that conceptual capacities are operative in perception is consistent with Dreyfus's phenomenological descriptions of everyday perceptual coping and exceptional expert bodily skills.

How might we resolve these competing accusations of a prejudicial mythologizing of experience? Can we appeal to phenomenology, transcendental arguments, conceptual analysis, a hermeneutics of everyday perception, scientific research in neuroscience, or some other source of philosophical authority to determine whether or how perceptual experience is already conceptually articulated? Fortunately, such tendentious philosophical appeals are beside the point. Dreyfus and McDowell each implicitly invoke fundamentally different accounts of conceptual understanding, so that each argues in ways that often seem utterly irrelevant to his interlocutor. By getting clearer about the dividing lines among prevalent approaches to concepts and conceptual understanding, we can understand what is at issue in the opposing accounts and arguments advanced by Dreyfus, McDowell, and others undertaking philosophical analysis of conceptual understanding and rational normativity.

Dreyfus's and McDowell's surface disagreement about the conceptual or nonconceptual character of ordinary perceptual experience marks a deeper disagreement between opposing ways of thinking about conceptual understanding as an operative process (Dreyfus) or a normative status (McDowell). Operative-process accounts take conceptual content to be actually present or operative in specific performances by concept users.[2] Jerry Fodor exemplifies this approach. He began his book on *Concepts* by saying: "The scientific goal in psychology is to understand what mental representations *are*. . . . Nothing about this has changed much, really, since Descartes" (1998, vii). To use a concept is to have something in mind or causally implicated in what one does; in Fodor's specific version, concept use involves token mental states that possess representational content.

Normative-status approaches to conceptual articulation, by contrast, identify the conceptual domain with those performances and capacities that are appropriately assessed according to rational norms. The issue is whether various performances are accountable to reasoned assessment and can stand up to it sufficiently. Whether comportments are accountable in this way is then itself a normative issue: the question is whether assessment according to conceptual norms is appropriate. Whether relevant kinds of representations or structures are present or causally implicated in a thought or action then does not matter, but only whether that thought or action is accessible and potentially responsive to conceptual assessment.

2. In most cases, operative-process accounts identify *causal* processes operative in producing and deploying conceptual understanding. Husserl (1982, 1970a) is nevertheless a notable example of someone who seeks to explicate intentionality and conceptual understanding by describing a process (acts of temporal, "noetic" synthesis that constitute ideal, "noematic" senses) that constitutes a meaningful experiential directedness toward the world that cannot be explicated causally. Despite significant differences in their accounts of experience, Dreyfus also follows Husserl in situating his phenomenological descriptions of skillful comportment outside of the causal realm.

Dreyfus's examples make clear that he takes philosophical accounts of conceptual understanding to characterize an operative process. Two examples are prominent in his original response to McDowell (Dreyfus 2005): blitz chess played at a speed that allows only two seconds per move and a brief period in the career of professional baseball player Chuck Knoblauch when he made frequent errors on simple throws even though he did well with more difficult plays when he had no time to think. These two examples respectively indicate to Dreyfus the absence of conceptual understanding from skillful coping with one's surroundings and the potentially deleterious effect on bodily skill when explicit reflection and conceptual articulation are brought into play. The examples highlight a supposed "mindlessness" to expert understanding, from which Dreyfus infers its nonconceptual character. With parallels to nonhuman animals who also are "experts" at negotiating their environment, expert players of blitz chess or baseball do not have conceptually articulated thoughts in mind but instead respond directly to the affordances or solicitations of a situation on the board or field. Expert chess players or second basemen need not, and perhaps cannot, have concepts explicitly or implicitly "in mind" and cannot take up a stance of reflective detachment while performing well.

John Haugeland's (1998) opposing use of chess and baseball examples highlights that an operative-process approach is not mandatory here. For Haugeland, chess at any speed involves conceptual normativity. No nonhuman animal can play chess, because no animal grasps the relevant concepts; they cannot recognize pieces and moves, the legality of those moves, or their strategic significance. Moreover, players' perceptual and practical skills at recognizing positions and making moves must be responsive and accountable to those concepts and norms. Otherwise they would not be playing chess. Haugeland would undoubtedly say the same of baseball, which he jokingly characterized as the "all-star" example of intentionality (1998, ch. 7). Knoblauch's grasp of a base, an out, and winning a game are on display in his fielding, even when "mindlessly" successful. The relevance of concepts is normative rather than operative. Nothing turns on whether a concept is in mind or in brain but only on whether one's performances are, or can be, held accountable to the relevant standards in the right way. Not surprisingly, Haugeland agreed that perception is permeated by conceptual understanding. For Haugeland, as for McDowell, if perception is not conceptual, it is not genuinely perception of objects in the sense in which object-perception normally plays a role in human cognition and action.

Haugeland's or McDowell's and Dreyfus's concerns are thus orthogonal. Haugeland's (1998) arguments against the possibility of a biologically based understanding of human intentionality make this mismatch especially clear, for his line of argument also provides a decisive consideration against treating expert chess play, and other forms of skilled perceptual-practical responsiveness, as nonconceptual. Haugeland argued that biological functioning can only differentiate the patterns in the world to which it normally responds, even if those patterns are gerrymandered from the perspective of conceptually articulated understanding. For example, a bird whose evolved perceptual responses are to avoid eating most yellow butterflies, except for one oddly mottled pattern of yellow, would not thereby be mistaken about the *color* of the mottled yellow ones. *We* identify the bird's responses as almost in accord with a conceptual category we endorse ("yellow"), but the bird's behavior itself provides no basis for concluding that it was striving but failing to accord with that classification. Moreover, even if the bird's response patterns were de facto coextensive with conceptually significant features of the world, as in always and only avoiding eating yellow butterflies, those patterns would not then display an intentional directedness toward the butterflies' color, for that coincidence would merely be a de facto contingency. For Haugeland, intentionality or conceptual understanding[3] must introduce a possible gap between what some comportment is directed toward and the manner or content of that directedness such that a mismatch between the two accounts for the possibility of error. The birds' pattern of behavior is only a complex pattern of response to actual circumstances. The single pattern of what the birds do in varying circumstances cannot then generate a dual pattern that could differentiate what they are responding *to* from how they *take* it to be.[4] Individual birds can malfunction with respect to species-normal patterns of discrimination and response, but there is no further basis for concluding that the overall response pattern within the population

3. Haugeland also makes a similar claim about intentionality, which he contrasts to the "ersatz intentionality" that can properly be attributed to nonhuman animals or machines running sophisticated computer programs. He thus identifies conceptual understanding with intentionality more generally. I take up the relation between intentionality and conceptual articulation in the next section.

4. Cummins (1996) makes a similar argument concerning representations, arguing that a representation can be erroneous only if the target of the representation and its content are determined independently. Haugeland's version of the argument does not depend upon understanding intentionality or conceptual articulation as representational.

aims for but falls short of something different than what its members actually, typically do.

Now consider a grandmaster playing blitz chess. The grandmaster's ability to recognize and respond rapidly to complex patterns on the chess board is the outcome of an extended "selective" regime (artificial selection of perceptual patterns via reflective study of past games rather than natural selection operating on a population). If grandmasters' play were simply a felt responsiveness to complex perceptual configurations experienced as tensions and solicitations, as Dreyfus insists, then nothing they did would amount to errors in play. Grandmasters playing blitz chess do make errors, of course. Dreyfus's account of skilled coping as a nonconceptual intentional directedness cannot recognize them as errors, however, but at most as responses that are abnormal for grandmasters. They could only be errors if the regulative and strategic norms of chess play already constitutively governed the pattern-recognition capacities involved. This point would be especially telling for any board patterns that frequently elicit mistakes even from expert players in blitz chess. Just as "there is nothing that the [bird's] response can 'mean' other than whatever *actually* elicits it in normal birds in normal conditions" (Haugeland 1998, 310), if Dreyfus were right that expert chess play were nonconceptual, then there is nothing that a "normal" grandmaster's blitz chess play would be "trying" to do apart from what grandmasters normally do in various actual board configurations. Any board patterns that trouble blitzing grandmasters could only be recognized—by conceptually reflective systems that actually understand and deploy chess concepts and standards—as design limitations in their trained cognitive orientation rather than errors in play (the counterfactual cases of obscure positions that might not be encountered in the ordinary run of play would be relevant here as well). Dreyfus takes for granted that grandmasters are playing *chess* at a rapid pace, but he is not entitled to that claim unless their play is informed by and accountable to the conceptually articulated norms of the game. Haugeland's arguments suggest that if Dreyfus were right about expert chess play, then the people we normally identify as expert chess players would not be playing chess but only an oddly gerrymandered simulacrum of the game.

Dreyfus's, Haugeland's, or McDowell's points about these examples are in fact compatible, however, because they rely upon different ways of thinking about conceptual understanding. For the reasons I just indicated, Dreyfus should agree with Haugeland and McDowell that grandmasters' play in blitz chess involves an understanding of the concepts of rooks, moves along ranks and files, and winning. He must likewise acknowledge that they take their play to be accountable to that understanding even if they normally need not explicitly think about such matters, having already brought them to bear through a nonreflective bodily capacity. Haugeland and McDowell could and do then agree with Dreyfus that such conceptually articulated abilities can be and often are executed without explicitly attending to or reflecting upon a concept or its application. Wayne Martin uses the example of blitz chess precisely to dissociate explicit or reflective application of concepts from conceptual normativity and the judgments that express it: "In [playing speed chess] I make judgments—I reach a conclusion that is in some sense responsive to evidence—even though I don't undertake any conscious deliberation and I experience my judgment as issuing more-or-less instantaneously" (2006, 2). Moreover, Haugeland does and McDowell can endorse a further component of Dreyfus's concern—namely, that many of the patterns actually recognizable by grandmasters and other skilled perceivers may have no higher-order expression than that constituted by the ability to recognize them, so that skillful recognition is irreplaceable by any rule-governed system.[5]

What matters for a normative-status account of conceptual understanding and judgment, such as those advocated by McDowell, Haugeland, or Martin, is not whether concepts are explicitly represented or employed in the course of actual performances. The issue is whether those performances are accountable and responsive to the relevant conceptual norms.[6] Conceptual understanding involves the possibility of reflection, with subsequent revision or repair of the associated practical-perceptual skills, but it need not be identified with any present component of the exercise of those skills as an operative process. In that context, Dreyfus's examples serve a different inferential role than he

5. Such cases nevertheless only count as "recognition" and as "skill" because of their conformity to the rules of chess and their conduciveness to successful play. They are conceptually responsive even though there are no extant concepts that express them generically. Cases of recognition skills that do not correlate with already-articulated linguistic terms or phrases are in this respect like colors that we can discriminate but have not named, which McDowell has often discussed (see McDowell 1994, lecture 3). For McDowell, the conceptual domain extends beyond the explicit classificatory concepts already at our disposal, which is why anaphoric, demonstrative, and indexical expressions are integral to the linguistic expression of conceptual understanding.

6. The point is not that whatever cognitive, bodily, or interactive processes are going on in conceptually accountable performances are of no importance but that they do not demarcate the conceptual domain; the very same kinds of processes may be involved in performances lacking any conceptual character. Whatever processes actually produce conceptually contentful comportments recede even further to the extent of the counterfactual stability of their responsiveness to norms. If one neural or bodily process for implementing such comportments were blocked, others might be recruited to fulfill its role.

proposed. They would not exemplify a domain of skillful practice in which concepts and rational norms are not yet operative. They would contribute instead to the phenomenology of conceptual understanding, as counterexamples to any claim that explicit representation or conscious deliberation is essential to conceptual understanding. They help rule out some *accounts* of conceptual understanding rather than limiting its scope as Dreyfus had proposed.[7]

Having recognized the extent of McDowell's and Dreyfus's agreement about their respective central concerns, the question still remains how to demarcate the domain of conceptual understanding. Why prefer either a normative-status or an operative-process account of conceptual understanding over the other? Disagreements about how to use the term 'conceptual' are not merely verbal, despite the extent of the implicit agreement that I have proposed. At stake here are which phenomena belong together in philosophically significant classifications and what tasks philosophers should undertake in thinking about conceptual understanding and intentionality. Moreover, the operative-process/normative-status divide is not the only fault line in recent philosophical demarcations of the topic of concepts and conceptual understanding. In the next section, I complement this distinction with other telling fault lines among philosophical approaches to conceptual understanding. My initial reflections upon these differences will help situate my own project for how to characterize scientific conceptual understanding within an appropriately naturalistic understanding of our capacities and achievements. In doing so, I will eventually return to the McDowell/Dreyfus discussion. I do not follow Dreyfus in thinking of practical-perceptual coping as a distinct, preconceptual "level" of intentional directedness. Yet I also do not simply endorse McDowell's therapeutic acceptance of conceptual normativity as pervasive even in perception. Despite insisting upon a normative account of the conceptual domain, I will draw upon considerations from Dreyfus's work to place conceptual normativity within a scientific understanding of nature.

7. Dreyfus (2013) challenges McDowell to show how the pervasiveness of conceptual norms in perception is actually experienced by perceivers. On the line I am suggesting, McDowell's response should be that Dreyfus himself has already described that experience on his behalf. In many cases, including Dreyfus's favorite examples, we experience our responsiveness and accountability to conceptual norms as a kind of "mindless coping" in which we are not thematically aware of concepts or engaged in reflective assessment. Nevertheless, we also understand that our performances are accountable to norms that could be applied reflectively and how to bring such norms to bear, even when we do not actually do so and have no concepts explicitly or implicitly in mind.

II—Mapping Philosophical Approaches to Conceptual Understanding

Philosophical treatments of concepts and conceptual understanding are complicated by their entanglement with more general discussions of intentionality. Intentionality has been perhaps the central topic of philosophical work on language, mind, and action over the past century. Attempts to characterize intentionality as a topic are themselves controversial, since alternative philosophical accounts of intentionality often bring with them competing descriptions of the phenomena to be understood.[8] Yet the central cases within the domain and some of their characteristic features can be readily identified. Propositionally articulable mental states, linguistic expressions and utterances, and meaningful behavior or action are the prototypical cases of intentional phenomena, although theorists often differ about which cases are primary and which, if any, are derivative. I will use 'intentional comportments' as a putatively neutral term for whatever states, performances, systems, capacities, or signs should properly be characterized as intentional. Intentionality then has several central features. First, intentional comportments are not self-contained but are directed toward or "about" an intended object. Second, this directedness is guided, mediated, or governed by an "aspect," a description, or some other partial mode of presentation or representation.[9] Third, this directedness is also intensional, such that in many contexts, one cannot straightforwardly replace the mode of presentation/representation with another mode of directedness toward the same object.[10] My belief that the author of *Origin of Species* is buried in

8. The ability to talk and reason about the same topic, even in the absence of any shared conception or description of that topic, is a pervasive feature of discursive practice that is often overlooked in philosophical work. Even semantic externalists (e.g., Putnam 1975; Kripke 1980) who allow for causal determination of reference usually only extend that capacity to causally efficacious objects. My account of how discursive performances are anaphorically interconnected in this way is developed in chapters 4–5.

9. The difference between operative-process and normative-status accounts is often most clearly manifested here. Does one understand the "aspectual" character of intentional directedness to be a representation or other de facto process or structure that mediates intentional relations to an object? Or does one understand it as a normative orientation "governing" such relations?

10. These contexts, in which substitution of one mode of presentation for another can fail to preserve some important feature of a comportment's directedness toward what was presented in the first "mode," are often labeled "intensional contexts." Central examples include "propositional attitudes" (such as "belief that . . . ," "desire that . . . ," etc.) and modalities ("possible that . . . ," "necessary that . . . ," "obligatory that . . . ," etc.). Philosophical strategies differ in whether one first identifies which contexts are intensional and then uses that determination to help clarify which

Westminster Abbey, for example, is different from a belief that the naturalist aboard the circumnavigational voyage of HMS *Beagle* is buried at Westminster Abbey, even though both beliefs correctly indicate and characterize Charles Darwin. Moreover, intentional comportments can be directed toward their "object" even if no such object exists or does not (or not uniquely) satisfy its characteristic mode of presentation. In this respect, intentional comportments and their modes of directedness are also open to assessment, since the intentional relation is in some sense "deficient" if its object does not exist or is mistakenly presented or represented.[11] Assessment applies holistically to whole groups of intentional comportments, such that they are intentional in significant part through systematic, normative relations to other intentional comportments. Relevant groupings of intentional comportments are appropriately assessed in broadly rational terms that involve, at a minimum, consistency and coherence but also instrumental efficacy. Intentional entities can sometimes fail to satisfy these norms, but any collection of putatively intentional comportments that always, or even mostly, failed to satisfy rational norms would thereby fail to be intentional (at least in that systematic grouping). As Davidson (1984) and others have long argued, errors and lapses in rationality make sense only against an extensive background of success.

One important fault line among philosophical accounts of intentionality and/or conceptuality then concerns the degree and character of continuity or discontinuity between human capacities or performances and those of nonhuman animals or other putative candidates for intentional directedness. Everyone recognizes that there are some important differences between human capacities and those of other intentional or conceptual systems. Daniel Dennett (1987) stands on one side of a continuum in claiming that intentional ascription can apply in much the same way to a range of systems from thermostats to human agents. Dennett still recognizes important pragmatic differences, since for thermostats, other explanatory stances more usefully supplant the attribution of rationally accountable beliefs and desires, whereas for human agents,

comportments are intentional or appeals to an independent determination of intentional relatedness to explicate the characteristic "intensionality" of the modes of presentation of intentional comportments.

11. This sense of deficiency is subtle, varied in its import, and context sensitive, since many intentional comportments (hopes, plans, suppositions, imaginings, tryings, etc.) are directed toward nonexistent or counterfactual situations as such, and quite appropriately so. Many of the most obvious apparent counterexamples dissolve once the holistic character of intentional comportments and their assessment is taken into account. Thanks to an anonymous referee for reminding me of the need for such qualification of the normativity of intentional directedness.

intentional explication is often indispensable. More strikingly, language dramatically expands the range and depth of possible intentional attributions to humans: "The capacity to *express* desires in language opens the floodgates of desire attribution. 'I want a two-egg mushroom omelette, some French bread and butter, and a half bottle of lightly chilled white Burgundy.' How could one begin to attribute a desire for anything so specific in the absence of such verbal declaration? How, indeed, could a creature come to *contract* such a specific desire without the aid of language?" (Dennett 1987, 20). For Dennett, we are nevertheless making the same sort of attribution in ascribing such a desire to a speaker (along with a belief that uttering that sentence to the waiter will help satisfy that desire) as in ascribing to a thermostat a desire to keep the room at 68° F (along with a belief that closing the circuit to the furnace will restore that temperature). In each case, we predict behavior from the intentional stance.

At the opposing end of this continuum, John Haugeland criticizes Dennett and others who understand intentionality as a similar phenomenon in human and other animals, despite differences in its explication and articulation. Haugeland (1998, ch. 11) acknowledges that real patterns in the world show up from Dennett's intentional stance but insists that these patterns are not what is tellingly manifest in human thought and action. Apparent continuity between human understanding and nonhuman animal behavior is achieved only by misdescribing the distinctive normativity and reflexivity of the human activities that allow entire domains of phenomena to be genuinely intelligible. For Haugeland, what seem to be parallels between the perceptual and behavioral capacities of humans and other animals show only that some animals display a kind of simulacrum of genuinely intentional comportment:

> As far as we know, the intentionality of animals is entirely ersatz. That is, we can understand animals as having intentional states, but only relative to standards that *we* establish for them. This makes animal intentionality exactly analogous to biological teleology. We say that the "purpose" of the heart is to pump blood, that it's "supposed to" work in a certain way, that functional descriptions are "normative," and so on. . . . But finally, of course, the heart does not have any *purposes* in the way that a person does, nor does it accede to any *norms* on its own responsibility. (Haugeland 1998, 303)

Ersatz intentionality is a genuine phenomenon that Haugeland recognizes as more than just "as-if intentionality" but utterly different from the fully human forms of intentional normativity that it only superficially resembles.

Tracking differences between continuist and discontinuist conceptions of intentionality or conceptual understanding is complicated by the shared recognition of both continuities and differences between human capacities and performances and those of nonhuman animals. Moreover, many endorse both evolutionary continuity and the gradual emergence of novel human capacities whose differences from their precursors amount to differences in kind. Further complications arise from divergences in the use of key terms. Some theorists identify "intentionality" as what we share with other animals while reserving "conceptual understanding" for a distinctively human capacity. Others use "concept" more liberally, suggesting for example that a clear, reliable, and appropriate behavioral distinction between what one does or does not try eating when hungry suffices as an implicit grasp of the concept of food. On the latter view, human beings can express concepts verbally and reason about them in ways that other animals cannot, but other animals also grasp some concepts. Still others, such as Haugeland or Sellars, place both (genuine) intentionality and a grasp of concepts firmly on our side of a significant divide while accounting for what nonhuman animals do in other terms.

Part of the difficulty in assessing disagreements over the extent of continuity between human and nonhuman organisms as intentionally directed is that at least two relevant distinctions are in play, although often conflated. The first distinction concerns the flexibility or inflexibility of an organism's responses to features of its environment. Some behaviors are quite reliably and rigidly responsive to environmental cues—moths fly toward light, bacteria move in the direction of a positive sugar gradient. Simply cued responses can still produce relatively complex but rigid patterns of behavior if a series of such responses are sequentially linked, exemplified by Wooldridge's (1963) classic description of the egg-laying behavior of the sphex wasp.[12] By contrast, many organisms can change their behavioral patterns in flexible, instrumentally rational responses to novel or conflicting patterns of multiple cues and can make further adjustments shaped by the outcomes of their own earlier efforts. Call this difference between rigid and flexible responsiveness to environmental cues (which may be a difference in degree rather than kind)

12. Sterelny (2003, 14) introduces the term "detection agent" for organisms whose behavioral repertoire is dominated by such directly cued responses to features of their environment. See chapter 3.

the "sphex/flex" distinction; to describe it initially in terms of causality, rationality, or intentionality may beg key questions.

A second distinction differentiates "taking-up" from "taking-as." Organisms "*take-up*" features or components of their surroundings by responding to their presence or absence. An organism's biological environment encompasses whatever it takes-up from its surroundings, including developmental as well as behavioral responses. "Taking-up" environmental features includes flexible and multivalent patterns of responsiveness along with rigidly cued detection and response. Some organisms also respond to some features of their environment in ways that support a distinction between merely *taking* them *up*, and taking them *as* relevant "under an aspect," "as meant," or "under a description," such that they can *mistake* them. Exactly what is involved in "taking-as" is contested. At a minimum, the organism's own behavioral repertoire must somehow differentiate correct from incorrect "taking-as." *Correcting* a mistaken prior response involves more than merely *changing* its response to some recurrent environmental feature, even if the change is beneficial. One can argue that the two distinctions coincide—if so, any flexible, nonsphexish response to some environmental feature would be an aspectual taking-as—but that coincidence should not be assumed.

Discussions about the scope of intentionality or conceptual understanding sometimes go astray due to lack of clarity about which differences are at issue. Neither the absence of standardized terminology nor acknowledgement of differences in degree should block recognition of significant dividing lines here. For some philosophers, the crucial target of philosophical explication is a distinctively human capacity; for others, human capacities are just elaborations or extensions of more basic capacities we share with nonhuman animals. The remainder of this section nevertheless sets aside the difference between continuous and discontinuous approaches to understanding intentional/conceptual phenomena. I will return to this issue later in this chapter, and in chapters 3 and 4, when I ascribe a continuous basis for what has subsequently evolved into a significant discontinuity between conceptually articulated intentionality and the flexibly rational responsive capacities displayed by many nonhuman animals.[13] Unlike most ascriptions of

13. One could reserve the term "conceptual" for what is (so far as we know) distinctively human while using "intentional" to characterize behavior on the more flexibly rational side of the "sphex/flex" distinction. My reasons for identifying "intentionality" with conceptually articulated behavior will become clear later, but in the remainder of this section, I use "intentionality" in a more undifferentiated way.

a distinctively human capacity, however, mine does not differentiate "levels" of ability on a common scale. Many nonhuman animals have capacities for discrimination and flexible responsiveness that we cannot match. More important, nonhuman animals live very different kinds of lives, and theirs and our capacities can only be assessed with respect to relevant goals.[14] Conceptually articulated intentionality is not something nonhuman animals *lack*; it would be irrelevant or even deleterious to their ways of life. For now, however, the consideration to keep in mind is a temporary need for some terminological flexibility in whether we are talking about intentionality, rationality, conceptual understanding, or some different or more finely graded phenomena. Questions of continuity between human and nonhuman animals have often guided the explication of these terms, but the differences in approach that I will now consider cut across supposed parallels or differences in how different kinds of organisms engage the world perceptually, practically, and cognitively. With these distinctions settled, we can then return to the continuist/discontinuist divide more constructively.

My initial discussion of Dreyfus and McDowell highlighted the difference between approaches that treat intentional or conceptual phenomena as operative-processes or as normative statuses (the remainder of the section will use the term 'intentionality' to refer to this entire domain, without regard to whether some subset of intentional comportments can be demarcated as conceptual). An operative-process approach to intentionality seeks to discern features of intentional comportments that are operative in producing their directedness toward and normative accountability to their objects. Salient examples of operative-process approaches include Fodor (1998), for whom intentionality results from representational structures that play a functional role in cognition; Husserl (1982, 1970a), for whom the structured correlations between noetic act and noematic sense constitute the meaningful directedness of consciousness; Searle (1982), for whom intentionality is a complex biological property of organisms; Millikan (1984), for whom representations acquire evolved proper functions; Carnap (1967) or the Marburg neo-Kantians on the logical structure of a language; Jackson (1998), for whom conceptually

14. Mark Okrent (2007) provides an especially clear and thorough explanation of why we should think of biological teleology in terms of goal-directedness rather than functional roles. In chapter 3, I discuss why 'goals' is the right term for expressing the teleological orientation and associated normativity of various animals' capacities, including ours. The grounds for both the continuities and discontinuities that I will then attribute will concern the character of the goal-directedness in question.

articulated intentionality is established by a partition of possibilities that is operative in mental life and manifest in intuitions; or Dretske (1981), for whom intentionality is constituted by the primary information-bearing features of cognitive states.

A normative-status approach to intentionality, by contrast, identifies its domain with those performances and capacities that can be held normatively accountable in the right way. There must be some way in which intentional performances or states can be held accountable to relevant standards, and they are intentional in virtue of whether they would mostly stand up to such accounting. Thus, for example, Dennett identifies intentional systems as those that are interpretable as mostly rational in context, while Davidson claims that they must be systematically interpretable as mostly speakers and believers of truth by the interpreter's own standards. Unlike operative-process theories, normative-status accounts can allow that the defining feature of intentional performances (e.g., normative accountability upon reflection or interpretation) need not be operative in all or even most cases. Thus to return to an example from the McDowell/Dreyfus debates, on a normative-status approach to intentionality, chess grandmasters playing blitz chess need not have a concept "in mind" when they respond to the board position with a rapid move. It suffices that they *could* and *would* hold what they are doing accountable to the regulative, constitutive, and strategic norms of chess play. These performances are intentionally directed toward rooks and knight forks rather than to plastic figurines on dark and light squares because the players understand these concepts and norms, and their play is responsive to and mostly accords with them. Some characteristic normative approaches to intentionality include Brandom (1994, 2000, 2008) on the game of giving and asking for reasons, Davidson (1984, 2005b) on radical interpretation, Dennett (1987) on the intentional stance, McDowell (1994, 2009) on conceptual understanding, Heidegger ([1927] 1962) on care and the existentiell possibility of *Eigentlichkeit*, or Haugeland (1998) on existential commitment to domain-constitutive standards.

A second dividing line among approaches to intentionality is most easily drawn in terms of Husserl's (1970b, investigation 6) distinction between empty and fulfilling intentional relations. An empty intending can be directed toward its object in its absence, including the modes of absence marked by the nonexistence of the object or by failure to satisfy its intentional manifestation under some aspect. By contrast, a fulfilling intending presents the object itself as directly given under some intentional aspect.[15] Husserl's distinction highlights opposing directions taken by two different approaches to understanding intentionality. The more traditional philosophical approach has been to start by understanding empty intending and then to ask what it is for an empty intention to be fulfilled. These approaches typically identify intentional directedness with some form of representation or other intralinguistic or mental pattern. The problem of how to understand nonreferring intentional states and erroneous presentations seemed to dictate beginning with what it is to have intentional content (when the intended object may be absent, misrepresented, or nonexistent), and only then to ask how some intentional comportments present their objects "directly." Such approaches have notoriously confronted problems of skepticism, among other difficulties.

An alternative approach begins with a system's actual relations to entities and then asks what it would be for those relations to be intentional (and thus meaningful, as conceptually or otherwise aspectually articulated).[16] The primary challenge is then to show how a pattern of engagement or interaction with one's surroundings opens a space of *articulated* engagement accountable to norms. The most common motivation for such an approach has been "baldly" naturalistic, in McDowell's phrase. Intentionality is taken to be a feature of some entities, states, or performances that are causally or otherwise physically interactive with their surroundings. One then asks what it is for such causal interaction to involve intentional directedness under an aspect, such that how the system comports itself toward its surroundings could be appropriately understood as an error, either by intending something *other* than what it actually interacts with or by intending it under an aspect that it might not actually possess or display. Dretske's (1981) appeals to information-bearing states or Millikan's (1984) teleosemantic functional norms are familiar examples of such an approach. Naturalists such as Dretske or Millikan are not alone in taking this strategic direction, however. Heidegger ([1927] 1962) also begins with intentional fulfillment (an understanding of being exhibited in an ability-to-be) without construing

15. The sense of fulfillment carried by fulfilling intentional performances need not be infallible. Some intentionally directed states, performances, or entities can turn out not to be fulfilling presentations, for example, even though they present objects in ways that are indistinguishable in the first person from a perceptual or other intuitive givenness of the object itself.

16. Sellars also highlighted the importance of this issue in *Empiricism and the Philosophy of Mind*: "The real test of a theory of language lies not in its account of what has been called (by H. H. Price) 'thinking in absence,' but in its account of 'thinking in presence,'—that is to say, its account of those occasions on which the fundamental connection of language with nonlinguistic fact is exhibited" (1997, 65).

Table 2.1. Philosophical Approaches to Understanding Intentionality—a Matrix

Accounts of intentionality	1: Primacy of **empty intending / intralinguistic holism**	2: Primacy of **fulfillment**: causality, perception, being-in-the-world, etc.
A: Operative-process account	Husserl: essential structures of consciousness Carnap: logical structure of language Jackson: a priori partitions Searle: intentionality as biological Minsky et al.: GOFAI	Dretske: information-bearing states Millikan: teleosemantics Fodor: cognitive representations Dreyfus: practical/perceptual coping Dennett: what "satisfies" the intentional stance
B: Normative-status account	Sellars: we-intentions? Quine: radical translation Dennett: the intentional stance? Davidson: radical interpretation Rorty: conversation of mankind Brandom: game of giving/asking reasons	Heidegger: Dasein's disclosedness McDowell: perception and action as rational second nature Haugeland: existential commitment Dennett: the intentional stance? Sellars: we-intentions?

fulfillment in causal or other narrowly naturalistic ways. Both Dreyfus's practical-perceptual coping as a preconceptual mode of intentionality and McDowell's "direct realist" account of perception as rational second nature also start with a fulfilling intentional comportment, despite each rejecting any "baldly naturalistic" construal of that engagement with the world.

These two distinctions together form a 2 × 2 array that locates various approaches to understanding intentionality within that space (see Table 2.1). Four distinct philosophical strategies stand out in these terms:[17]

A1: operative-process accounts of the constitutive structure of some domain of possible intentional comportment (e.g., the logical structure of a language, the constitutive presuppositions of a "worldview," or the essential structure of transcendental consciousness)

17. Dennett's intentional stance is ambiguously placed on the chart. If intentionality is identified with the gerrymandered properties of a system that allow it to be sensibly interpreted from the intentional stance, then it describes a form of what Haugeland calls "ersatz intentionality" and belongs in A2. If instead, intentionality incorporates and ultimately depends upon an interpreter's *ascription* or the ascribability of the pattern of rationality-in-context to those systems by an interpreter, then we need to know more. Depending upon how those explanatory ascriptions are themselves normatively accountable, Dennett's theory may belong in B1 or B2, but it would then take our biological understanding of nonhuman animal behavior rather than that behavior itself as exemplary of "original" intentionality. Sellars's account of the social normativity of intentional content is similarly ambiguous, depending upon how one interprets its relation to the anticipated fusion of the manifest and scientific images.

A2: operative-process accounts of the causal, functional, or practical patterns of a system's interaction with its surroundings, which suffice to open a possible gap between what the system interacts with and how the system's performances "take" it be

B1: normative-status accounts of how the performances of a system or group of systems as a whole mostly conform to a systematically construed ideal of rationality in context, such that the goals with respect to which it would be rational are appropriately taken as authoritative for it

B2: normative-status accounts of how a system's actual engagement with its surroundings is articulated in a way that renders it accountable to something beyond its own actual performances or those of its larger community of intentional systems

One revealing feature of this classification is that it highlights the difference between two kinds of philosophical disagreement about intentionality. Some of the most focused but narrow philosophical disagreements about intentionality differ concerning what plays a more or less agreed upon philosophical role. We can recognize such "intramural" disagreements between Husserl and Carnap (A1) over whether logical syntax or essential structures of consciousness constitute meaning; between Dretske and Millikan (A2) over whether evolutionary history is directly relevant to intentional content; among Quine, Dennett, Davidson, and Brandom (B1) over how best to characterize rational interpretability in context; between Dreyfus and Fodor (A2) over whether practical-perceptual engagement with the world should be understood in terms of "coping skills" or causal-functional representations; or between Minsky and Searle (A1) on whether syntactic structure or conscious awareness constitutes intentional content. On the other hand, in some critical philosophical encounters, what is primarily at issue are more far-reaching differences between operative-process and normative-status accounts, or between taking empty or fulfilling intentions as the point of philosophical departure for understanding intentionality. Here we find the considerations that lead naturalists like Dretske, Millikan, or Fodor (A2) to reject Carnap's formalism or Jackson's two-dimensionalist possible worlds semantics (A1); Quine's or Davidson's (B1) criticisms of traditional theories of meaning and the "idea idea" (A1); or McDowell's and Haugeland's (B2) criticisms of Davidson's or Brandom's (B1) coherentism. Heidegger's (B2) criticism of Husserl (A1) in the first part of his 1925 Marburg lectures (Heidegger 1985) takes up both issues at once: in arguing for the primacy of categorial over eidetic intuition, Heidegger takes Husserl to task for starting with empty intending, while his criticism of Husserl's supposed naïveté about the being of consciousness points toward his own normative conception of Dasein's relation to its own ability-to-be (as concern for its own being rather than self-awareness).

The most important reason to classify strategic approaches in this way is to highlight important differences in philosophical approach that are often taken for granted and consequently misunderstood, exemplified by the exchanges between Dreyfus and McDowell. This classification nevertheless also encourages explicit reflection upon which strategy to pursue in seeking to understand intentionality and why. In the remainder of the chapter, I indicate what I take to be compelling reasons for undertaking a B2 strategy: a normative account of the intentional domain that begins with a system's actual involvement in the world. I also call attention to the most important challenges confronting a naturalistic account of B2 intentionality. The ensuing chapters consider how those challenges can be met.

III—Why Intentionality Should Be Understood as a B2 Phenomenon

An important achievement of John Haugeland's (1998) *Having Thought* is a series of arguments that show why some of the most prominent philosophical strategies for understanding intentionality are broken-backed. Although Haugeland does not explicitly identify these strategies under my classifications, his arguments track those distinctions. Haugeland not only develops a general line of argument against the A2 and B1strategies, but his arguments show that their failures are reciprocal. Broadly speaking, his point might seem to be that the A2 positions (characteristically represented by Dretske, Millikan, or Fodor) can account for how intentional comportments are engaged with the world at the cost of being unable to show how that engagement is meaningfully articulated by genuinely normative intentional/conceptual content. Similarly, it may seem as if the B1 positions (characteristically, Brandom and Davidson, but also those who would seek normative bedrock in the accepted comportments of a community) can account for a richly articulated space of meaning and normative authority at the cost of losing its accountability to and thus directedness toward the world. Yet this formulation overlooks the interdependence of world-directedness and conceptual-articulation. Because the socially constituted articulations appealed to in B1 cannot be held accountable to anything beyond their own responsive dispositions, they do not actually achieve an articulated *conceptual* space either. Reciprocally, because the causal or biological systems from which Dretske or Millikan start cannot articulate conceptual differences in the causal chains or lineages of descent in which they are implicated, they also fail to identify the *intentional* relation as causal or etiologically functional. Thus Haugeland's arguments show that truth and meaning, or objectivity and aspectuality, are constitutively intertwined in the form of "norms of objective correctness" (1998, 317).

Haugeland's arguments and their targets can be organized in a sequence from physical to social accounts of intentional relations, in which differences between philosophical approaches (A2 and B1) fall out as differences in the normativity of the systemic interactions involved. The sequence starts with Dretske's effort to understand the intentional relation between a perceptual system and its perceived object in physical, information-theoretic terms. The problem Dretske sets is how to discern the intended object of a perceptual state from its long and involved causal chain that culminates in, for example, hearing someone at the door.[18] Dretske's account turns on considerations of perceptual constancy. The object of a perceptual state is whatever it carries information about in a "primary" way (rather than *via* some more proximal way or *of* some more distal object). Perceptual intermediaries such as a vibrating eardrum or oscillations in the air are not perceived as objects because different intermediaries could have produced a qualitatively indistinguishable experience, and hence the experience does not inform us which intermediary was involved. More distal causes, such as the button being depressed or the person pushing the button, are also not primary information bearers because their involvement is indicated only via the more proximal stimulus of the bell. Without hearing the bell ringing, we would obtain no auditory information about visitors.

18. In the text, Haugeland follows Dretske in the latter's assumption that the proper object of this perceptual experience is the ringing of the doorbell. Dretske's account, if it succeeded in its own terms, would indeed identify the proper object of auditory perception in that way. Haugeland's own constructive arguments, which proceed from the constitutive standards that govern the perceptual situation, ought to indicate (rightly, I think) that what we hear is someone at the door (via the ringing of the bell) rather than hearing the bell (the role of the bell is more comparable to that of the resulting vibrations in the air). Even when we speak of "hearing the doorbell," we don't literally mean the bell (which is usually located physically away from the door, in order to communicate the information about someone at the door to someone elsewhere in the building). If we were to say to someone else under normal circumstances, "I hear the doorbell, could you go take care of it?," we would look askance if they used a towel to muffle the bell in the hallway rather than answering the door. Moreover, if our perception were in question ("Was that the doorbell?") we would go check for a person in the doorway rather than seek evidence that the bell had recently vibrated.

Haugeland argues that Dretske is able to pick out the bell as the supposedly proper object of perception only because he is already committed to its unity and perceptual significance as an object on other grounds. A perceived object remains constant across variance in the visual or auditory stimuli that can be experienced as presentations of *it*. These presentations differ from one another in many respects, and Dretske's aim is to identify objects as the underlying locus of constancy. Haugeland points out that Dretske's account of primary information bearing must therefore rule out *any* relevant similarity among sensory presentations of one and the same object, and that he cannot do:

> If there were *any* single kind of stimulus that mediated all and only the constant perceivings (same kind of perception of the same kind of object), then the perception would carry information that the stimulus was of that kind [rather than about the object]. . . . What's worse, it seems that there *must* be such kinds, if sensory perception is to be possible at all. For if one *can* reliably recognize the squareness of the table from varying perspectives, then there must be *something*—something higher order, global, relative to context, or whatever—normally common to all and only the stimuli from such objects, on pain of rendering perception magical. (1998, 245–46)

Haugeland could go on to point out, but does not, that such higher-order stimulus kinds are not unique in blocking direct information conveyance from perceived object to perceptual state. A similar argument could be posed for *any* of the relevant causal intermediaries: a common object as cause of recognized auditory similarities must equally well present higher-order similarities among the vibratory patterns in the air that cause the higher-order stimulus pattern or the electrical excitations of the nerves that convey it. The grounds for picking out one stage of the sequence of causal intermediaries and precedents of perception as the common "object" of the perceptual effect cannot be causal or information theoretic. Causal or physical interactions as such cannot meaningfully articulate the world.

Biological evolution does introduce normative considerations into organisms' interaction with their surroundings in contrast to the merely physical or causal conveyance of information. An organism is a functional complex, and functional and evolutionary biology work together to explain the typical presence of certain components or operations in organisms of that lineage: "The normative force [of biological functioning] is part of and integral to a larger account of how individual organisms of that kind work as a whole on the whole. . . . The understanding is holistic and statistical: the norms governing the component functions are intelligible together in terms of their interdependent roles in enabling the whole system to succeed—that system-level success being understood in turn in reproductive and evolutionary terms" (Haugeland 1998, 308). The question here is whether biological-functional normativity *suffices* to both pick out the object of biological comportment and articulate its intentional aspect or mode of presentation. Haugeland argues that it does not suffice because biological systems cannot distinguish between proper functioning and objective correctness or truth. The conclusion to his argument, however, is that biological function cannot pick out a determinate "content" or "way" in which an organism's functional performances *take* things to be. Organisms' functional involvement with their surroundings takes up or targets some features of their physical surroundings and not others as part of a holistic pattern of organismic functioning in its environment. What they cannot do is to take them up in a way that also takes them *as* having some intended sense or aspect, such that the intended aspect could be mistaken about what was targeted in that way.[19]

To see why this is so, consider again Haugeland's central thought experiment for this argument, which I discussed briefly earlier in this chapter. Haugeland asks us to imagine a species of bird whose normal functioning connects its behavioral response to its perceptual input by refraining from eating most yellow butterflies while eating most other butterflies when it can. These perceptual and behavioral mechanisms evolved in response to an environment in which most yellow butterflies are poisonous and most others are not. Haugeland's claim is that biological functions can genuinely account for the failures of those birds whose mechanisms do not make the normal discriminations (they are abnormally deficient in this respect, rather than merely variant, to the extent that the variance is functionally relevant). These functional norms cannot, however, articulate the birds' normal behavior aspectually. Specifically, under no circumstances can they show normal function to have been *mistaken* in either direction. A (normal-functional) avoidance of rare, nonpoisonous yellow butterflies does not involve

19. As I noted earlier, Cummins makes a similar argument about the conditions for attributing representational errors, and he shares Haugeland's sense of a pervasive failure to fulfill these conditions among contemporary theories of representation: "It is precisely the independence of targets from contents that makes error possible. If the content of a representation determined its target, or if targets determined contents, there could be no mismatch between target and content, hence no error. Error lives in the gap between target and content, a gap that exists only if targets and contents can vary independently. It is precisely the failure to allow for these two factors that has made misrepresentation the Achilles heel of current theories of representation" (1996, 7).

mistaking them for poisonous ones. What the birds' behavior differentiates does not quite map onto a conceptual distinction between what is and what is not poisonous. Consequently, in the more common case of poisonous yellow butterflies, the birds' evolved mechanisms are also not intentionally directed toward avoiding them *as* poisonous butterflies. Significant extensional overlap between the two differentiations, one conceptual and one not, lets us explain why this response pattern contributes to the birds' relative fitness, but it does not justify identifying the two. In the other direction, if normal functioning does not discriminate certain odd combinations of other colors from yellow, or certain odd patterns of yellow from nonyellow, that behavior likewise cannot be mistaken about the butterflies' color. *We* understand color and poison as possible features of entities rather than just cues or solicitations to respond to circumstances in one way rather than another. Such normal-functional responses in another organism are design limitations but not errors, because "correcting" those limitations has no role in the organism's behavior, development, or its selective history.[20]

Biological functions only articulate the patterns in the world to which they actually respond when functioning normally, even if those patterns are gerrymandered from the standpoint of our conceptualization of relevant features of the world. Moreover, even if those response patterns were de facto coextensive with conceptually significant features of the world, such that the birds always and only avoided yellow butterflies, those patterns would still not display an intentional directedness toward the butterflies' color, for that result would merely be a de facto contingency. Haugeland does not spell out the underlying principle, but the point is clear enough: intentional directedness must introduce a possible gap between what is meant and what is actually encountered, such that there is a *possibility* of error, even when no errors actually occur. In cases where the birds' avoidance response is coextensive with the butterflies' color, a counterfactual query would be telling: if there *were* to be a shade or pattern of yellow that the birds' normal functioning would not lead it to avoid, would it have any means of self-correction to accord with a

20. Strictly speaking, they would only indicate design limitations if they resulted from a lack of relevant variance in detection or response mechanisms on which selection could operate or from the intrusion of nonselective "forces" such as drift. In other cases, the actual patterns exhibited could be functional and even adaptive, if a more fine-grained selective mechanism were too costly to the organism in energy requirements, discrimination time, or some other selectively relevant feature. The important moral here is that what is a selectively relevant biological trait or a selectively relevant feature of an organism's environment is holistically determined by the organism's overall pattern of behavior and selection history and not simply by local correlations between perceptual input and behavioral output.

conceptual norm?[21] If organisms (individually, collectively, or via prosthetic phenotypic extensions) cannot hold their own performances to account in some respect, then their behavior cannot be properly understood as intentionally directed in that respect.[22] Haugeland concluded, "The trouble with the insectivorous birds is that there is no definition of that to which they are *supposed* to respond except as that to which they *do* respond when everything is functionally in order. . . . The colors of the butterflies have no normative status at all apart from their involvement in that normal functioning" (1998, 314).

A constructive claim is embedded in this characterization of the limitations of any selective-functionalist approach to intentionality. Haugeland is arguing that the kind of intentionality characteristic of human understanding requires reflexive self-directed comportments that would constitute a standard to which they are accountable and the ability to self-correct according to that standard. Without such a capacity for corrective response in accord with a standard, the standard could have no determinacy. How the system takes things to be could never be pried apart from how they normally are. Moreover, openness to self-correction cannot be limited to interaction with actual entities but must also encompass a modal "space" of possibilities and impossibilities. Intentionality cannot just involve a pattern of response to actual surroundings but must somehow constitute a more comprehensive pattern in which the actual response pattern is situated. We can now recognize

21. Haugeland (1998, ch. 10) himself uses such a counterfactual thought experiment to a similar end by imagining a dog who seems able to recognize and distinguish different members of the same family but is incapable of responding appropriately to the (impossible) counterfactual situation in which the family members' individual physiognomic properties were redistributed among them. So long as an "ersatz intentional" system is exposed only to the actual conditions to which its development is already adapted, it can seem intentionally directed, but its inability to respond appropriately under extraordinary circumstances exposes the illusion. Haugeland's strategy here parallels Dreyfus's (1979) earlier objections to the alleged intentional directedness of AI programs like Roger Schank's (1975a, 1975b) restaurant scripts, which cannot handle counterfactual circumstances for which they were not already designed (e.g., Schank's restaurant scripts' inability to answer questions about whether the waiter is wearing clothes).

22. The notion of an extended phenotype (Dawkins 1982) originated in Dawkins's genic-selectionist program but has since been adapted to a more biologically adequate conception of natural selection as operating upon "developmental systems" (Oyama, Griffiths, and Gray 2001) or through "niche construction" (Odling-Smee, Laland, and Feldman 2003). Extended phenotypes incorporate such characteristic features of a biological life pattern as beaver dams, bird nests, spider webs, or cities as part of the organism's phenotype. I include the possibilities of collective and phenotypically extended means of correction within the scope of Haugeland's argument because just as we cannot assume in advance what is a biologically relevant environmental feature or organismic trait, so we cannot assume in advance what is the relevant intentional "system." Biological intentionality could be a property not of individual organisms but of communities of organisms or of organism/environment complexes.

why Haugeland characterizes nonhuman animals' involvement in the world as a kind of "ersatz" intentionality. Such involvement would be an ersatz "imitation" or simulacrum of its genuine counterpart, in the sense that a system could be "designed" (by natural selection for organisms or by programming for artificial intelligence programs) to produce *actual* patterns of comportment that are nearly coextensive with a subset of the patterns that a genuinely intentional system would accept as *possible* or *correct*. Those normal or "designed" response patterns are nevertheless not accountable to anything beyond their own functional success. They can establish a de facto correlation, but cannot constitute a space of possibilities within which, and a standard to which, the correlated behavioral patterns are accountable.

We can now sum up Haugeland's criticisms of those positions that occupy A2 on the table. These positions account for how intentional comportments are directed toward the *world* by starting from a system's actual engagement with its surroundings, whether as causally interactive or selectively adapted. They then try accounting for that system's engagement with the world as *intentional* and hence aspectual in terms of its characteristic patterns of perceptual constancy or evolved functionality. Yet these patterns of causal transmission of information or natural selection of functional performance cannot articulate a "taking-as" in some determinate respect or aspect that is independent of what those performances "take up" or target in their surroundings and what actually or normally characterizes that target. The constancy that Dretske hoped would pick out an object as a unique target of recognition must recur at every stage of causal transmission. *We* readily recognize its location at the object, and not at other stages, because the mediating patterns of stimulation or vibration are gerrymandered from our conceptual perspective. Yet the difference between gerrymandered and coherent patterns is not recognizable at the level of information flow but only via standards imposed upon that process from elsewhere. Dretske's account thus takes for granted what it seeks to explain: the difference between perceptual relations to perceived objects, and perceptual relations to the object's causal descendants or antecedents. Similarly, biological functionality (or artificial design, in the case of sophisticated AI computer systems) can mimic patterns of intentional directedness by creating patterns of normal or programmed response that under typical conditions are similar to patterns that a genuinely intentional system could recognize conceptually. We recognize the conceptually specifiable pattern (e.g., a difference in color or toxicity) as a standard and thereby treat extensional similarity as accord to the standard and any marginal deviations as errors.

"Ersatz intentionality" cannot distinguish between correctness and error because whatever pattern it actually picks out under normal conditions *is* the target it is "aiming toward." The normal functional pattern is a standard against which some behavior is abnormal. No comparable gap can open between normal function and correct identification, however, since biological normality is the only normative consideration in play.

Consider now the B1 accounts, which first characterize intentional content in a holistic way, and then use that characterization to understand how intentional norms could be fulfilled causally or perceptually. Haugeland describes a key feature of these approaches to intentionality as "interrelationist" (1998, 207–8). The B1 approaches (including Dennett's, Davidson's, Brandom's, or Rorty's, among others) each begin by specifying a holistic pattern of comportment that, if all works out, would collectively constitute meaningful, intentionally directed performances and states. In some cases (e.g., Dennett), the pattern is only recognizable as such from the outside; in others (e.g., Brandom and, I would argue, Davidson), the recognition of the relevant pattern is itself part of the pattern. Such patterns, however, whether construed in terms of the intentional stance, radical interpretation, the game of giving and asking for reasons, or the like, would only be genuinely intentional if causal or perceptual encounters with the world have a normative grip upon the patterned comportment. Advocates of this approach differ on how to account for the normative force of the world's grip upon putatively intentional comportments. Dennett relies upon predictive success or failure, for example, whereas Davidson resorts to the token identity of rational patterns and causal interactions, and Brandom appeals to the ongoing adjudication of differences in speakers' conceptual perspectives. Haugeland's arguments are intended to show that the world could never actually get a grip upon such patterns, because the only consideration these accounts can acknowledge is the internal coherence or incoherence of their constitutive holistic patterns. The entities that such patterns of comportment are supposedly directed toward can exercise no external constraint upon them at all.

Haugeland develops this line of argument against B1 views from two complementary directions. First, he argues that they explicate a "mere coherence" among the comportments they recognize as rational rather than any genuine accountability to objects that could constrain what we say and do. As one example of this objection, he endorses McDowell's criticism of Davidson for illegitimately "help[ing] himself to the idea of a body of beliefs" (McDowell 1994, 68) and he develops a parallel objection that Brandom's supposed objectivity proofs only "show that there is no legal move, in Brandom's system, from 'Everybody believes *p*' (or 'I believe *p*') to '*p*.' But they don't show anything at all about what *could* legitimate '*p*' instead; in particular, they don't begin to show how '*p*' could 'answer to how things actually are'" (Haugeland 1998, 358). Dennett's intentional stance does not even purport to constrain its own application: nothing in Dennett's account (apart from the institutional constraints imposed within normal scientific practice that a charitable reading might extract from Dennett's naturalism) provides a check on an interpreter merely claiming predictive success for her interpretations or, in the other direction, dismissing a putatively intentional system as irrational after only the most desultory failed efforts to make sense of its performances.[23] Rorty (1979, 1982, 1989, 1991) even celebrates the abandonment of any "hankering after objective truth," with the result from Haugeland's perspective that Rorty's (1979) conception of inquiry as conversation would be governed solely by curiosity and a hankering for novelty.

Haugeland's second complementary argument against the B1 strategies emphasizes the inadequacy of any constitutive appeal to social conformity as a normative constraint upon individual deviance.[24] He thinks that social conformity can successfully produce complex patterns of social *institution* (paralleling how biological evolution can produce patterns of normal function):

> It isn't only the norms as such that are socially instituted, but also the respective behaviors and circumstances that those norms "connect." Thus, what it is to greet someone, and what it is to be a circumstance in which greeting is appropriate, are nothing other than what the community members accept and deem as such— . . . they are themselves instituted along with the normative practices in which they occur,

23. The real force of Haugeland's argument against Dennett, therefore, is his argument in "Truth and Rule-Following" (1998, ch. 13) that the social-institutional constraints of scientific practice are insufficient to secure accountability to the world without existential commitment. I discuss Haugeland's objections to social-institutional strategies beginning in the next paragraph.

24. While Sellars, Brandom, and Rorty explicitly make appeal to "we-intentions," socially recognized normative statuses, and community agreement, Davidson and Dennett develop a B1 strategy that eschews any explicitly social references. Haugeland nevertheless clearly finds social conformity at work implicitly in their account. Davidsonian radical interpretation relies upon a postulated "massive agreement" in beliefs as criterial for radical interpretation of what a speaker means and closes the circle of social conformity with the reciprocal application of this criterion "at home" (and also in interpreting others' interpretation of oneself). Dennett does not explicitly apply his account reflexively in this way, but I argued above that his account of predictive assessment thereby tacitly relies upon acceptance by a scientific community.

> and by the same socializing process. This allows for considerable intricacy and interdependence . . . making it possible for current behavior and circumstances to incorporate not just "manifest" recognizabilia, but instituted statuses and roles, accrued over time. Social rank and office can be instituted in this way, as well as finer-grained actions, rights, and responsibilities, such as those contingent on whose turn it is, who owns what, which water is holy, or how the teams stand in the league. (1998, 311–12)

What social conformity or "agreement" cannot do, Haugeland argues, is to allow objects themselves to serve as authoritative standards to which community agreement is itself accountable. An entire community can establish subtle norms for whether a folk dancer has or has not "grocked" in circumstances that were or were not grockworthy. Moreover, in Haugeland's more telling example, a community can distinguish between appropriate responses to what this community authoritatively designates as the sacred jonquil and the quite different responses called for by the profane jonquils normally found in gardens and florist shops. The community can also institute connections between this difference and other socially instituted distinctions and standards (so that, for example, only the nobility are permitted to see the sacred jonquil, those who see it impermissibly are punishable, those who legitimately see it are then permitted to marry, and so forth). What that community cannot do is to confer any further, extrainstitutional significance upon this socially instituted designation. In Haugeland's telling example, the community can indeed determine that the socially appropriate response to the sacred jonquil is to utter 'scarlet' rather than 'yellow'; it cannot thereby determine that the sacred jonquil differs from its profane counterparts in color. For a socially appropriate utterance of 'scarlet' to be about an object's color, conformity to that social practice must be accountable to a standard not subject to the same social authority.

Haugeland's conclusion from each line of argument is that the supposedly normative conceptions of intentionality proposed by the advocates of B1 approaches cannot actually account for intentional directedness and accountability. These approaches can show how norms of correct performance are instituted within a community of persons, whether those are standards of correct behavior, utterance, or even "inference."[25] What they cannot do is to show how the de facto practices

25. I put 'inference' in scare quotes because Haugeland would have to conclude that inferentialist accounts of semantic content such as Brandom's or the truth-theoretic accounts of inferential relations within a speaker's idiolect that result from Davidsonian radical interpretation can only produce a simulacrum of inferential normativity. Brandom or Davidson can account for what further utterances or actions a linguistic or other community (including the fleeting discursive

of such communities could ever be accountable to anything beyond what they thereby *take* to be authoritative. Individual performances can be deviant with respect to the community's authorization, but no further standard could be authoritative over the collective performances of the community as a whole.[26] At most, a community's actual performances could be subject to retrospective revision, but the revision could not correct a prior error, for there would be no norm with respect to which prior performance could have been mistaken. In this respect, this inability parallels the inability of biological functionality to articulate the *dual* normativity of conceptual understanding. In each case, the performances or states of individual organisms or agents can be abnormal with respect to what is typical for their biological taxon, or deviant with respect to what is appropriate within their community, but they are not thereby accountable to anything beyond biological function or social institution. Neither process can distinguish how some features of its surroundings *are*, as distinct from what is the normal functional or socially appropriate response to them. As a result, they respond to their surroundings merely as cues for a biologically functional or socially appropriate response and not as objects understood *as* being in one way or another. To be intentional requires of a system (or group of systems) that its comportments be directed beyond itself toward an intended "object," in the sense that the success or failure, correctness or incorrectness, appropriateness or inappropriateness of the system's performances can be held accountable to the *object*. At their best, A2 or B1 approaches can show how a putatively intentional system could be accountable to norms established by the system's ongoing biological maintenance and reproduction or its institutional applications of precedent but not to the objects of any supposed intentional directedness.

Haugeland does not argue specifically against the A1 strategies, but similar lines of argument would be telling and indeed have often served as primary motivation for the more widespread contemporary pursuit of A2 or B1 strategies. Advocates of A1 strategies face a dilemma. If the logical, transcendental, or presuppositional structures they posit as

"community" composed of a Davidsonian radical interpreter and the individual speaker of an idiolect being interpreted) normally *takes* a speaker to be committed or entitled to given her prior performances in context, but they cannot account for why those are norms that constitute semantic relations between propositional contents rather than just social norms of appropriate behavior.

26. Haugeland talks about the "general telling" of the community as what cannot be mistaken on such an account to allow for the possibility that "on isolated occasions, all or most members of the community (by an amazing coincidence) happen to mis-perform in the same way at the same time" (1998, 315). Such an occasional collective error could still be assessed by reference to the community's more general telling of differences.

constitutive of meaning are simply psychological or sociological generalizations about what intentional systems actually do, then they do not constitute a space of meaning and intentional directedness. Frege's or Husserl's well-known criticisms of psychologism long ago showed why such conceptions cannot account for intentional or conceptual content. Yet if these structured relations instead diverge from what the putatively intentional system actually does, one would need to identify how these idealized structures were nevertheless authoritative over the system's actual performances so that divergence from the supposed ideal amounts to an error. Claims that these structures are logically, transcendentally, or contextually "necessary" ring hollow in the face of widespread nonconformity to these supposedly necessary relations. This line of argument has been variously developed by the later Wittgenstein, Heidegger, Neurath, and Quine, among many others, and it rightly directed most subsequent philosophical work away from the predominance of A1 approaches during the early twentieth century.[27]

IV—Meeting Haugeland's Challenge to Naturalism

I regard these lines of argument from Haugeland and others as decisive against the viability of A1, A2, or B1 strategies for a philosophical explication of intentionality or conceptual understanding. Much more detailed working out of the arguments would be needed to adapt these broad argumentative approaches to respond to specific, sophisticated philosophical developments of each strategy. Since my primary concern is constructive rather than critical, I leave the dialectical development of these lines of argument to others or other occasions. A more telling concern from my perspective is that the import of Haugeland's critical arguments may seem to go too far. Haugeland's arguments against the A2 and B1 strategies make it initially difficult to see how biological

27. The one prominent recent exception to this widespread rejection of appeals to logical, transcendental, or presuppositional necessity as the basis for understanding intentional normativity has been the various attempts to utilize the resources of modal logic rather than first-order logic to account for how and why some intentionality-constitutive structure is necessary. In some cases, these technical resources have been harnessed to conceptions of nomological rather than logical necessity, which yields an A2 rather than an A1 approach (Dretske, Millikan, and Fodor all incorporate modal considerations within their accounts of intentional content). In other cases, Jackson (1998) and Chalmers (1996) have drawn upon the additional resources of two-dimensional-possible-worlds semantics to supplement the inadequacy of a one-dimensional necessity for this task. My reasons for regarding this kind of A1 approach as also broken-backed have been developed in my previous book (Rouse 2002, especially chapters 1 and 9).

functioning or social interaction could play *any* constitutive role in understanding intentionality. Any viable account of intentionality and conceptual understanding should be consistent with understanding ourselves as evolved, functioning organisms who participate in complex, iterated social practices. Haugeland's arguments may seem to rule out biological functioning or social relations as making any contribution to understanding intentionality and its normative accountability to the world, and Haugeland himself endorsed that interpretation of their significance. Moreover, although Haugeland defended a strongly discontinuous account of intentionality, as a distinctively human phenomenon, he rejected any appeal to language as a decisive factor in constituting intentional normativity and conceptual content. It is difficult to understand how one might explicate intentionality in ways consistent with a broadly naturalistic philosophical orientation without acknowledging some constitutive role for human biology, social practices, or linguistic capacities.

Three further steps are needed to endorse Haugeland's critical arguments against A1, A2, and B1 strategies, as I do, and yet still plausibly proceed to understand intentional and conceptual normativity in broadly naturalistic terms grounded in human biology and social life. The first two steps involve rethinking some widely accepted philosophical conceptions of biology and social life, respectively. An important reason philosophers have failed to explicate intentionality in biological or social terms is the inadequacy of familiar conceptions of the biological and social domains. Haugeland's arguments do indeed show that these familiar conceptions of evolutionary and functional biology and of social institutions cannot provide the basis for a more adequate account of intentional and conceptual normativity. I argue in chapters 3 and 4 that familiar ways of thinking about biology, and about social life, are the root of the difficulty. The naturalistic project of understanding our conceptual capacities and norms as explicable within nature and history can and should be separated from its usual linkage to these familiar conceptions. The third step, already prepared for in the first two, is to recognize the artificiality of separating the biological from the social dimensions of human life and understanding. Even with appropriate revisions, neither a biological nor a social conception of intentionality by itself will do, if these are regarded as alternative or even opposing approaches. We need to grasp human social life as integral to human biology if we are to see how it opens a "space of reasons" of intelligible possibilities. A central part of such an account, I argue, must be a better understanding of the biological evolution and development of language and conceptual understanding more generally. Language and other conceptually articulative practices enable human organisms to relate to one another and to our surroundings in ways that are neither possible nor useful for nonlinguistic organisms.

The possibilities for such a naturalistic reconstruction of intentionality as a biological and social phenomenon emerge from a more careful reconsideration of the import of Haugeland's critical arguments. What Haugeland's arguments against A2 strategies show is that intentional normativity cannot be explicated by evolved norms of human biological functioning.[28] While a functional-teleological conception of living organisms and their behavior has been widely influential in biology, psychology, and philosophy, it is not the only available alternative. Indeed, I argue, it is not the best available alternative. Yet even a better understanding of human biology, if narrowly construed, is still not sufficient for an adequate naturalistic explication of intentionality and conceptual understanding. We also need to rethink our understanding of human beings as social animals.

Haugeland's dismissal of attempts to understand intentionality as a social phenomenon similarly relies upon a familiar but problematic conception of the social domain. Social science and social philosophy have been dominated by two opposing conceptions of human beings as participants in social life. The more prevalent and influential approach within American social science has been an individualist conception of social life grounded in neoclassical economics and decision theory. On this conception, human agents are best understood as rational marginal-utility maximizers whose interactions can be modeled mathematically using the tools of decision theory, game theory, and microeconomics. Any divergences from rationality can be accounted for by appeal to explicable forms of nonrational behavior. Even if these individualist models were adequate for other aspects of social life, however, the general difficulties confronting the A1 strategies for explicating intentionality make them unpromising approaches to intentionality and conceptual content. Indeed, rational individualist theories of social life normally presuppose some other conception of intentionality and conceptual content to establish the domain of possibilities available for rational choice. The primary recognized alternative to rational individualism within social theory has been some form of communitarian conception of the social world. These conceptions identify social norms with patterns of social

28. Nor would it help appealing to functional characteristics of some other taxonomic clade or analogical group if conceptual understanding is also a capacity of some nonhuman animals.

conformity or agreement as background for the intelligibility and assessment of individual performances and commitments. Communitarian conceptions of social life were in fact the target of Haugeland's arguments against a social explication of intentionality.

Haugeland's arguments against understanding intentionality in social terms only apply to communitarian conceptions of social normativity, however, just as we saw that his criticisms only challenge functionalist conceptions of human biology. Here too, he has telling arguments against that way of construing the "social" character of intentional normativity. Communitarian social theory is likewise neither the only nor the best way of thinking about how human social interaction makes conceptual understanding possible. Chapters 4 and 5 show how communitarian conceptions of social norms ignore or idealize away some of the features of social life that enable accounting for conceptual normativity. When integrated with a better understanding of the relevant aspects of human biology, a more adequate conception of the normative accountability constituted by social practices enables a B2 account of intentionality and conceptual understanding that is both naturalistic and untouched by Haugeland's objections to A2 and B1 approaches.

This approach to intentionality and conceptual understanding will be developed in greater detail below. Several important themes will nevertheless locate the project at the outset among some more familiar alternatives. First, this approach fits clearly within B2 on my classification of philosophical approaches to intentionality. It is a normative approach that considers how intentional comportments are accountable within a "space of reasons" rather than characterizing natural-selective, cognitive, behavioral, institutional, or other processes that supposedly operate in producing intentional comportments. The approach begins with an intentional system's involvement in its surroundings and shows how that involvement becomes articulated conceptually, rather than beginning with empty (representational or holistically intralinguistic) intendings and then explicating the difference between fulfilled and empty intentional directedness.[29] The space of reasons is not an idealized or theoretical construction, however, but is the practical configuration of the world we live in as a discursively articulated environmental niche. Our way of life as human beings interacts responsively with some components

29. The 'surroundings' of the system cannot be specified physically or by spatial proximity because the relevant environment of an intentional system is instead defined in relation to its own patterns of activity. What matters is interaction between the system and its surroundings, and not everything in proximity to it is involved in relevant interaction or involved in the same way. See chapters 3–4.

or aspects of our material surroundings and not others; salient among these are our patterns of utterance and inscription and their incorporation within our other expressive and responsive activities.

This approach distinctively addresses the continuity or discontinuity between the forms of conceptual understanding central to human life and the perceptual, cognitive, and practical capacities of nonhuman animals. Its overall orientation emphasizes continuity, since intentionality and conceptual understanding are capacities that are part of our lives as evolved organisms. These biological capacities take distinctive form within human social life, however, as social practices make possible a form of conceptual normativity with no evident place in the lives of nonhuman animals. That outcome reflects a long history of coevolution between human organisms and the discursive practices that have become increasingly central to a human way of life.

On this overall conception, our intentional/conceptual involvement in the world is biological as organisms whose behavior and physiology aim to maintain and reproduce the dynamic pattern of our simultaneous belonging to and differentiation from our environment. This boundary between organism (as a reproducible, self-maintaining pattern) and its environment is ambiguous in several ways. As a "way of living," we are individual organisms who also participate in a larger pattern that constitutes that way of living as human.[30] Other human organisms belong both on "our" side of the boundary between our (shared) self-maintaining way of living and the environment we share and as part of the environment in and against which we sustain our individual existence. The same is true of the various "companion species" (Haraway 2008) whose life patterns are significantly intertwined with ours. The specific "biosocial" character of human life that results is crucial for grasping the distinctive character of conceptual understanding. It opens a space of possibilities for self-understanding in which the maintenance of our way of life is at issue for us within that way of life rather than being fixed biologically.

Language is also both a pervasive, salient aspect of the environment in which human beings develop and an expressive capacity incorporated into our bodily repertoires. Normal human neural and cognitive development only occurs within a linguistic environment surrounded

30. A similar point has played a prominent role in Haugeland's (1982, 2013) own exposition of Heidegger's *Being and Time* ([1927] 1962) in the form of Haugeland's insistence that Dasein is a singular entity, which is nevertheless also articulated into individual "cases." Whether or not that point is correct as an interpretation of Heidegger (I think it is correct if suitably qualified), I regard it as indispensable to an adequate understanding of intentionality.

by spoken and written words; the characteristic capacities and way of life that develop in that context in turn serve to sustain and reproduce that linguistic environment for subsequent generations. The developmental uptake of spoken and written language as integral to the human environment is thus a preeminent example of the evolutionary phenomenon of niche construction (Odling-Smee, Laland, and Feldman 2003). That language is integral to human biological development and evolution is now a widely recognized (although still controversial) claim in evolutionary biology. We are linguistic/discursive beings and not merely animals with an evolved capacity for language.[31] What languages are, and how they are situated within the broader domain of intentional/ conceptual comportment, has nevertheless often been misunderstood. In this book, I emphasize language as a public phenomenon. It encompasses patterns of expressive utterance responsive to circumstances but also includes the normal human developmental response to those patterns that maintains and reproduces them in the next generation and the evolutionary selection pressures for easier, faster language acquisition and use. The neurological and anatomical evolution that occurred in the human lineage under selection pressure for language acquisition is an important part of any full story about intentionality and conceptual understanding, especially for a naturalist, although I keep this point mostly in the background.

Language as a public phenomenon is also an exemplary case of a socially interactive practice. Social practices only exist through human beings continuing to reproduce them in mutual responsiveness to one

31. The claim in question is not the widely accepted view that a general capacity for language evolved at some point in the differentiation of *Homo sapiens* as a species and was then a consequential factor in our species' survival and demographic/geographic expansion. It is instead the stronger and more controversial claim that languages, human neural and anatomical development patterns, and our distinctive patterns of neotenous development all coevolved. The functional/ anatomical patterns of human bodies are shaped by their development and evolution amid discursive practice, while languages take the shape they do by selective reproduction suitable for human bodies and patterns of living. On such a conception, language and human cognitive functioning are still continuing their coevolutionary dance. For discussion, see Dor and Jablonka (2000), Jablonka and Lamb (2005, especially ch. 6, 8), Deacon (1997), Bickerton (2009, 2014), and chapters 3–4 of this book. Bickerton (2014) plausibly argues for an evolved neurological reorganization that allows for rapid, subconscious assembly and recognition of longer strings of linguistic units, which has remained relatively stable since its initial emergence and which implements what remains of the Chomskian program of Universal Grammar, the "Minimalist Program" laid out in Chomsky (1995). That neurological structure nevertheless significantly underdetermines the grammars of the various specific languages, which continue to change on more rapid time scales. Bickerton then proposes to distinguish the biological evolution of language, which supposedly culminates in this neurological reorganization, from its subsequent cultural evolution in the various natural languages, but once one recognizes niche construction as integral to biological evolution, this distinction between biological and cultural evolution cannot be sustained.

another in partially shared circumstances; such practices can change very quickly despite maintaining their continuity. I emphasize how these aspects of language (its perceptually accessible public character, its developmental coevolution with human organisms, and its socially interactive mutual responsiveness) belong together as a biosocial phenomenon. I then argue that the coevolution of these features of our discursive, biosocial way of life *together* enable conceptually articulated intentional directedness.

This account distinctively emphasizes the practical-perceptual aspects of language use. The ability to recognize and produce articulated linguistic performances is itself a biologically evolved capacity, which also materially changes the environment in which humans develop and to which we respond.[32] Anyone who undertakes immersion in a newly acquired language after childhood is all too familiar with the perceptual and practical challenges. Acquiring the abilities to recognize the spoken and written patterns around you, and to contribute to them fluently in real time, can be frustratingly difficult. Similarly, in our language-pervaded environment and responsive way of life, any impairments of linguistic capacity, including the ability to respond appropriately to other speakers, are debilitating "abnormalities."[33] Dreyfus's treatment of skilled practical-perceptual coping makes an important contribution to my account at this point.[34] Chess players see and respond to meaningful chess situations rather than meaningless data or even preconceptual solicitations and repulsions; I argue that "native" speakers and listeners are in a similar way perceptually and practically responsive to the semantic significance of their own and others' discursive performances.

A second important feature of my account is to recognize that languages are what I call "partially autonomous" practices. On the one hand, language use has developed over time in ways that enabled ex-

32. In chapter 3, I discuss the coevolutionary adaptation that both enhanced human children's capacities to acquire linguistic ability through normal exposure to spoken language and also adapted languages themselves to fit those patterns that were more readily learnable via these genetically and neurologically assimilated capacities for ease and speed of acquisition in childhood.

33. A striking example of the intertwining of the social and the biological dimensions of discursive practices are the widespread, ongoing efforts to transform discursively articulated social life so as to incorporate as full participants those persons whose biological development impairs their ability to engage in linguistic interaction in the same ways most others do. Deafness, dyslexia, and autism are among the biological variations that take on new significance within our discursive way of life yet might be better assimilated through changes in social practices.

34. Dreyfus himself resists this way of expanding upon his account because it undercuts his efforts to sustain a distinction between conceptual understanding and practical-perceptual coping skills as supposedly preconceptual. See Dreyfus (2000) for his response to my earlier efforts to appropriate his account of coping skills within a more comprehensive account of intentionality.

traordinary capacities for "intralinguistic" articulation. For many specific utterances, and some entire domains of linguistic articulation, an occasioned utterance's relation to "extralinguistic" circumstances is extensively mediated by other linguistic expressions. This mediation takes multiple forms: words are normally understood as iterable expressions whose prior and subsequent uses bear upon the current use; words and sentences have extensive grammatical and inferential relations to other words and sentences; and utterances often belong to a conversational context, in which the most relevant circumstances for interpreting them are the preceding linguistic utterances and the possibilities they offer for further linguistic response. On the other hand, linguistic performances also remain holistically connected to other forms of involvement in the world and are only intelligible amid these more extensive capacities for practical-perceptual interaction with our surroundings. Speaking and hearing language is itself a subtle and complex perceptual and practical capacity, as we have seen, but linguistic exchanges are also directed toward, responsive to, and ultimately accountable to our worldly circumstances. These two sides of the only partial autonomy of linguistic expression function together to allow speakers not only to pick out and respond to aspects of their surroundings but to do so in ways that can "take them as" other than they are. The mostly internally articulated pattern of discursive practice enables a "taking-as" that is distinct from other practical and perceptual responses to our surroundings. These patterns nevertheless remain directed toward and accountable to our practical and perceptual engagement with the world because they are not entirely free-floating or disconnected from ongoing involvement with our surroundings. The combination of a substantial degree of intralinguistic autonomy situated within a broader biologically grounded engagement with our environment (including other people as ambiguously part of that environment) is the most distinctive feature of the capacities for articulation and accountability that are constitutive of conceptual understanding.

The sense that linguistic expression constitutes a partially self-contained domain is supported by the somewhat specialized character of linguistic capacities. Defenders of the notion of "nonconceptual content" are rightly attentive, but mistaken in their response, to the fact that many aspects of human practical-perceptual responsiveness to our surroundings are not themselves linguistic abilities and enable discriminations that are often difficult to characterize verbally. Language is not a general-purpose expressive capacity. Its more limited and specialized repertoire is biased toward some domains of human expressive and responsive activity and always functions alongside other capacities.[35] Despite this apparently limited functional role, however, language is not entirely self-contained. Linguistic capacities open onto and "incorporate" other sensory/cognitive/performative capacities. That happens not only because linguistic utterances and their uptake are always situated within a wider practical and perceptual context. "Recognitive," demonstrative, anaphoric, and indexical locutions also serve to bring utterances into specifically linguistic engagement with their surrounding circumstances.[36]

Philosophers have often assigned special importance to perception in constraining our otherwise free-floating capacities for conceptual spontaneity. Even Sellars, Davidson, Brandom, McDowell, and others who reject the Myth of the Given (the effort to ground conceptual content in the mere "having" of some nonconceptual input) typically treat perception as the primary locus for the objective accountability that is needed for discursive performances and commitments to have content. In doing so, they also emphasize the passive receptivity of perceptual encounters. I argue that receptivity is not enough. Perception can only play the role of constraining and thereby enabling conceptual content in concert with our characteristic forms of activity, vulnerability, and sociality. Our capacities for active exploration of the world are themselves partly constitutive

35. The coexistence of language with other expressive capacities that are not readily expressible linguistically is part of what lends mistaken plausibility to the notion of "nonconceptual content." At least three mistakes must nevertheless be combined to make this notion plausible. The first is to treat language as a self-contained practice rather than one that depends upon both its incorporation within and its semantic inclusion of the whole of human bodily intra-action with our surroundings (including other discursively articulate human beings). The second mistake is then to conflate the conceptual domain with what is readily expressible in language. Language enables a distinctive capacity for conceptually articulated normativity, but it is not coextensive with it. We can articulate and express conceptual understanding through nonlinguistic activities, although those activities are transformed by being caught up in discursively articulated normative accountability. The third mistake is to extract the *result* of their entanglement with discursive practice and identify it as an inherent feature of our various nonlinguistic capacities in isolation. Here, the metaphorical connotations of "content" (suggesting distinctions between container/contained and inside/outside) undoubtedly contribute to the ease of making this error.

36. Kukla and Lance (2009) introduce the pragmatic category of "recognitive" expressions for those speech acts whose function is to express the speaker's recognition of something. They include perceptual recognitives that express perceptual uptake of some aspect of one's surroundings ("Lo! a rabbit!"), vocative expressions that recognize and call other speaker/agents ("Hey, Alice!," "Hey, you!," "Hello"), or acknowledge and respond to such calls, among other forms of social or practical/perceptual recognition. Many of the explicit semantic markers of this pragmatic role, such as "Lo," seem archaic, and the pervasive role of recognitive performances in discursive practice is masked by the absence of explicit grammatical or semantic markers for many contextually recognitive speech acts.

of perceptual receptivity,[37] but I argue that perception is more thoroughly entangled with our activities in the world. We are also vulnerable to what happens around us, which are not merely occasions for a disinterested receptivity to objects but possible threats to the continuation of our life patterns. Conceptual capacities are responsive to a surrounding environment that both enables and threatens our lives as organisms and our projects as agents.

This account of conceptual understanding also emphasizes how perceptual responsiveness and vulnerability to the surrounding world are caught up within social life. Human beings are social animals. In this respect, our species is hardly alone; as primates, we are attentive and responsive to our fellows' activities and affects. To that extent, other primates' lives are salient within any primate's biological environment. Language nevertheless enables but also depends upon a further kind of mutual responsiveness among human agents. We address or call one another individually, recognize ourselves as so called, and understand such calls as imposing default expectations and/or obligations to respond. Kukla and Lance (2009) show how language centrally involves this "vocative" dimension of call and response. We will see in later chapters how "second-person" accountability to others and first-personal responsibility for one's own performances and commitments, both of which emerge in characteristic form through the vocative dimensions of language use, also help establish the normativity that is constitutive of conceptual understanding. The point to emphasize now is that the social-vocative and perceptual-practical aspects of conceptual abilities are not independent. The ability to recognize and respond to one another's calls, after all, requires perceptual-practical skills. Less obviously, perhaps, vocative abilities are integral to the discursive significance of what is perceptually accessible. We do not merely report what we see but also call attention to its perceptual presence for us and invite others to see the same things. We help one another learn to see new things and call one another to account for perceptual encounters whose presence, relevance, or significance might otherwise have gone unnoticed. We do not thereby merely enhance individual perceptual capacities by drawing upon those of other perceivers. Those capacities take on new dimensions of normative significance for us as organisms through being caught up in the resulting patterns of mutual responsiveness, or so

37. Nöe (2004) provides an especially clear account, well grounded in recent work in the neurophysiology of perception, of how our perceptual responsiveness to the world depends upon our characteristic forms of movement and agency within it.

I will argue in chapters to come. Only in part by calling others, and recognizing and responding to such calls, do we become normatively accountable in ways that go beyond merely functional normativity. The vocative capacities that language both draws upon and helps constitute do not merely invoke a de facto propensity to respond to one another's hails but also institute an accountability for responding to them and thereby make conceptual normativity possible.

Understanding conceptually articulated intentionality in this way as an integrated social-biological phenomenon has at least three further important philosophical consequences. The first is a shift in the primary locus for the normative accountability of intentional comportments. A long and influential tradition gives philosophical primacy to the epistemic assessment of claims to knowledge and/or truth. In that context, the articulation of conceptual content is mostly taken for granted as a prerequisite to epistemic assessment. On the account of intentionality and conceptual understanding that I develop, epistemic assessment remains important but subordinate to conceptual normativity. Moreover, epistemic assessment remains contextual in a way that blocks general formulation of epistemological questions, including the concerns with skepticism that have long preoccupied philosophers. Such a shift was already proposed in recent work by Davidson, Brandom, McDowell, and Haugeland, among others, with important historical precedents in Kant, Hegel, and Heidegger. In chapter 5, I argue that recent efforts along these lines, which emphasize the objectivity of conceptual understanding, have not sufficiently freed themselves from familiar problems in the tradition they criticize.[38] The naturalistic account of conceptual understanding developed here can help free us from these problematic vestiges of the epistemological tradition. In part 2, I then consider how this shift in focus constructively reformulates our predominant cultural and philosophical images of science.

This challenge to recent conceptions of the normativity of conceptual understanding in terms of objectivity also points toward a second further consequence of this social-biological approach to intentionality. Emphasis upon conceptual objectivity rightly acknowledges our need for responsive deference to the world. How the world impinges upon our inquiries nevertheless in turn depends upon the goals of those inquiries and their place within larger patterns of social practice. Just as an organism's environment is partly defined by its own life patterns, so

38. This critical side of the project extends my earlier work on Davidson, McDowell, Brandom, and Haugeland in my previous book (Rouse 2002), especially chapters 2 and 5–9.

normative accountability to the world in our practices and life activities depends upon what we are up to, and that is not already settled. Conceptual normativity therefore has an ineliminable temporal dimension. This temporality can be expressed in the anaphoric terms of what is at issue and at stake in the future development of the practices that we find ourselves already in the midst of. The partial openness of who we are to become, and thus of how different possible trajectories of social practice matter, accounts for the element of freedom in social life long recognized as essential to conceptual normativity. To be normatively accountable is to be bound, but not determined, to respond to what is at issue and at stake in our ongoing involvement in these patterns of social and biological life. How we do respond then partially reformulates the issues and stakes in subsequent performances.

A third consequence of this approach shifts the focus of philosophical explication from what concepts are to what conceptual understanding and conceptual articulation are. If conceptual understanding were an ability to deploy and recognize mental, neural, or linguistic representations, it would seem natural to begin with an account of concepts. Concepts would be the elements in such a representational structure, and that structure would be the principal target of philosophical and empirical psychological explication. Even those who understand concepts in this way now typically recognize their holistic interdependence. As Wilfrid Sellars influentially noted, "One can have the concept of green only by having a whole battery of concepts of which it is one element. . . . While the process of acquiring the concept green may—indeed does—involve a long history of acquiring *piecemeal* habits of response to various objects in various circumstances, there is an important sense in which one has *no* concept pertaining to the observable properties of physical objects in Space and Time unless one has them all—and, indeed, a great deal more besides" (1997, 44–45). The view of conceptual understanding developed here is holistic in a much stronger sense, however. The claim is not just that having one concept requires having many interdependent ones; the resulting form of conceptual holism would still then be too static and unified. The performances that constitute conceptually articulated practices are both socially differentiated and dynamically responsive to that differentiation through ongoing efforts to sustain the coherence of a common discursive practice. Conceptual understanding is then not the grasp of a static holistic structure but an active capacity to track, adjudicate, and respond appropriately to the more or less divergent performances within social practices (of which expressive speech is a paradigmatic example). This tracking and adjudication takes place in two registers simultaneously: for a performance's appropriateness and significance within a practice as a partially autonomous context and for the broader practical and perceptual significance of both the performance and the *only* partially autonomous practices to which it belongs. The results of such adjudication in both registers are also continually reintegrated into ongoing practice in ways that reverberate through the practice as a whole. Conceptually articulated practices sustain a shifting, uneasy equilibrium between these competing pulls toward unity and divergence.

This strongly holistic account of conceptually articulated practices makes the dynamics of entire domains of conceptually articulated practice (from natural languages to empirical sciences, artistic traditions, work domains, games, and many more) more philosophically basic than is any specific component of those practices. In the paradigmatic case of language we can, as a first approximation, identify concepts with words as articulable, iterable, and recombinable elements of linguistic practice. Linguistic practice encompasses much more than words, however, and what any word contributes to the practice or specific performances within it depends upon how it is interconnected with other "elements" of linguistic practice, including other words. Ongoing use both builds upon and changes those interconnections, with consequent effects throughout the entire domain. What the significant concepts and conceptual relations are within a practice, and how they matter to the practice and its practitioners, are always to some extent at issue within the practice. They therefore are not and cannot be predetermined elements or units of which the practice is composed. It nevertheless matters crucially to conceptual understanding that the interconnected performances that it tracks and adjudicates are articulated into recombinable elements that function together to constitute a partially autonomous domain of practice. These considerations provide an initial indication of why "concepts" are not the book's primary focus. Part of the argument it develops is that a naturalistic account of the relevant human capacities should be focused instead upon conceptually articulated practices and conceptual understanding as the ability of some organisms to track, adjudicate, and contribute to such conceptually articulated domains of practice.

The remainder of the book develops this account of conceptually articulated understanding as integral to our natural life as social, discursive organisms. This first part of the book undertakes a more extended account of how conceptual understanding is grounded in human biology and social practice in ways that can enable grasp of scientific practices and scientific understanding as themselves scientifically explicable natural phenomena. The second part turns to the question of how to understand scientific practices as exemplary cases of the normativity of conceptual understanding and as the basis for the understanding of nature and of philosophical naturalism that frames my entire inquiry. The ordering of these two parts of my argument was pragmatic, but the sequencing partly masks their mutual dependence. In place of the shopworn architectural metaphor for understanding or knowledge as a structure built upon an argumentative foundation, I present its two parts as composing an arch. Neither part stands entirely on its own, but together they exhibit a complementary stability and mutual support.

THREE

Conceptual Understanding in Light of Evolution

I—Refining the Issue of Linguistic and Conceptual Evolution

Conceptual understanding is at least a capacity of human beings and perhaps also of organisms of other species. Not all such capacities of organisms are themselves directly evolved, since many capacities are by-products of other evolved traits or capacities. Part of understanding a capacity exhibited by organisms is to recognize the evolved capacities and evolutionary constraints that enable it to be exhibited. Such evolutionary understanding is dependent upon how a trait or capacity is characterized, however. There are many ways in which we might describe our capacities for conceptual understanding and conceptually articulated description of and engagement with the world. There is a strong reciprocal relation between the description of a trait or capacity and an understanding of that trait or capacity in the light of biological evolution. An adequate evolutionary account of how a trait or capacity arose presupposes a correct characterization of the trait or capacity as it emerged developmentally and ecologically so as to be directly or indirectly responsive to natural selection. The correct biological characterization of the trait nevertheless significantly depends upon how it emerged and was retained and modified in the course of evolution. Breaking into this circle is a serious challenge for scientific and philosophical understanding of evolution.

Accounts of the evolution of humans' or other organisms' capacities for conceptual understanding thus involves decisions about how to characterize the explanandum of those accounts. Perhaps the most widespread approach, at least since the neo-Darwinian evolutionary synthesis, has been guided by the often unstated assumption that conceptual understanding is a capacity that would always emerge in an organism with sufficient general intelligence. Such intelligence in turn reflects overall cranial capacity and specific patterns of representational inner wiring and connection or externally interpretable behavior (Sterelny 2003, 4; Godfrey-Smith 2002). This assumption leaves no place for an evolutionary explanation of conceptual capacity as such. It only allows for an explanation of changes in brain size and in relevant patterns of development—such as neoteny and its associated requirements for parental care and social life or the differential growth and connectedness of various regions of the central nervous system. The selective advantage conferred by the resultant capacities may then play a significant role in the explanation of these neurological developments, although on some accounts, neurological and developmental changes that occurred for other reasons produced conceptual capacities as a by-product. In either case, evolution could account for the emergence of bodily and behavioral conditions that enable conceptual understanding, but conceptual understanding itself would have to be characterized in more general terms that are independent of its specific evolutionary history in any particular species.

The prevalence of this assumption that conceptual understanding is a general capacity shows itself in various places in our intellectual culture. Most work in philosophy of language and mind in the past century has proceeded on the assumption that the characterization of these key features of our cognitive capacities need not be deeply rooted in evolutionary biology.[1] More striking is the widespread acceptance of the idea that "intelligence" and the associated capacity for grasping symbolically articulated concepts might emerge in different forms in other terrestrial species or in radically different organisms and environments on other planets.[2] Such commitments to the "multiple realizability" of intelligence, conceptual understanding, and symbolic expression suggest that

1. Millikan (1984) and Dennett (1987, 1995) are among the notable exceptions.

2. Smocovitis (1996, 172–88) emphasizes that the first generation "architects" of the neo-Darwinian evolutionary synthesis were committed to this generalizable conception of intelligence as an adaptive trait and shows how this conception played a role in extending "evolution" to incorporate cosmological as well as biological processes.

it would be parochial and misleading to ground an account of conceptual understanding in our particular evolutionary history.

The prevalent assumption has been that evolution only displays the contingent history through which a more general biological capacity for conceptual understanding acquired particular instantiation in *Homo sapiens*. Several prominent alternative approaches admittedly have treated conceptual understanding as a more directly evolved trait or set of traits. Ruth Millikan's (1984) teleosemantics introduced a novel approach to understanding mental representation as determined by the causal-functional role for which types of mental states were originally selected and retained. A different approach to the evolution of conceptual understanding emerged from Chomskian linguistics. Chomsky influentially argued that the capacity for understanding and producing linguistic expressions must be hardwired into human brains, given the extraordinary rapidity, flexibility, and productivity with which human infants, seemingly alone among biological species, learn deep, complex, grammatical structures from a very impoverished evidence base. Chomsky himself did not explicitly address the evolution of linguistic competence until late in his career (Hauser, Chomsky, and Fitch 2002). His insistence that linguistic competence depends upon recursive, transformational-syntactic structures that require a hardwired language module nevertheless already suggested a specific configuration of human cognitive evolution. The evolution of linguistic understanding would become a separate problem from the evolution of other cognitive capacities that may be more closely parallel to the capacities of nonhuman animals. Moreover, Chomsky's emphasis upon syntactical transformations and the recursive productivity of linguistic competence focused explanatory attention upon syntax and recursion as the central issues to explain as novel and characteristic features of human conceptual understanding.[3] Finally, for some psychologists, biologists, and philosophers, Chomsky's postulation of a relatively autonomous "language module" in human brains served as a model for reconceiving conceptual understanding as "massively modular." The idea was to disaggregate conceptual understanding and other forms of cognition into a variety of distinct capacities that evolved separately, as adaptations to specific conditions of early

3. Chomsky's conception of Universal Grammar has undergone several important conceptual shifts over his career. Bickerton (2014) argues that Chomsky's (1995) later, more minimalist, derivational conception of Universal Grammar can be understood differently in an evolutionary context as the outcome of neurological adaptation for more efficient acquisition and use of the capacities to produce and interpret extended strings of linguistically significant units. Bickerton's proposed reconceptualization of Chomsky's Minimalist Program is discussed further below.

hominid life or as responses to specific evolutionary problems (Barkow, Cosmides, and Tooby 1992).

These familiar approaches to the evolution of human cognitive capacities, whether seen as distinctively human capacities or as more continuous with the capacities of other animals, have received extensive critical appraisal.[4] Many of the specific difficulties confronting each approach are by now well documented in the literature. I find these specific critical arguments convincing overall but will not review the details. I do want to call attention, however, to a central problem confronting all these influential ways of thinking about the evolution of human conceptual understanding due to important recent developments in evolutionary biology. All these familiar approaches have been formulated with reference to versions of the neo-Darwinian synthesis that first emerged in the middle decades of the twentieth century. Although they generally acknowledge that some inherited traits are nonadaptive, neo-Darwinians usually make the default presumption that evolutionary novelties, including language and articulated conceptual understanding, result from gradual, adaptive changes in gene frequencies in populations under selective pressure. The available variation in phenotypes in that population is acted upon by their "external" environment in ways that shift the distribution of genetic variants in the next generation, with cumulative effects over longer spans of time.

During the past several decades, neo-Darwinian orthodoxy has been challenged and augmented by new or newly resurrected alternative mechanisms of evolutionary change that are relevant to characterizing and understanding conceptual capabilities in an evolutionary context. This "extended synthesis" (Müller and Pigliucci 2010) encompasses multiple revisions and expansions of neo-Darwinism: efforts to reintegrate development into evolution (developmental evolution, ecological-developmental biology, and developmental systems theory); to consider the interactions of multiple "dimensions" of evolutionary change (epigenetic, behavioral, and cultural as well as genetic), including renewed attention to how phenotypic plasticity can become genetically assimilated (Jablonka and Lamb 2005); to recognize the evolutionary importance of niche construction and the coevolution of organisms and their environments (Lewontin 2000; Odling-Smee, Laland, and Feldman 2003); to recognize the nature and extent of the plasticity of neural development; to pin down more carefully our behavioral and cognitive similarities

4. Important examples include Cowie (1999), Tomasello (1999, 2008, 2014), Sterelny (2003), Okrent (2007), and Bickerton (2009, 2014).

and differences from our most closely related species; and to rethink many familiar assumptions about language and its evolution in that context. I regard these developments in evolutionary biology as sufficiently far-reaching and relevant to shift the burden of proof toward proponents of more traditional neo-Darwinian explanations of the evolution of language and conceptual understanding. Advocates of more traditional conceptions need to either accommodate their views to these more recent developments or show more convincingly why no accommodation is called for.

My aim in this chapter and its immediate successor is not to undertake such an evaluation of the prospects for familiar approaches to the evolution of language and/or conceptual understanding in light of recent empirical and theoretical developments in evolutionary biology. I instead begin to articulate a different way of thinking about language and conceptual understanding that draws upon these new developments in biology and philosophy. Some recent work in philosophy, linguistics, and biology has already taken important steps in this direction (Deacon 1997; Dor and Jablonka 2000, 2001, 2004, 2010; Sterelny 2003, 2012; Lloyd 2004; Okrent 2007; Bickerton 2009, 2014; Odling-Smee and Laland 2009), although I have yet to see a comprehensive overview of the emerging perspective on conceptual understanding in light of these biological developments.

An important initial issue in accounting for the evolutionary emergence of articulated conceptual understanding is to understand the relation between language and conceptual capacities more generally. A capacity for producing and consuming linguistic expressions, at least in anything resembling its mature, developed human phenotype, has long seemed to be unique to humans.[5] While many nonhuman animals have acquired sophisticated communicative abilities of various sorts, the differences between human language and other forms of animal communication have become increasingly clear (Hauser 1996; Deacon 1997; Bickerton 2009, 2014). That recognition leads to a strategic choice from the outset. Should we initially approach language as an evolutionary novelty within the human lineage, whose emergence under specific

5. Radick (2007) provides an informative history of scientific investigation of whether other primates have their own language or the capacity to acquire ours. Radick explores repeated shifts in scientific consensus concerning whether linguistic abilities in other primates are relevant to understanding our linguistic capacities. His narrative concludes in the early 1980s, with renewed efforts by Peter Marler, Robert Seyfarth, and Dorothy Cheney to use audio recording to perform the "primate playback experiment," alongside renewed efforts to teach rudimentary language to captive chimpanzees and bonobos.

selection pressures accounts for much of what is distinctive about human cognition, including perhaps the difference between genuinely conceptual understanding and other forms of cognition? Is conceptual understanding then first and foremost a linguistic phenomenon or at least a capacity intimately involved with the capacity for language use? Or does the human capacity for language instead arise from traits that are homologous with other species? In that case, the continuities between human cognition and the cognitive capacities of other species would stand out, especially in the case of our nearest ancestors in the primate lineage. Language then might be derivative from more fundamental cognitive capacities, including a capacity for conceptual understanding, even though the emergence of language facilitates and dramatically expands the articulation of concepts and their importance within human development, reproduction, and evolution. Needless to say, these are not the only options, including more complex relationships among these possibilities. Moreover, multiple aspects of language would need to be considered as possible sites for its evolutionary emergence: the emergence of syntax and recursive grammatical transformations, of capacities for phonological expression and their auditory (or other sensory) recognition, of displacement and symbolic understanding that detach cognition from its orientation toward current practical and perceptual circumstances, of the social relations that allow recognition of and response to individual speakers and respondents and consequently novel forms of social cooperation, of the complex abilities at imitation that make possible iterated social learning, and/or some other distinctive and putatively critical capacities embedded in the learning and reproduction of languages as both social practices and cognitive achievements.

Elisabeth Lloyd (2004) shows that the fortuitous success of the bonobo Kanzi in acquiring a rudimentary linguistic capacity has changed the terms in which these issues should be addressed (Savage-Rumbaugh, Shanker, and Taylor 1998). Kanzi inadvertently participated in experiments on language acquisition because his mother was a research subject, and he was too young to be separated from her. While his mother struggled with the experimental protocol, Kanzi did much better despite not being initially targeted for instruction. Eventually, Kanzi acquired not only a substantial vocabulary of symbols but also the ability to produce novel, intelligible syntactic recombinations. The experimenters plausibly characterized his eventual linguistic capacities as in some respects comparable to those of a thirty-month-old normal human child. The interpretation of these data is controversial (see Pinker 1994;

Savage-Rumbaugh, Shanker, and Taylor 1998; Lloyd 2004; Bickerton 2009), but I follow Lloyd in her insistence that Kanzi's achievement shows that the neurological capacity for linguistic understanding is homologous between humans and bonobos and probably extends further to common ancestors.

A frequent criticism of the Kanzi or Washoe experiments (e.g., Bickerton 2009, 78–83) is that whatever symbolic capacities apes acquire in these settings is still used almost exclusively in ways that engage or manipulate their immediate surroundings, without actually displaying a grasp of symbolic significance comparable to that of human children. That claim may well be correct, but it is also beside the point. Lloyd emphasizes that beginning with the fully developed human phenotype, rather than considering more basic capacities from which current forms of human language subsequently developed, is a crucial mistake in many assessments of whether language is an evolutionary novelty in humans. It would indeed be surprising if individual apes used such newly and artificially acquired capacities in ways disconnected from those organisms' evolved way of life. The importance of the Kanzi experiments is not that some apes have capacities for language understanding and use that are comparable to those that humans have evolved through natural selection in discursively articulated environments. The experiments only show that bonobos have the neurological capacities for initial responsiveness to such a selective regime. Yet once the homology is acknowledged, the critics are right that Kanzi's limitations and distinctive circumstances of acquisition are also instructive and that any serious reflection upon the evolution of human linguistic capacities must take both Kanzi's abilities and his limitations into account.

What conclusions should be defeasibly drawn from these experiments? If we take Kanzi's abilities at face value, as I think we should, then the capacity for producing and consuming linguistic expressions is not uniquely human and did not emerge as a novel capacity in the *Homo* lineage. Other species in the primate lineage who share this capacity have nevertheless not developed language on their own, even in rudimentary forms, despite having the neurological basis for producing and understanding symbolic expressions with syntactic structure. This capacity for linguistic expression and understanding has only been expressed in experimental settings that bring other apes into contact with an analogue to human language adapted to their perceptual and expressive abilities. This fact strongly suggests either a lack of selection pressure in other lineages for linguistic communication or substantial barriers to the realization of this latent capacity.

The difference between Kanzi's ability and the rather more limited or even nonexistent linguistic capacities of apes who encounter language later in their neurological and social development is also clearly important. The circumstances of Kanzi's development indicate that the ancestral capacity to understand language shares with human linguistic development the need for early exposure to already extant uses of symbolic expressions. Presumably this developmental window results from changed patterns of neural development due to linguistic exposure, but it is also likely from the role of immature development in facilitating symbolic displacement from the perceptual circumstances of symbol use (Deacon 1997; Lloyd 2004; Bickerton 2009). The availability of an extant public linguistic practice within an organism's developmental environment thus appears to be crucial to the realization of linguistic ability; moreover, as several commentators have recently argued (Deacon 1997; Dor and Jablonka 2000, 2001, 2010; Bickerton 2009, 2014), an extant discursive practice then contributes to selection pressures for the coevolution of languages along with the neurological capacity for language. The existence of even a rudimentary discursive practice can be a resource for the scaffolding of linguistic development and linguistic innovation. These considerations highlight the importance of recognizing language as one of the most salient and powerful forms of niche construction (Odling-Smee, Laland, and Feldman 2003; Odling-Smee and Laland 2009). They also strongly suggest that in retrospect, a crucial issue for the evolution of language is overcoming initial barriers to the realization of linguistic capacities within a species or a population of that species. If language is a social practice whose continuing existence depends upon its early exposure to developing infants, then it would be difficult to establish even a rudimentary discursive practice in the first place. Once such performances are entrenched in a lineage and integral to its ongoing way of life, selection pressure might well arise toward easier and faster acquisition, and then for the "ratcheting" of the practice in its successive iterations (Tomasello 1999). Yet very steep barriers may block the initial establishment of a protolinguistic practice that would then be salient in the subsequent development of infants. If *prior* exposure to an extant adult practice is a developmental prerequisite for the acquisition of linguistic competence, then developing an initial adult competence and performance is a serious challenge.

The difference between Kanzi's capacities and his mother's also reinforces a substantial body of evidence suggesting that the primary evolutionary-developmental challenge for language acquisition is displacement (Bickerton 2009, 2014) or symbolic understanding (Deacon

1997), even apart from the catch of needing already to have an extant linguistic practice as developmental background for the formation of the relevant neural capacities in infants. This barrier is not a *deficiency* that would merely require cognitive innovation to overcome. The difficulty is instead that symbolic understanding requires partial *suppression* of the richly articulated perceptual discernment and practical significance of an organism's involvement in its current circumstances. Symbolic displacement involves perceptual recognition and practical responsiveness to a perceptually salient feature of one's surroundings that serves as a symbolic, protolinguistic expression. Yet grasping the symbolic significance of that expression displaces it from its immediate practical/perceptual involvement in those circumstances. The more refined and developed an organism's capacities for perceptual discrimination and practical responsiveness to those discriminations are, the more substantial is the barrier to displacement. As Bickerton (2009, ch. 1) points out, all nonhuman animal communication forms seem to be strongly tied to the circumstances of utterance. A vervet monkey's distinctive warning cries in response to different dangers from predators, for example, serve to focus attention on the possible locations of those dangers and the appropriate response within current circumstances (Seyfarth, Cheney, and Marler 1980; Cheney and Seyfarth 1990). In the absence of such circumstantial indications, the warning cries would have no point. The very capacities for perceptual discrimination and practical response cultivated in those animals that develop effective communicative capacities thus work directly against the appropriation of those capacities for displaced symbolic communication.

Emphasis upon symbolic displacement rather than syntactic recombination and recursion counters the predominant orientation of many linguists toward the problem of understanding the evolution of language.[6] The Kanzi phenomenon, reinforced by several important arguments developed in recent literature, nevertheless strongly suggests that primary emphasis upon the achievement of syntactic structure and re-

6. Bickerton (2009, see 49–51) is a prominent case of a linguist who once took syntax and recursion to be central but has now changed his mind in favor of the emergence of symbolic displacement, although Bickerton (2014) still reserves a prominent place for the evolution of syntax and recursion to accommodate and support the behavioral and neurological significance of linguistic understanding and expression. Dor and Jablonka (2000, 2001) present evidence that some of the syntactic structures that linguists take to be sui generis, notably island constraints, can instead be understood as at least partially semantic. Tomasello (2008) argues that a usage-based account of the development of language can account for the gradual "grammaticalization" of combinations of symbols and the articulation and refinement of an initially simple syntax toward more complex forms.

cursion as the key challenge for the evolution of language is mistaken. Although Kanzi's syntactical grasp is quite limited compared to the adult human phenotype, he shows sufficient understanding of the significance of combining symbols and the possibility for semantically significant recombination to indicate that the ancestral, unrealized capacity in bonobos includes the ability to grasp rudimentary syntactical structure and the differential significance of judgments rather than names or tags. Moreover, as Deacon (1997) has argued, however difficult the problem of attaining the capacities for rapid learning of combinatorial syntax, it remains secondary and derivative. Once even a rudimentary protolanguage is under way, it will only survive if it is readily reproducible through what the next generation can discern from the available evidence. Languages themselves evolve, and only those grammatical structures that are readily learned will be reproduced. So the central problem for an evolutionary understanding of syntax is not how human beings could ever have achieved the syntactical complexities of the languages we now speak (as if these were requisite for any language whatsoever), but why, given the ongoing evolutionary refinement of an extant discursive practice, these syntactic forms happened to be so readily discernible and reproducible within our lineage. The neural realization of these particular syntactic predispositions is an interesting and surely complex issue, but it occupies an intermediate place in understanding how and why linguistic capacities arose between the initial emergence of extensive protolinguistic expression and the subsequent diversification of the world's many extant languages.[7] These syntactic capacities are in any case presumably derived modifications of earlier forms of language, and they could be scaffolded by those earlier forms.[8] As one further consideration, Dor and Jablonka (2000, 2001) argue that many of the subtle syntactic transformations that linguists nowadays emphasize to justify the preeminence of syntax may also have a more basic semantic significance (a point to which I will

7. The emergence of *some* way to produce and disambiguate strings of linguistic symbols in real time to permit fluent, effective communication was surely essential to the evolution of anything like current human capacities, but the currently extant forms or mechanisms for doing so do not seem essential.

8. See Bickerton (2014) for extensive discussion of how to grasp the evolution of minimalist forms of syntactical combination as a likely neurological reorganization to accommodate more efficiently the production and consumption of protolinguistic expressions while recognizing the importance and range of subsequent patterns of variation and change within specific linguistic traditions. Useful background on this issue is also provided by Bickerton (1975) on the formation of pidgins and creoles and Tomasello (1999) on "grammaticalization" as telling phenomena that indicate how more complex grammatical structures can emerge within more rudimentary linguistic or protolinguistic practices.

return in the next chapter for its importance in locating language in relation to other perceptual and expressive capacities).

The final initial lesson from Kanzi's acquisition of rudimentary symbolic understanding concerns the limitations of his capacities. Kanzi shows not only that the underlying capacity for linguistic understanding is ancestral but also that linguistic ability has had extraordinary articulation and modification in human beings over our evolutionary history. Kanzi's capacities, though significant, are quickly surpassed by any normal human child and are utilized differently by human children from early in their development (Tomasello 2008, 2014). This difference highlights the significance of niche construction, neural plasticity, and the partial genetic fixation of phenotypic plasticity in shaping the historical development of linguistic and conceptual capacities at multiple levels. At one level, the relevant changes incorporate substantial evolutionary and developmental transformations in neurology and anatomy. Deacon (1997) provides an initial sketch of some of the relevant changes to the organization and development of human neurology and anatomy under ongoing coevolutionary selective pressure for language and the social and cultural practices that it enables. Changes in childrearing and other social practices are also a crucial component of linguistic niche construction. Sterelny (2003, 2012) and Bickerton (2009, 2014) both emphasize the ways in which the development of mature linguistic ability is now shaped by the scaffolding available for language learning. Such support arises not only from the social pervasiveness of spoken and written language but also from the ways in which adults initiate children into language by explicitly modeling discursive interaction and engaging them linguistically in ways that encourage development of mature human linguistic facility.

Linguistic practices themselves have also changed considerably in human history and prehistory. The emergence of written language and the associated changes in discursive practice and linguistic development, with further elaboration through various aspects of print culture, have had important consequences for human cognitive development (including the manifestation of novel forms of cognitive "limitation" in these new environments, such as dyslexia). The earlier associations of linguistic expressions with various kinds of musical expression presumably had comparably dramatic effects in human prehistory. As has been widely recognized, the historical emergence, social incorporation, and elaboration and articulation of linguistic practices and capacities have also interacted extensively with other forms of cultural elaboration and material niche construction. The development of languages both facilitated and was transformed by their contribution to the associated development of articulated and reticulated tool use and social role differentiation, including tools that make a difference to cognitive development, such as new forms of visual representation.[9] Linguistic and other forms of social niche construction, neural plasticity and its partial genetic fixation, and the consequent interaction among cultural, behavioral, epigenetic, and genetic aspects of evolutionary change have had crucial roles in the development of language as both a central feature of our evolutionary niche and an extraordinary human capacity.

II—Language, Perception, and Conceptual Understanding

The preceding remarks draw upon recent developments in evolutionary theory as well as insistence upon the significance of the Savage-Rumbaughs' work with Kanzi (Savage-Rumbaugh, Shanker, and Taylor 1998). These considerations let us reframe the underlying question about the relation between linguistic competence and conceptual understanding given the arguments in chapter 2. Evolutionary theory rightly emphasizes continuity within lineages among different organisms and their traits and capacities. Derived modifications of common ancestral forms or behaviors can nevertheless take strikingly different forms and functions (think of the manifold variations on the tetrapod limb, for example) and can be co-opted in ways that later seem remote from their evolutionary origins. The possibility of radically different evolutionary trajectories from common origins is heightened by the forms of cumulative niche construction that have become dominant in so many aspects of human evolution. I argued in chapter 2 that performances or states exhibit conceptual understanding only if they express a content that might be objectively mistaken. Such two-dimensionally normative performances are distinct from mere evolutionary design limitations or satisficing compromises in maintaining an organism's ongoing way of life. I will argue that conceptual understanding in this sense is a possibility that only emerges through a divergence within the human lineage from our nearest

9. Bickerton's (2009, 2014) specific hypothesis for the evolutionary context in which there were strong selection pressures for symbolically displaced protolanguage—the establishment of a new nutritional niche based upon territorial power-scavenging of the carcasses of megafauna—includes the emergence and proliferation of Acheulian hand axes as integral to this hypothesized way of life, both for cutting through carcass hides before other scavengers could do so and as projectiles useful in defending scavenging sites against other predators.

common ancestors and their other descendants. To understand how and why that is so, however, requires some further ground clearing.

Comparative consideration of the cognitive capacities of humans and nonhuman animals has long been plagued by unwarranted claims for human exceptionalism and "superiority" but also by anthropomorphic projections of human traits, capacities, and needs onto nonhuman animals. Consider two influential but problematic alternative forms these assumptions have taken. First, the ability to speak a language has been assumed to be a general cognitive capacity, which in principle might be both attainable by and functional for other organisms. If that were so, not having achieved a capacity for or an actual grasp of even a rudimentary language would seem to be a cognitive deficiency in some sense, at least for organisms whose bodily organization and metabolism would be sufficient to support its physical prerequisites, whatever those may be. This sense of deficiency then can become the basis for an all-too-familiar self-congratulation; we are the one species, or one of some small number of species, that has actually developed and realized capacities for linguistic expression and understanding. In the other direction, this same assumption that language belongs on a scale of progressive cognitive accomplishments can encourage too facile an assimilation of the communicative capacities and behavior of various nonhuman organisms with human languages. Precisely in order to avoid pointless or groundless self-congratulation on our part, one may insist upon attributing manifold forms of "language"—in a more general sense of expressive, communicative behavior—to various species. The specific and peculiar features of human language can thereby recede from view.

A second problematic assumption, also motivated partly by concern to avoid problematic forms of human exceptionalism and self-congratulation, has been that animal cognition involves language-like internal representations that mediate between perception and action. Even organisms without language might conceivably engage in similar kinds of representation and processing of representations as part of their cognitive response to their surroundings in perception and other bodily activity. As Peter Godfrey-Smith (2002) and Kim Sterelny (2003, 4) have noted, this assumption takes different forms: representational content is sometimes ascribed to an organism's neural "wiring-and-connection" and sometimes attributed by a more global interpretation of its overall responsiveness to different environmental circumstances. In either form of the assumption, language would only be a further articulated, socially interactive expression of the kind of cognitive activity that already goes on in a wide range of organisms. This assumption is a substantive commitment that needs to be argued for, however, and it may well distort and diminish the cognitive capacities of nonhuman organisms. John Haugeland nicely highlights the underlying assumption that is all too often taken for granted:

> The number of "bits" of information in the input to a perceptual system is enormous compared to the number in a typical symbolic description. So a "visual transducer" that responds to a sleeping brown dog with some [internal representation or linguistic] expression like, "Lo, a sleeping brown dog" has effected a huge data reduction. And that is usually regarded as a benefit, because without such a reduction, a *symbolic* system would be overwhelmed. But it is also a serious bottleneck in the system's ability to be in close touch with its environment. Organisms with perceptual systems not encumbered by such bottlenecks could have significant advantages in sensitivity and responsiveness. (1998, 220)

I shall argue that nonhuman animals are indeed perceptual/practical systems unencumbered by symbolic bottlenecks. Not surprisingly, human perceptual responsiveness also then needs to be understood similarly, although our niche-constructive adaptation to development within a public discursive practice complicates the role of perception in our overall cognitive engagement with the world. If this line of argument is correct, then the familiar strategy in cognitive science and cognitive ethology of understanding animal cognition in terms of inner processing of representations would be doubly mistaken. The first mistake would be to ascribe internal cognitive-representation processing to other organisms to account for their often rich, subtle, and complex perceptual, practical, and affective responsiveness to their surroundings. That mistake would be a form of anthropomorphism that mistakenly projects language-like contents onto the cognitive repertoires of nondiscursive organisms. The perceptual/practical articulation of an experientially significant environment in nonhuman organisms may not map onto our conceptually articulated understanding in the ways that such projections suggest. The mistake may go all the way down to the postulation of an internal "central processor" intervening between perception and action at all rather than seeing what goes on in the central nervous system as part of a more complex bodily responsiveness to the organism's surroundings. Second, as we shall see, evolutionary considerations strongly suggest that we not think of conceptual understanding in terms of internal representation even in our own case. Conceptual understanding instead emerges as part of our own responsive engagement with, and evolutionary history within, an environment reconfigured by the salience of language and discursive practice more generally. If that suggestion is correct, representationalist conceptions of nonhuman animal minds would be a form of pseudoanthropomorphism projecting a mistaken conception of human cognition onto nonhuman animals.

To see why these two assumptions—that language is a general cognitive capacity and that cognition is a form of language-like representation processing—might be mistaken, we need to understand better the character of most organisms' practical/perceptual interaction with their environment. Consider first the oft-noted point that an organism's "selective environment" is not composed of everything in its immediate physical surroundings but is instead defined in relation to the characteristic way of life of that organism (Brandon 1990; Brandon and Antonovics 1996). Richard Lewontin incisively captured the significance of this point about the organism/environment relation:

> Every element in [an ornithologist's] specification of the environment [of a bird species] is a description of activities of the bird. As a consequence of the properties of an animal's sense organs, nervous system, metabolism, and shape, there is a spatial and temporal juxtaposition of bits and pieces of the world that produces a surrounding for the organism that is relevant to it. . . . It is, in general, not possible to understand the geographical and temporal distribution of species if the environment is characterized as a property of the physical region, rather than of the space defined by the activities of the organism itself. (Lewontin 2000, 52–53)

Less often remarked is the important converse of this constitutive coupling of organism and environment: every biologically significant trait of the organism is a mode of responsive interaction with some aspect or aspects of its environment. The biologically significant properties of an organism's sense organs, nervous system, and metabolism are patterns of interconnected responsiveness to its environment—indeed, that responsiveness is what determines them as sensory, nervous, and metabolic.[10]

Kathleen Akins (1996) has argued persuasively that understanding organisms' sensory systems as responsive to a selective environment requires thinking differently about what organisms register perceptually. Sensory systems do not register objective properties of the environment but instead detect differences that matter to the organism's physiological or behavioral response to those conditions. Akins strikingly characterizes

10. For more extensive discussion of the close entanglement of organism and environment in development and evolution, see Sultan (forthcoming).

these organism-relevant features as "narcissistic properties," although I prefer to think of them as "intra-active properties." The neologism acknowledges that the organism for which these environmental properties are "narcissistically" defined is not a separately definable entity but instead only exists in sustaining these characteristic forms of environmental intra-action.[11] In Akins's primary example, an animal's thermal detection system does not register continuous changes in the ambient temperature but instead is only sensitive to discontinuous thresholds for different behavioral or physiological responses to high or low and increasing or decreasing temperature:

> This [complex interrelation of static and dynamic thermoreceptor response properties] seems somewhat strange on the traditional view of sensory processing, of thermoreception as a system that disinterestedly records temperature facts. Just how inept could this system be? Viewed as narcissistic, however, the system makes perfect sense. What the organism is worried about, in the best of narcissistic traditions, is its own comfort. The system is not asking, "What is it like out there"?—a question about the objective temperature states of the body's skin. Rather, it is *doing* something—informing the brain about the presence of any relevant thermal events. Relevant, of course, to itself. (Akins 1996, 348–49)

Akins's point does not just connect already-determinate sensory input with behavioral output, however. A sense organ such as an eye is not just a passive receptor but is instead embedded within characteristic patterns of bodily movement in ongoing response to changes in the organism's surroundings.[12] Those movements, from saccades to focal changes, eye

11. I take the term 'intra-action' from Karen Barad (2007), who coined it to call attention to those phenomena in which the supposedly interacting objects only have definite boundaries *as* objects through their involvement in these constitutive patterns of intra-action. In this case, the relevant "objects" are organisms and their environments. An unfortunate association of Akins's term "narcissistic" is with conceptions of evolution that emphasize "selfishness" as motivating all animal behavior (rather than goal-directedness). Apart from the difficulties with isolating motivation from a selective regime that sustains various patterns of environmental response, such conceptions tend to isolate the organism, identified with a kind of "self-interestedness," from its entanglement with its developmental and selective environment. In part 2, I will explore the relations between the natural or laboratory phenomena that allow the conceptual articulation of scientific domains and the biological phenomena I am considering here. Part of my argument is that scientific understanding is itself a further extension of the underlying biological phenomenon of discursive niche construction.

12. I shift examples to consider vision rather than Akins's example of thermoreception because of the pervasive but highly uneven and differential distribution of thermoreception throughout the body. The mistake that identifies vision with the impact of light on the rods and cones rather than with characteristic patterns of bodily movement in responsive intra-action with what happens in the rods and cones thereby stands out more sharply. Yet the same point applies in both cases. Temperature sensitivity is not a capacity of thermoreceptor cells in isolation but instead of

and head movements, or bodily repositioning, are partially constitutive of visual and other forms of perceptual experience. As Alva Nöe (2004) and others going back at least to Merleau-Ponty (1962) have insisted, perception is a mode of enactive skill rather than the passive registration of an image or other feature. Moreover, the exercise of these skills is not simply a matter of taking in information from the environment and formulating a bodily response. Perceptual awareness extends beyond what physically impacts sense organs to incorporate what is perceptually accessible, such that relevant patterns of bodily response are already built into what is registered perceptually. In this respect, Haugeland notes that "the very distinction between perception and action is itself artificially emphasized and sharpened by the image of a central processor or mind working *between* them, receiving 'input' from the one and then (later) sending 'output' to the other. The primary instance is rather *interaction*, which is simultaneously perceptive and active, richly integrated in real time. . . . There is little reason to believe that symbol processing has much to do with it—unless one is already committed to the view that *reasoning* must underlie *all* flexible competence" (1998, 221).[13]

There is, however, a wide range of characteristic ways in which organisms respond to their surroundings. Recognizing the real differences among these forms of intra-action of organisms with their biological environments has mistakenly seemed to support the idea that the more flexible and robust forms of interaction must involve something like symbolic representations decoupled from the organism's immediate practical/perceptual engagement with its environment. Kim Sterelny (2003) highlights the most important difference in play here, although we should remember that we are considering a distinction along a continuum rather than two discrete kinds of practical/perceptual intra-action with an organism's environment. At one level, there are what Sterelny calls detection systems, which are "mechanisms that mediate a *specific adaptive response* to some feature (or features) of their environment by registering *a specific environmental signal* that tells the organism of the presence of that feature" (2003, 14). Many organisms ("detection agents") have a behavioral repertoire largely composed of such tight couplings of signal and response, including sequential cascades

characteristic patterns of movement and physiological function in response to differences and changes in temperature. As Akins herself notes, the differential distribution of thermoreceptors throughout the body "is a fact that will strike you as immediately plausible if you imagine wading into a cold lake. As a matter of fact, some steps *are* harder to take than others" (1996, 346).

13. For reasons noted above, I would prefer to speak of "intra-action" rather than 'interaction' as Haugeland does.

of such couplings that can generate more complex forms of behavior, from courtship rituals to the oft-cited egg-laying behavior of the sphex wasp (Wooldridge 1963). Many of the cues prompting such detection cascades are themselves generated by previous performances by the organism or its conspecifics and companion species. These include chemical signals like pheromone trails, physical markings or constructions, and behavioral responses that function as signals generating the next response, including many forms of animal communication. The result can be highly complex and (mostly) adaptively appropriate responses to the organism's environment. Detection systems offer many evolutionary advantages: they can be reliable, metabolically cheap, environmentally appropriate, articulated when combined in more complex arrays, and appropriately geared to the comparative costs to the organism of "false positives" and "false negatives."[14]

The limitations of detection systems become clear, however, when one recognizes that the features of the environment that would be most relevant to the organism's continual survival and reproduction might be partially disconnected from the features the organism can reliably detect. Sometimes the organism's informational environment can become translucent or opaque because the causal sequences that cue the organism's characteristic response pattern are only indirectly and contingently connected to the features of its environment that matter to its way of life. Sterelny emphasizes, however, that informational translucence or opacity most commonly arises when other organisms play a prominent role in the organism's perceptual and behavioral environment: "Action can be safely based on a single cue only when it is directed toward indifferent or cooperative features of the environment. Thus adaptive behavior targeted on the inanimate world (and biologically indifferent parts of the animate world) can often be controlled by simple cues of

14. The scare quotes around the expressions 'false positives' and 'false negatives' reflect Haugeland's earlier point that such organismic systems can be abnormal, and can even embody design limitations, but cannot be mistaken. The reason is that the only norms in play here are normal-functional (or, as I shall argue shortly, normal-purposive). Whatever the system actually detects when it develops and functions normally is what it "aims" to detect, including whatever seemingly false positives or false negatives could be identified by redescribing the system's aims in terms drawn from our conceptual understanding. Such redescription can usefully explain why the actual system functions as well as it does in its normal environments but does not thereby show that the system is somehow aiming for something "better" but rather is falling short. This recognition is heightened when one realizes that the adaptive functionality of a detection system cannot be assessed in isolation but must instead be understood as integrated within the organism's entire cognitive and metabolic economy. A more subtly or flexibly discriminatory system might be counteradaptive depending upon its metabolic demands, its effects on the speed or reliability of response, or its maladjustment to the costs of different forms of insensitivity to difference.

environmental structure. . . . In contrast, biological agents pose far more difficult epistemic problems. An animal's predators, prey, and competitors *are* under selection to sabotage its actions" (2003, 25). Under such conditions, simple detection systems are targets for deception, mimicry, and disruption by other organisms.

In response to environmental translucence, organisms typically develop more robust tracking capacities, combining information from multiple aspects of their surroundings, arriving through multiple channels. They also develop more flexible behavioral repertoires in response to these more complex, ambiguous, and often novel combinations of potentially relevant environmental features. Such repertoires can develop differentially by tracking and responding to a particular individual organism's history of perceptual encounters and the sequential outcomes of their own previous past responses to those encounters. Although such complex and flexibly responsive perceptual-behavioral repertoires are readily distinguishable from simple detection systems, they are not fundamentally different in their normative character. Sterelny expresses the point with admirable clarity:

> In translucent worlds there is a complex relationship between the incoming stimuli that the organism can detect and the features of relevance to it. When no one cue is sufficiently reliable, selection can favor the evolution of the capacity to make use of multiple channels. . . . Agents with robust tracking—with the ability to use several cues either built-in or learned—have islands of *resilience* in their behavioral repertoire. The cues that control behavior have become flexible and intelligent. . . . Robust systems, like detection systems, are [nevertheless] behavior-specific. Their function is to link the registration of a salient feature of the world to an appropriate response. (2003, 27–29)

As such robust and flexible response systems become more complex, they enable differential responsiveness to quite high-level, global patterns of environmental features. The combinatorial possibilities also allow for novel responses to unprecedented stimuli, especially when the relevant features take the form of other organisms' behavior in response to the organism's prior behavior and the organism is capable of tracking those sequential interactions. Relatively robust social relationships such as dominance hierarchies, intermittent interactions with other conspecifics in fission-fusion social groupings, or the ability to track and respond differentially to the past behavior of individual organisms can then be incorporated into what is at base a fairly direct coupling of environmental configuration and perceptual-behavioral response. These high-level, multiple-channel, flexibly responsive couplings of behavioral repertoires with perceptual uptake include what J. J. Gibson (1979) famously described as organisms' perceptual responsiveness to the global "affordances" of a situation for its own possible, significant behaviors.

These patterns of responsive organism-environment intra-action have a characteristic normativity. Philosophers and biologists have commonly but mistakenly articulated this normativity in proper-functional terms: the function of the heart to pump blood through the circulatory system, for example, or the function of thermoreceptors to inform the organism of life- or comfort-threatening conditions or changes of condition. On Millikan's (1984) influential formulation, the "proper function" of some biological trait in a population of organisms is the function it served in their ancestors in a lineage such that the trait was selected for and consequently maintained in organisms of the current generation. As Mark Okrent (2007) notes, such accounts could at best explain the statistical prevalence of the trait within a population rather than its presence within any individual organism. The more fundamental problem, however, is that such functionalist accounts presuppose, but cannot account for, the goal of maximizing the fitness of the organism: "Millikan attempts to explicate the teleological notion of a function by appealing to what she takes to be the unproblematic notion of what an item with a function was selected to do by natural selection. . . . *That* some item was selected for satisfying some function implies that it contributed to the goal of maximizing the fitness of some containing system. But which containing system that is, and what it is for an item to serve some goal, is not thereby defined" (Okrent 2007, 103). Okrent then argues that the only plausible way to rectify this deficiency is to understand organisms in terms of goal-directed rather than functional teleology. The goals in question, however, are not something extrinsic to organisms' various goal-directed activities but are instead the maintenance of the organism's own life-constitutive patterns of intra-action with its environment. Organisms are processes rather than substantial entities, and their various performances make sense (defeasibly) as contributions toward the maintenance of those processes within a changing environment. That environment in turn is only picked out as such by its relevance to the possible satisfaction of that goal:

> Organisms are, essentially, agents that act on [i.e., intra-act with] their environment in order to realize ends that are intrinsic to and necessary for their continuance. And, insofar as organisms [intra-]act, that [intra-]action *itself* amounts to the organism taking features of its environment *as* serviceable or detrimental to its interests. . . . Since [these] meanings of things are revealed only in light of the context of significance established by the interlocking interests of the organic agent, each of those meanings

are defined only in relation to the meanings of the other things in the world of the agent. The world of the organism is not a collection of independent things. It is a context of significance, where that significance is relative to the organic interest and ends of the organism [and vice-versa].[15] (Okrent 2013, 135–37)

Okrent introduces this account of organismic teleology as the setting for a philosophical account of intentionality and conceptual articulation. He explicitly contrasts his naturalist-pragmatist approach to the more philosophically common strategy that identifies intentional directedness with some form of "mental" representation (whether the underlying representational structure is supposedly grounded in the organism's inner neural wiring and connection, in other organisms' habitual strategies for interpreting their performances, or both).[16] Sterelny (2003) explores the possibility of splitting the difference, by understanding much of animal cognition in terms of a robust and flexible goal-directed responsiveness to environmental affordances, while considering a possible further evolutionary transition in the hominid lineage and perhaps elsewhere that incorporates representational intermediaries. Such a transition would allow for the registration and recall of internal representations of features of an organism's environment decoupled from any specific behavioral responses to what is thereby represented.

To assess Sterelny's proposal, it is important to recognize that, unlike the capacities for detection and robust tracking he discerns in other organisms, accounts of decoupled representation are not descriptions of what an organism does. What is proposed is instead an explanatory hypothesis to account for a higher level of perceptual robustness and responsive flexibility in what the organism does. Sterelny (2003, 45–50) points out that the evidence for any form of decoupled representation in nonhuman organisms is limited and equivocal, for very good reason. Given the richness and flexibility of robust tracking systems, it is difficult to know how to discriminate tracking of multiple, conflicting environmental indications from decoupled representation. Behavioral responses to actual circumstances in all their complexity provide the

15. I have interpolated into the quotation the [intra-] of intra-action and the concluding acknowledgment that environmental significances and organismic interests are reciprocally determined. This entanglement of the organism with its environment in determining what is the goal of its various intra-actions is recognized throughout Okrent's account, but it is not explicitly marked in his terminology of organisms acting on or interacting with their environments.

16. I am indebted to Godfrey-Smith (2002) for this way of both distinguishing and amalgamating familiar representationalist accounts of intentionality and to Sterelny (2003) for expanding Godfrey-Smith's reference to "inner wiring" to incorporate "wiring-and-connection."

only evidence for what an organism tracks or represents. Much of the information that one might imagine as internally represented can instead be reconstructed by the organism in response to sequentially appearing indications, including using the organism's own response to one perceptual feature as a component of the multiple signals that collectively generate subsequent responses.[17] Thus even Sterelny, who is inclined to defend the claim that folk psychological categories map onto an evolved representational structure in human cognition, nevertheless finds reason to doubt that spatial orientation, ecological information, or even social intelligence among primates has provided clear evidence of the presence of decoupled representations in nonhuman organisms. This lack of evidence for mental representation in other organisms is not due to any supposed simplicity, rigidity, or linearity in their behavioral responses, however. On the contrary, decoupled representations may simply be unnecessary to explain the intelligence or instrumental rationality of animal behavior given that capacities for robust tracking and flexible responsiveness to high-level combinations and sequences of multiple environmental features can account for extraordinary complexity, subtlety, and flexibility in animal behavior.

There are stronger reasons, however, for doubting whether decoupled representation plays a role in the cognitive repertoires of nonhuman animals, including our closest evolutionary relatives among the primates. First, the very organisms with more developed and sophisticated cognitive capacities, such as nonhuman primates, marine mammals, or birds, which might otherwise be presumed to be the organisms most likely to develop sophisticated representational capacities, are precisely the ones whose developed cognitive capacities provide the greatest barrier to decoupled representation. If their prior cognitive capacities consist in more robust tracking of situational information through multiple channels, combined with more flexible motor and behavioral responsiveness to different perceptual configurations, then their evolved patterns of cognitive development would make them all the more attuned and sensitive to the rich detail and subtle differentiations accessible perceptually and practically in their surrounding circumstances.[18] Bickerton (2009)

17. A phenomenologically familiar parallel to this pattern within our own cognitive repertoires is manifest in how we recall memorized verbal patterns like songs, poems, or speeches. Typically, we do not have "random access recall" but instead use the recurrence of one word or line as part of the basis for producing its successor.

18. I describe this attachment to circumstances as perceptual and practical because a crucial part of what perceptual uptake leads to is further exploratory movement to acquire relevant circumstantial information. Such movement ranges in scale from saccades of the eyes, to turning the

points out that the known communication systems of nonhuman animals are similarly situational, functioning as perceptual cues rather than symbolic expressions: "[Animal communication systems] aren't just a cheap substitute for language, but something completely different. Their users, in the process of reacting to situations, provided clues as to how other animals should react to those situations; interpreting such clues correctly improved those animals' chances of survival" (2009, 18). Moreover, as Marc Hauser (1996) has argued in his detailed overview of the evolution of animal communication, the information conveyed in and picked up through nonhuman animals' communicative signaling is almost exhaustively concerned with three fitness-relevant features of the animals' circumstances: individual survival, mating and reproduction, and interactions within a group of social animals. Communicative expressions are thus all the more tightly integrated into perceptually salient circumstances and generally serve to reorient the receiving organisms' response to those circumstances rather than to provide occasions for symbolic displacement.

Under those circumstances, one should expect that the primary barrier to the acquisition of linguistic abilities and repertoires in an organism previously lacking them would be the difficulty of acquiring a capacity for symbolic displacement from what is perceptually salient. The development and evolutionary enhancement of more subtle, robust, and flexibly responsive capacities for integrating perceptual tracking and behavioral responsiveness requires a multiple-channel, "broadband" receptive engagement with situated perceptual configurations. These sensitive and robust capacities would need to be blocked or suspended for an organism to attend to something perceptually present as symbolically significant, thereby turning its orientation away from rather than toward intimate involvement in its surrounding circumstances. As I noted earlier, that expectation is precisely what one finds in the experimental literature on animal language learning in both directions. Adult organisms with developed practical/perceptual capacities are too strongly attuned to their current circumstances to grasp the symbolic rather than perceptual/practical significance of signs introduced to them by experimenters, except with great difficulty and in the most rudimentary ways. For these adult animals, the symbolic expressions introduced and learned within these

head or pricking up the ears, to tracking a scent, to the seemingly counterintuitive phenomenon of predator-tracking, in which animals move toward a predator to keep it in view rather than fleeing its vicinity.

experiments continue to function more as perceptual cues rather than displacing symbols (Deacon 1997; Bickerton 2009). On the other hand, the organisms that do more readily acquire some grasp of symbolic displacement, notably Kanzi, do so when extensively exposed to the symbolic uses of perceivable markers during a relatively immature stage of development, before their capacities to attend, track, and respond to multiple entangled perceptual features have been fully developed. Immature organisms, including young human beings, more readily discern and respond to symbolic expressions precisely because they encounter fewer barriers to displacement.

Haugeland's arguments for distinguishing the cognitive capacities of nonhuman animals from full-fledged intentional or conceptual capacities take on a new significance in this context. Sterelny's attempt to graft folk-psychological accounts of decoupled representation onto a broadly naturalist-pragmatist strategy for understanding most animal cognition and Deacon's and Bickerton's alternative proposal that appeals to discursive niche construction in the hominid lineage are also relevant here. Haugeland had concluded that we should interpret nonhuman animal behavior as "extensionally" picking out patterns of normal response to environmental circumstances; we should not regard such behavior as intentionally and intensionally directed toward objects under an aspect, which might be mistaken in either respect. The resulting implicit classification of those circumstances is then often gerrymandered from our perspective.[19] Any lack of conceptual coherence to their behavioral responses does not signal a mistake on their part, however. There is no basis in their behavior itself for ascribing a striving toward, but falling short of, a differentiation of objects or their features other than their actual patterns of normal response.[20] Haugeland thereby treated

19. I now interpolate "mostly-successful goal-directed activity" for what Haugeland himself described as "normal functioning," following Okrent (2007) for the reasons discussed above.

20. An important reason for Haugeland taking this stance, although he does not put it in these terms explicitly, is the holism of an organism's behavioral response to multifaceted circumstances, also discussed in chapter 4. As biological theorists attempting to explain the evolution of these changing patterns of responses, we might regard them as expressing a satisficing compromise among multiple, distinct, conceptually coherent classifications. For example, there might seem to be evolutionary trade-offs among capacities to discriminate some fitness-relevant feature of the organism's environment, the energetic costs of maintaining and utilizing more fine-grained discriminatory capacities, the enhanced risks of predation due to more sustained exploration of available information, and structural constraints upon the development of multiple capacities together. The differentiation of such conceptual categories that are then recombined in a supposedly satisficing compromise would, however, be accomplished within the biological explanation rather than in the behavior being explained. The organismic lineage incorporates various actual responses to actual

nonhuman animals' cognitive repertoire as exhibiting what he called "ersatz intentionality." Such cognitive and behavioral repertoires could at best partially mimic the genuinely intentional normativity of human activity, including the perception of objects as objects.

We can now see why Haugeland is right to insist upon this difference and yet is mistaken from an evolutionary perspective in therefore characterizing animal cognition as an "ersatz" mimicry of any genuine intentionality that displays conceptual understanding. His implicit comparison to a conceptually intelligible pattern is simply irrelevant to the organism's responsiveness to its environment. What we see in nonhuman animals is a robust capacity to discriminate and respond flexibly and mostly appropriately to subtle, often disguised aspects of their actual circumstances that matter to their species-characteristic way of life, including novel behaviors by other organisms. Moreover, these perceptual discriminations and motor-behavioral responses are not distinct but correlated subsystems of the organism's overall way of life. They instead constitute an integral entanglement of the organism's physical and behavioral repertoire with its selective environment. As Haugeland himself recognized in other contexts, these practical/perceptual capacities gain extraordinary sensitivity, robustness, and flexibility precisely through a receptive openness to relevant aspects of their immediate surroundings. Yet it is that openness and sensitivity to their immediate surroundings that also binds them cognitively and affectively to those circumstances. Although something akin to displaced representational content can sometimes be mistakenly read into such performances, their characteristic feature is instead the tight, sensitive coupling of their perceptual, practical, and cognitive capacities to salient features of

patterns of circumstances, some of which are then reproduced in later generations, and some of which disappear. It is no objection to the biological explanation that it draws upon conceptual divisions that the organism itself cannot differentiate. The explanation of why the organism's actual pattern was successful draws upon counterfactual considerations that we introduce as part of biological theorizing, whereas the organisms only undertake holistic patterns of response to actual circumstances.

The role of the counterfactual considerations in establishing conceptual classifications is nicely illustrated by Daniel Dennett's (1991) argument for taking noisy patterns (i.e., patterns that seem to incorporate some "errors" in the patterning) to be real patterns. The crucial consideration behind Dennett's argument is that a pattern that contains "noisy" interruptions of the pattern *is* a real pattern if the pattern would persist even if the noise were randomly revised. Yet the counterfactual revisions presuppose a prior distinction between signal and noise. From the organism's perspective, that is precisely the distinction that is in question, since what marks noise for a conceptually articulated, counterfactually revisable classification for explanatory purposes is, from the organism's perspective, an integral part of the overall signal to which the organismal lineage is responding.

the selective environment defined by its relevance for their way of life. Language and conceptual understanding must then be recognized as derived, partial modifications of these evolved, "nonintentional" capacities for robust tracking and flexible response to the organism's ongoing intra-action with its environment.[21]

Both Bickerton (2009) and Deacon (1997) emphasize in this context that the central challenge for a more adequate grasp of the evolution of human language is to understand the circumstances and selection pressures in the hominid lineage that initially loosened that tight coupling of mind and environment so as to permit the emergence of even rudimentary forms of language and symbolic displacement.[22] Each offers a hypothesis for the critically distinctive selection pressure that first overcame the cognitive barriers to displacement.[23] If one's aim is instead

21. Part of what is at issue in this entire discussion is how to think about intentionality and conceptual capacities, including the kinds of continuity or discontinuity exhibited by humans' and nonhuman animals' perceptual and active responsiveness to their surroundings. Other organisms' conjoined perceptual and active capacities are clearly directed toward and responsive to their environments, so to that extent they are intentional. The question implicit in the use of scare quotes is whether the "extensional" determination of what belongs to an organism's selective, developmental, or ecological environment, in part through its physiological and behavioral response to its surroundings, is also aspectual and normative in the ways usually characterized as both intentional and intensional. At this point in the book's argument, this question remains open.

22. Okrent (2007) proposes an alternative approach that more closely amalgamates human and animal cognition. The difference between human understanding and most animal cognition is that the latter ways of "taking-as" are only vaguely articulated, whereas human intentional comportments are capable of much more extensive and multifaceted articulation. This difference in turn is explained by augmenting the instrumental rationality characteristic of organisms that embody robust tracking systems (rather than simply correlated detection systems) with the practical rationality involved in the use of tools (including language) in socially appropriate ways in the context of stable, socially articulated roles. In chapter 4, I shall argue that while Okrent is right to see indefinitely extensible articulability of concepts as characteristic of human discursive niche construction, he is mistaken in postulating *this* form of continuity between vague animal concepts and articulated human ones. Haugeland is right in (retrospectively but nonteleologically) taking the highly flexible and robust forms of animal intra-action with their environments as a simulacrum of conceptual understanding rather than as a more vaguely articulated version of it. Okrent's and my accounts converge in most respects but differ over whether nonhuman organisms' environmental intra-actions involve (vaguely articulated) taking-as rather than just robust tracking and flexible responsiveness that has no symbolic or conceptual content.

23. In order to explain the emergence of language and symbolic displacement in the hominid lineage, Deacon appeals to hypothetical problems of hominid social and reproductive relations. Bickerton develops a more detailed hypothesis, grounded in ecological considerations that emphasized changes in food sources and vulnerability to predation and also paleontological evidence for early hominids' extensive production and use of the Acheulian hand axes found in very large numbers at early hominid sites. The key part of his explanatory account is that language emerged as part of the exploitation of a new food source: scavenging the carcasses of megafauna. Hand axes would enable hominids to both cut the skin of decaying bodies several days before other scavengers could do so with only claws and teeth and also defend the site against other scavengers and predators. Scavenging large carcasses would both support relatively large groups and require large

to understand the resultant human cognitive capacities, however, it matters less which, if either, of these explanatory hypotheses is correct. What does matter is the form of the problem, due to the conjoining of two key points: first, that symbolically displaced conceptual capacities would normally be selected against, so that even if the capacity for symbolic displacement is latent, it won't actually develop; and second, that once these initial cognitive and selective barriers to symbolic displacement are overcome, in whatever way, the selection pressures could easily reverse direction in favor of easier and more rapid protolanguage learning. Early hominids somehow did develop a first rudimentary linguistic practice, whichever hypothesis most adequately accounts for it. That early protolanguage could then coevolve with human anatomy, physiology, and neural organization under selection pressure for rapid and reliable acquisition of linguistic competence and whatever subsequent abilities were cumulatively "ratcheted" from those base features of discursive social life.

The philosophical issue is then to characterize the resulting capacities and achievements more adequately and to understand how they relate to and affect humans' continuing capacities for perceptual and practical responsiveness to our circumstances. If the perception and behavior of nonhuman animals is not fully "intentional" in the traditional sense of expressing a conceptual content that might mistakenly characterize aspects of their environment, then we need to understand the difference between these two ways of being directed toward or "about" something. Moreover, since we remain animals whose perceptual and motor capacities are largely continuous with those of other animals, we need to understand how our continuing perceptual and practical abilities contribute to and are affected by our developed capacities for linguistic expression and conceptual understanding.

---

groups to defend the site. Since carcasses are both widely and randomly distributed, however, such a way of life would require the ability to forage widely in smaller groups and then to assemble large groups rapidly when significant food sources were discovered. This hypothesized need to exchange information within dispersed fission/fusion groups and mobilize collective action at distant sites suggests a significant selection pressure for symbolic displacement in communication. I find this hypothesis reasonably plausible, but my own argument does not turn on its empirical adequacy, for reasons discussed in the main text. In chapter 4, I also argue that Bickerton's hypothesis is not yet sufficient to account for the emergence of symbolic displacement, even if correct. In later writings, Bickerton (2014) explicitly recognized that his hypothesis could only explain the emergence of a protolanguage that falls short of what is needed for the kinds of linguistic and conceptual capacities later displayed by humans. Bickerton's account of the transition from protolanguage to language is discussed below.

III—Niche Construction and the Coevolution of Language

Most accounts of the evolution of language and conceptual understanding have started from the presumption that novel capacities of the hominid brain or central nervous system are what need to be understood. The relevant capacities have typically been identified as involving decoupled symbolic representation, especially the capacity to understand and produce syntactic combinations and recursive recombinations of symbols. These would indeed be the relevant considerations if the primary issue for the emergence of language were the abilities to understand and produce an indefinitely large range of novel, syntactically well-formed and semantically contentful expressions. Posing the issue in these broadly Chomskian terms has strongly suggested treating language as a relatively self-contained cognitive module. Additional empirical support for modular conceptions of linguistic competence has come from the evidence that various forms of aphasia (cognitive impairment of linguistic ability) are relatively independent of other forms of cognitive impairment. The more basic underlying conception has nevertheless been that the capacities for language and for symbolic understanding (however distinct they may be) are "internal" genetically coded capacities of human brains, which evolved under "external" selection pressure for general intelligence, behavioral flexibility, and socially coordinated action. The challenge then seemed threefold: to understand the representational and computational demands of linguistic competence, to understand how those capacities are realized in the human brain, and to understand the original variance and selection pressures that led to the evolution of the developed phenotype.

This framing of the issue now looks increasingly questionable. Two considerations, one critical and one constructive, pose especially important challenges to the assumptions underlying internal-representationalist accounts of linguistic competence. The critical consideration stems from recognition of the importance of robust perceptual tracking and flexible responsiveness to complex environmental configurations. As Haugeland once noted, "Perception is cheap, representation expensive" (1998, 219). In both metabolic and cognitive terms, behavioral reliance upon decoupled representations could be difficult and dangerous for an organism in several critical respects: maintaining a substantial representational storage, sustaining sufficient real-time updating to enable those representations to remain responsive to perceptual inputs and relevant to action in diverse settings, and providing relevant access to what is then effectively a large representational database. The difficulties come from the need for fairly comprehensive representation of relevant circumstances and the problems of scale and framing that consequently confront the organism's capacities for storage, updating, and real-time access. The danger comes, in turn, from the possible costs of relying upon outdated, irrelevant, or inaccessible representations in the wide range of situations relevant to fitness. If some nonhuman animal behavior already did utilize decoupled representation to a significant extent, then it would plausibly be adaptive whenever an organism could "off-load" to its perceptual capacities the problems of storage of and relevant access to representations of its environment. From an evolutionary point of view, however, that formulation expresses the issue backward. These problems of storage, access, and updating would lead to significant selective pressure against reliance upon decoupled representations in the first place. As a result, some of the very same considerations that have prompted the postulation of relatively hardwired modular cognitive capacities may instead indicate a selection pressure toward circumventing internal, decoupled representation altogether.

The constructive consideration is a recent development in evolutionary theory that emphasizes the significance of "niche construction" as a mechanism of evolutionary change (Odling-Smee, Laland, and Feldman 2003; Odling-Smee and Laland 2009; see also Sterelny 2003; Bickerton 2009, 2014). The concept of a niche originated in ecology, originally understood as a property of an organism's environment (e.g., Grinnell 1924, Elton 1927) but later reconceived (Hutchinson 1957) as an attribute of a population in relation to its environment (Colwell 1992). Whereas Hutchinson identified the ecological niche of a population with the environmental factors acting on these organisms, niche construction theory revises the concept for evolutionary biology by redefining it as the sum of selection pressures acting upon the population. This concept has a dual character relating to both the organism and its circumstances: "[An evolutionary niche] refers to natural selection pressures relating to the 'lifestyles' of organisms, and therefore to the many different ways in which different organisms survive by actively interacting [intra-acting—JR] with their environments. . . . It also refers to the real habitats of organisms in real space and time, . . . from which [the population] is actually earning its living, from which it is not excluded by other organisms, and in which it is able either to exclude other organisms or to compete with other coexisting organisms" (Odling-Smee, Laland, and Feldman 2003, 40).

The crucial recognition underlying niche construction theory is that the selection pressures bearing upon various organisms are significantly modified by the cumulative effects of these and other organisms' intra-actions with their more or less shared selective environments.[24] The result fundamentally changes how biologists conceive of evolution: "Niche construction should be regarded, after natural selection, as a second major participant in evolution. . . . Niche construction is a potent evolutionary agent because it introduces feedback into the evolutionary dynamic. Niche construction by organisms significantly modifies the selection pressures acting on them, on their descendants, and on unrelated populations" (Odling-Smee, Laland, and Feldman 2003, 12). The associated patterns of ecological inheritance then constitute a distinct but intra-active mode of inheritance alongside the genetic inheritance patterns that have previously been conceived as the evolutionary legacy that organisms bequeath to subsequent generations. Along with the genes they receive from their parents, organisms inherit a transformed environment exerting different selection pressures due to the cumulative effects of other organisms' activities on their selective environment.

Niche construction theory also draws upon renewed efforts to integrate developmental and evolutionary biology. Genetic inheritance is itself only expressed in the organism through developmental processes that involve environmental intra-action (Oyama, Griffiths, and Gray, 2001). The selectively relevant outcomes of genetic transmission are thus also exposed to the feedback effects of niche construction. These nonlinear relations between biological lineages and their developmental, ecological, and selective environments were long thought to have no evolutionary significance because of the changes from generation to generation in individual organisms' developmental environment. To the extent that populations of organisms inherit persistent changes in their normal developmental environment, however, even the evolutionary significance of genetic inheritance will be affected by developmental intra-action with that persistently transformed environment.

These intra-active consequences of niche construction can have sustained evolutionary impact in two further, widely recognized ways. The phenotypic plasticity exhibited in different developmental outcomes in different environments can then become genetically fixed if it is sufficiently advantageous under sufficiently stable features of the organism's developmental environment (Jablonka and Lamb 2005; Kirschner

24. See Brandon (1990, especially ch. 2) and Brandon and Antonovics (1996) concerning the difference between the physical surroundings of a population of organisms and the population's selective environment (defined in relation to those organisms' way of life, as the configuration of factors that influence the continuation and reproduction of that way of life).

and Gerhart 2005). This consideration has special significance in understanding cognitive evolution. Animals' neural "wiring-and-connection" is well known for its plasticity, as neural connections are established and reinforced or allowed to decay in the course of the organism's subsequent activity (Edelman 1987, 1992; Deacon 1997; Dor and Jablonka 2010). Genetic fixation of selectively important patterns of neural organization, including those that reduce learning time for critical tasks and skills, are therefore likely to be especially common.

In human beings and any other organisms that are capable of imitation of advantageous behaviors, a second locus for the evolutionary significance of niche construction is the cumulative "ratcheting" effect through which behavioral changes within the lifespan of an individual organism can be passed on to others: "The process of cumulative cultural evolution requires not only creative invention but also, and just as important, faithful social transmission that can work as a ratchet to prevent slippage backward—so that the newly invented artifact or practice preserves its new and improved form at least somewhat faithfully until a further modification or improvement comes along" (Tomasello 1999, 5). The feedback effects of niche construction can therefore lead to significant evolutionary change through cumulative ecological inheritance from the effects of organisms' activities, the genetic fixation of phenotypic plasticity, and the imitative stabilization or "ratcheting" of learned patterns of behavior, including cumulative effects of iterative behavioral niche construction.

Niche construction theory most obviously suggests physical changes in organisms' abiotic environments as salient ways in which organisms reconstruct the environments passed on to subsequent generations so as to change the selection pressures affecting them. Beaver dams and ponds, loosened soil from earthworm activity, birds' nests, the fungal farms of leafcutter ants, atmospheric oxygen from the cumulative respiration of cyanobacteria, or in our own case, cities, cleared agricultural land, technological devices, and increased atmospheric and dissolved oceanic CO_2 exemplify this obvious kind of case. Many influential aspects of the developmental environment of organisms are behavioral, however. Moreover, if salient behavioral patterns of an organism become influential components of the developmental environment of subsequent generations, in ways that reliably reproduce similar behavioral patterns from generation to generation, then those patterns will also function as part of those organisms' ecological inheritance.

One of the most pervasive, salient, and reliably reproduced features of the environments in which human beings develop into their normal adult phenotypes has been the presence of spoken and now also written language. These patterns of vocal utterance are normally coupled with facial expressions and expressive gestures and postures and situated amid patterns of action and interaction in partially shared circumstances. Moreover, linguistic expressions are not simply part of the ambient environment of developing human beings but are instead produced in ways deliberately designed to initiate children into discursive practice. Linguistic expressions are frequently produced by adults in ways that make them more salient for infants and toddlers. Adults do not merely speak in the vicinity of children but explicitly address them. Utterances thus directed at children are often presented initially in simplified forms for ease of recognition and uptake and frequently correlate with shared activities. As children begin to produce their own phonemically articulated expressions, adults often respond differentially in ways intended to help mold the developing child's behavior and skills into those of a speaker of a natural language.

The emergence of a linguistically expressive developmental niche was also facilitated from the outset by other dimensions of human niche construction. Human beings are social animals whose developmental pattern is highly neotenous. The dependence of human infants upon the caretaking of adult conspecifics produced significant selective pressures for the perceptual and cognitive salience of recognizable and trackable individuality, facial expressions and expressive gestures, perceptual attention, affects, and other behavioral aspects of human life that matter to the survival of young, dependent hominids. Many of these supportive features are shared with other primates but reinforced and further articulated within human development. The evolutionary emergence of language was undoubtedly scaffolded by these already extant features of early hominids' reliable ecological and then genetically fixed inheritance. Inherited abilities to focus perceptual attention upon the expressive performances of others, and to respond differentially to various individual performances, would also be substantially enhanced by the early emergence of what Kukla and Lance (2009) call the vocative and recognitive dimensions of language.[25] The vocative capacity to direct a vocal expression at someone in particular, and to recognize an expression as directed at me and as thereby calling for an appropriate response from me, makes possible an extensive further scaffolding of

25. Kukla and Lance (2009) show the fundamental, constitutive importance of both the vocative and the recognitive dimensions of discursive practice for facilitating discursive practice and more fundamentally for their contribution to opening and sustaining a normative "space of reasons."

linguistic learning. Similarly, the ability to express one's recognition of some aspect of partially shared perceptual circumstances and thereby call others' attention to it helps establish the ongoing connection between vocal performances and perceptual recognition (including the reflexive recognition of other perceivable vocal performances). So far, however, we are still talking about perceptual and practical capacities not fundamentally different from the communication systems developed by many organisms. A behavioral repertoire that is responsive to, and therefore also indicative of, specific circumstances is a salient feature of the developmental environment for many organisms—especially if one recognizes that among the most prominent "circumstances" indicated by other organisms' expressive performances are their own affective orientation and further behavioral dispositions. If the circumstances to which such expressions are normally responsive are selectively important for the organism, there will be consequent selective pressure toward abilities for more reliable and efficient learning to produce and recognize the relevant performances.

The crucial contribution niche construction theory makes to understanding the evolution of conceptually articulated language is to change the form of the problem. The issue is no longer how human ancestors could develop an internal capacity for decoupled symbolic representation, predication, and recursive recombination of the constituent symbolic expressions in place of the familiar perceptual capacities for robust tracking and flexible responsiveness to complex environmental configurations that are characteristic of many organisms. Capacities for robust perceptual tracking bind an organism all the more tightly to attentiveness and sensitivity to the salient features of its environment, whose perceptual salience and connection to appropriate behavioral responses are enhanced and reinforced by evolution. In that context, a direct transition to decoupled symbolic understanding seems inconceivable. Niche construction theory offers instead a multiple-stage process. What comes first is protolinguistic behavior, a more extensive and articulated pattern of gestures, posture, and eventually vocal expression.[26] Robust tracking of and flexibly appropriate responses to these expressions become part of these organisms' broader perceptual attentiveness

26. Tomasello (2008) argues from comparative primatological and human-developmental research that gestural rather than vocal expression provided the initial locus for the evolutionary trajectory that produced human language. My own argument does not depend upon whether this claim is correct.

and responsive behavioral repertoire. Moreover, a much more richly articulated protolinguistic repertoire can develop as communicative performances are increasingly keyed to other communicative performances as well as to their surrounding circumstances.

Once more articulated forms of expressive communication and response are in place, however, they can be recruited and adapted for symbolic displacement. Strictly speaking, these two stages need not be simply sequential. The eventual selection pressure for articulated vocal expression, uptake, and response could have come from its utility for limited forms of symbolically displaced expression. Bickerton's (2009) more detailed hypothesis about the ecological context for language evolution exemplifies such a proposal. The underlying idea is that symbolic displacement emerged in response to the need to recruit larger bands of hominids quickly to act together at distant locations.[27] What initially emerged would not be full symbolic displacement but instead expressive behavior that functioned as indirect perceptual indications of more distant circumstances. Such capacities are not so different from an animal's warning cry that alerts other animals to hidden or unnoticed rather than distant circumstances.[28] As social behavior becomes increasingly oriented toward responsiveness to less proximate circumstances, these expressive and responsive capacities can become more articulated. Moreover, as the coordination of collective action in response to distant conditions becomes integral to a population's way of life, there will be selection pressure supporting the expansion of communicative and coordinative capacities. The crucial point is that protolanguage thereby gradually emerges as a practical/perceptual capacity for expression and response that becomes integral to a social organism's reconstructed selective niche. The organism's neural capacities and organization are then subject to selection for more effective learning and performance within this communicative setting. The expressive repertoire can expand and diversify in response to the changing demands of a social life in which protolinguistic coordination of behavior is increasingly significant.

Language thus initially emerges not as the product of enhanced internal capacities of a larger hominid brain but instead as a perceptually

27. See note 23 of this chapter for a more extensive summary of Bickerton's hypothesis.

28. The best-known example of such warning cries is provided by vervet monkeys' use of three distinctive utterances correlated with the presence of three different kinds of predators that threaten vervets from different directions (Seyfarth, Cheney, and Marler 1980). Bickerton argues that extensive vulnerability to predation was likely an evolutionary consideration common to vervet monkeys and early hominids but not to most nonhominid apes.

salient, developmentally effective, and selectively important behavioral dimension of the developmental and selective environment of some hominid apes.[29] Vocal expressiveness and its behavioral integration into a transformed way of life persisted as an integral part of these organisms' ecological heritage only through its development and reproduction in each succeeding generation. Understanding this process as cumulative niche construction allows us to recognize that language is not a general capacity for symbolic representation that may happen to have emerged in only one species. Language is instead the outcome of a historically specific trajectory of niche construction that is consequently a particular trait of that species.[30] Moreover, this trajectory is one of coevolution between language and *Homo sapiens*. Language did not and could not initially emerge in this way as anything resembling the highly articulated, recursive symbolic system now in place but instead as a communicative, perceptually responsive, and expressively constrained dimension of early hominids' behavioral repertoire. As protolanguage became more

29. There is no clear evidence at this point to differentiate between two different scenarios for the place of language in hominid evolution. One possibility is that such an articulated vocal-expressive behavioral repertoire, and even its eventual transformation into a language marked by symbolic displacement, emerged as a common feature of one or more hominid species. As I noted earlier, there is now evidence that a common primate ancestor already had the capacity for rudimentary symbolic expressive behavior (even though the anatomical modifications that permit highly articulated vocal expressive behavior do not exist in nonhuman primates). A second, possibility, already suggested by Bickerton (2009), is that the gradual emergence of protolinguistic behavior and eventually language in some hominid populations was itself integral to speciation within the hominid lineage. The emergence and behavioral integration of a highly articulated vocal expressive repertoire in some populations of social, neotenous primates could well serve as an effective form of reproductive isolation of those populations from other conspecifics. The transformed selective pressures within such a vocally expressive behavioral niche could then hasten genetic differentiation in ways that reinforced and intensified the prior behavioral differentiation among populations of vocally articulated and less articulated hominids.

30. A comparison may help illustrate what it would mean to understand language as a historical development within a specific lineage rather than as a more general trait that happens to be instantiated only in that lineage. Arthur Fine makes a similarly antiessentialist proposal about science under the heading of the "Natural Ontological Attitude" (NOA):

> NOA thinks of science as an historical entity, growing and changing under various internal and external pressures. . . . The description of science as an historical entity was intended precisely to undercut . . . the idea that science has an essence. If that were our picture, then indeed one could imagine a sort of chemistry of science which seeks for regularities in the phenomena, the laws covering that, and then looks for even deeper structures that may lie behind those—the very molecules and atoms of science! If science is an historical entity, however, then no such grand enterprise should tempt us, for its essence or nature is just its contingent, historical existence. . . . As an historical entity science is an individual, like a particular species—the horse, for example. Many sciences contribute to our understanding of the horse, but there is no "science of the horse." From an evolutionary point of view, there is only a natural history. (Fine 1986b, 173–75)

If language is the product of niche construction within the hominid lineage, then there is likewise no general science of language but only a natural history, which is itself ongoing and may lead to significant changes in what language is and how it works.

integrated within a changing way of life, human beings changed anatomically, neurally, and behaviorally in response to selection pressures distinctive to that vocally expressive way of life.[31] Language itself was also changing in that context, as the patterns that were reproduced from generation to generation adapted to accommodate what was more readily learnable and responded to the place of vocal expression and uptake within transformed patterns of human activity.[32]

Understanding language as a form of niche construction also foregrounds the perceptual and practical-performative aspects of linguistic competence. That standpoint strikingly departs from most philosophical theorizing about intentionality and conceptual understanding, which tends to work from a very thin conception of language. Language is typically identified with some relatively abstract or formal structure or an equally abstract interpretive activity such as Davidsonian (1984) radical interpretation or Brandomian (1994) discursive scorekeeping. This structure must be concretely realized in the actual situated production and consumption of token utterances and in their ultimate accountability to aspects of the world encountered through perception and action. Its material realization is nevertheless not usually regarded as integral to a philosophical understanding of language and the conceptual relations it can express. The practical-perceptual skills of speakers and listeners, their bodily involvement in the world, and the social-institutional settings in which their skills are exercised are often taken for granted as philosophically unproblematic and as distinguishable from the logical and semantic relations that they embody. Even the semantic relations

31. For detailed discussion of how the structure and connection patterns in human brains evolved under the selection pressures generated by discursive niche construction, see Deacon (1997, pt. 3). Bickerton (2014) argues that the "Minimalist Program" of Chomsky's (1995) later work in transformational linguistics readily maps onto such neural plasticity and its genetic assimilation in order to enhance the efficiency and capacity for processing and combining linguistically significant expressions, both perceptually and expressively.

32. Among the more readily trackable patterns of linguistic change that result from the need to reproduce behavioral niche construction anew in each generation are those that result from functional reanalysis (Tomasello 2008, 299–308). Because linguistic communication is situated within a larger conversational context, speakers can often use more compact expressions that rely upon that context to be understood. Listeners (including language learners) who do not share that context must interpret the contribution of each component of what is said to the overall meaning, and in the absence of shared contextual considerations, will parse those contributions differently: "Children hear utterances and just want to learn to do things like adults—they do not know or care anything for any 'natural' roots of these [linguistic conventions]. Thus, when they hear utterances whose constituent parts are hard to hear or absent (or they do not yet know them), they may understand how that utterance works in a different manner from the adult producing it (i.e., which parts of the utterance are serving which communicative functions)" (Tomasello 2008, 304). Examples of such functional reanalysis in English include the shift of "will" from a volitional verb to a generic future tense marker or the adoption of "better" as a simple modal auxiliary (as in "I better go").

between words and aspects of the world are often tacked on at the margins as perceptual or practical entrances and exits to language proper, which is reduced to "intralinguistic" relations among types of expression (e.g., Brandom 1994, ch. 4).

In the preceding chapter, I called attention to the possibility that Hubert Dreyfus's account of skillful practical/perceptual coping with one's surroundings, which he mistakenly took to display a preconceptual and presumably prelinguistic level of bodily intentionality, might nevertheless play an important role in understanding language. We are now in a better position to see why this is so. Recognizing that language emerged as a form of niche construction requires that we understand it first and foremost as a practical-perceptual capacity for robust tracking of protolinguistic performances in their broader circumstances and for flexibly responsive performances (both linguistic and nonlinguistic) motivated by them. Instead of taking discursive practice as merely interrelated with practical-perceptual skills for coping with one's surroundings, we would have to take linguistic competence as "intimately" embedded in our practical/perceptual involvement in the world, in Haugeland's sense of, "the term 'intimacy' [which] . . . suggests a kind of commingling or integralness, that is, to undermine their very distinctness" (1998, 208).

Recognizing the integration of language with perceptual-practical involvement with the world highlights several features of language that are rarely foregrounded philosophically. Here we need not undertake imaginative reconstructions of the protolanguages of early hominids, since the practical-perceptual dimensions of language remain pervasive despite their philosophical marginality. As one obvious consideration, linguistic performances only take place through the acquisition and exercise of subtle and difficult practical-perceptual skills. Anyone who visits another linguistic community with little or no grasp of the language knows the difficulty of learning to perceive the semantically significant phonemic articulation of a spoken language and to produce it fluently. Wittgenstein famously remarked that if a lion could talk, we couldn't understand him (1953, part II, 223); more important, we couldn't even *hear* what he was saying, in the sense of perceptually discriminating semantically significant differences, let alone be able to roar back intelligibly. Acquiring language is inseparable from acquiring a complex orientation and set of ear, tongue, eyes, and body. Moreover, semantics and phonemics are not so readily separable, since part of what enables recognition of highly variant reproductions as instantiating the same phonemic pattern is grasping its semantic relevance.[33] This point cannot be reduced to the theory-ladenness of perception without begging the question, since the "theory" in question *is* this practical-perceptual pattern.[34]

As a second consideration, understanding language use as a form of niche construction also thickly embeds it in ongoing social practices and relations. Such involvement is sometimes acknowledged for a limited range of cases, such as Austinian performatives, but not for language more generally. Consider, however, the familiar philosophical treatments of names, often thought to be among the simplest discursive phenomena. Even in twentieth-century philosophy, from early Wittgenstein (1961) to Kripke (1980), names have often been understood as akin to tags conventionally connected to objects. Hanna and Harrison (2004) remind us that naming requires much more intricately articulated social and material practices: "To give a name is . . . to reveal, in the ordinary way of things, a label that has been used for many years, through occurrences of tokens of it in the context of many naming practices, to trace, or track, one's progress through life. Such [mutually referring practices include] the keeping of baptismal rolls, school registers, registers of electors; the editing and publishing of works of reference of the *Who's Who* type, the inscribing of names, with attached addresses, in legal documents, certificates of birth, marriage, and death, and so on" (Hanna and Harrison 2004, 108). Google searches are only the latest twist on the intricate and intertwined practices through which names are reliably attached to individual persons and even partially constitutive of what it is to be a reidentifiable person. Naming and understanding names cannot and does not make sense apart from its embeddedness in such a "name-tracking network."[35] Hanna and Harrison highlight

33. A similar point has recently been highlighted by Gary Ebbs (2009) in his rejection of what he calls a "token-and-explanatory-use" conception of words, although he does so in pursuit of a different philosophical project. The crucial point that Ebbs rightly rejects is the notion that our perceptual identification of word tokens as instances of word types is independent of our semantic understanding. Davidsonian radical interpretation, for example, presumes that interpreters can correlate independently identifiable word types (e.g., orthographically or phonemically identifiable marks or sounds) with the circumstances of utterance of tokens of those types in order to assign semantic significance to utterances in a speaker's idiolect. Ebbs counters that our semantic understanding of words as embedded in our own discursive practices is integral to the criteria we use in practice for identifying words and regularly overrides orthographic or auditory similarities and differences. If language is a form of ecologically heritable niche construction, then such public and practical criteria for identifying words would be indispensable.

34. That is because grasp of the relevant "theory" presupposes the practical-perceptual capacity.

35. Names also only function as names when they are usable in sentences, which involve predication. As Davidson (2005a) points out, predication cannot be understood via a representational theory since one cannot account for the unity of the proposition in those terms: a sentence is

that point by contrasting the social contexts in which people are recognizably tracked by names with mere occasional baptisms, such as "call me Captain Midnight" or the sequential occupants of the role of "the Dread Pirate Roberts" in the film *The Princess Bride*. Even with elaborate practices of name tracking in place, there are residual spaces for ambiguity and transformation, exemplified by Natalie Zemon Davis's renowned historical study of a contested case of personal imposture (Davis 1983). Apart from the institutional infrastructure of naming, Rebecca Kukla and Mark Lance (2009) argue that names are also caught up in the essentially second-person indexical character of discursive practice. John Perry (1979) famously argued for the "essential indexicality" of location and orientation, without which third-person descriptive facts are free-floating. Kukla and Lance insist in turn upon the essentially indexical call-and-response of discursive practice. If I cannot grasp that you are talking to *me*, with a defeasible obligation to acknowledge and respond, I am not a competent discursive practitioner. Yet recognition of such vocative expression is often perceptually contextual without being marked semantically in an explicit way. The vocative dimension of discursive interactions is a central part of their ineliminably practical-perceptual character.

As a third consequence, we should also recognize that learning a first language *is* learning to get a distinctive practical-perceptual hold on circumstances. We do not first recognize a certain class of circumstances and then attach words to them. The ongoing practice of using the word is instead part of the circumstances that we learn to negotiate in picking up on a discursive practice and acquiring linguistic competence. Wittgenstein highlighted this point in an important passage in the *Investigations*: "What's it like for him to come?—The door opens, someone walks in, and so on.—What's it like for me to expect him to come?—I walk up and down the room, look at the clock now and then, and so on.—But the one set of events has not the smallest similarity to the other! . . . It is in language that an expectation and its fulfillment make contact" (1953, part I, 444–45). We overlook this entanglement of an understanding of language and a practical-perceptual grasp of circumstances in part by implicitly equating language learning with learning a second language. Davidson provides an especially telling example here. In some ways he seems to be in accord with the broader point I am

not just a list of names. Both forms of embeddedness, of singular terms in sentential predication and of names in a network of name-tracking practices, are integral to language as a form of niche construction.

making, since he takes himself to have "erased the boundary between knowing a language and knowing one's way around the world generally" (2005b, 107). Yet Davidson does not adequately acknowledge the practical-perceptual basis of such knowing one's way around the world. Davidson blocks off any explicitly perceptual aspect to semantic understanding because he takes perception to be a causal process distinct from the anomalous, rational understanding constitutive of language and knowledge. Davidson's methodology of radical interpretation also presupposes the possibility of a prior perceptual discrimination of what only then can be interpreted as semantically significant types. Ebbs (2009) points out that Davidson thereby commits to the empirically implausible claim that the perceptual discrimination and reidentification of linguistic expressions is independent of their semantic significance. Both points are often overlooked because Davidson's analysis takes for granted that one already speaks a language. Second-language learning, along with the interpretation of speakers of other languages, can be relatively "thin" at first, because the world already has acquired a discursively articulated grip upon us as perceivers and agents. We discover how much a thin conception of a language overlooks only when we attend to subtleties. As just one revealing example, for many differences between circumstantial uses of prepositions in different languages, it is hard to disentangle a grasp of which circumstances call for one word rather than another from a sense of which word sounds right in context to fluent speakers. It is difficult to imagine language-independent differences that could triangulate the relationships marked by the English prepositions "in," "on," "at," "by," and "with" with the quite different patterns expressed by the French prepositions "à," "en," "dans," "dedans," "de," "près de," "vers," "par," "sur," "avec," and "pour," and the German "in," "auf," "an," "aus," "bei," "nach," "von," "zu," "um," and "über." A history and ongoing practice of uttering one word rather than another is integral to the identification of the relevant circumstances in each case. To paraphrase Wittgenstein (1953), at a certain point the explication of semantic differences stops with "this is what we say."

Abilities to discriminate the relevant circumstantial and phonemic similarities in uses of the "same" linguistic expression are thus mutually interdependent. So much of ordinary conversational practice is reliant upon a partially shared grasp of how the circumstances are relevant to what we say, and vice versa. Yet the prior course of a conversation itself belongs to the relevant "circumstances" to which utterances are connected indexically, deictically, gesturally, and vocatively. Philosophers have tended to overlook the integration of linguistic articulation within the practical-perceptual circumstances that frame conversation because of an implicit commitment to the philosophical primacy of written, nondeictic, and nonindexical assertions, but understanding a sentence as an assertion requires grasping the pragmatics of claim-making.[36] It is only because we are already situated within conversational practice that we can comprehend written sentences, whose placement on the blank, impersonal background of the page makes them seem freefloating and disconnected from any specific situation. Understanding what it means to be addressed by someone in conversation is the background that allows us to read written texts as implicitly making a claim that can address *anyone* who reads it. Kukla and Lance (2009) argue forcefully and rightly against the primacy and autonomy of assertions within discursive practice to accommodate the crucial role of vocative and recognitive uses of linguistic expressions in establishing and sustaining discursive normativity. Even Kukla and Lance do not sufficiently emphasize the perceptual and practical skills that are constitutive of discursive practice, through which semantic content is articulable and discernible via the intertwined abilities to correlate utterances with circumstances (including other utterances).

These phenomena that highlight the practical-perceptual concreteness of discursive practice have also been overlooked or marginalized within philosophy in part because we philosophers have been rightly impressed with the extensive expressive resources provided by logic and linguistics. The insights provided by these more formal disciplines encourage a misleading reversal in the order of understanding. Philosophers have tended to see logical and linguistic formal relations as a framework to which the concrete bodily, circumstantially embedded,

36. Derrida (1967a, 1967b) famously takes the opposite direction in criticizing philosophical "logocentric" conceptions of meaning as reflecting an imagined primacy of speech over writing. Derrida takes this stance, however, through a critical response to Husserl, for whom speech is not a public practice of signification and indication but an immediate, silent, "auto-affective" presence of the word spoken to oneself. This conception of speech is not even convincing as a foil for Derrida's argument. Apart from the recognition that silent, "inner" speech is a historical achievement, there would be no way to track the phonemic articulation of differences except on the basis of a prior practical-perceptual skill in discriminating and producing phonemically structured overt speech (Saenger [1997] provides a useful historical account of the gradual development of silent reading and the development of written texts to replace oral articulation of word patterns). Derrida similarly resists acknowledging the role of perceptually accessible circumstances in semantic understanding because he fears something like an appeal to what Sellars called the Myth of the Given to ground meaning in something not subject to further articulation. Once one realizes that neither the recognition of phonemic similarity nor the perceptual recognition of circumstances in the context of partially shared projects could take place independently, however, then speech takes over the role Derrida ascribed to writing as "pure indication," and writing against the abstract background of the blank page looks like a dependent form of only seemingly self-contained expressiveness.

and socially interactive aspects of discursive practice are attached as this formal structure is deployed in language use. These formal relations nevertheless only exist over time via their ongoing uptake and reproduction within discursive practice. Instead of seeing syntactic and semantic structures as overarching frames that govern language use, we should instead see them as emergent from an ongoing process of "grammaticalization" (Tomasello 2008, ch. 6) in which discursive niche construction is stabilized and further articulated.

Logic and linguistics are powerful expressive resources that are abstracted from and presuppose immersion in a natural language that is itself an integral part of the evolved, reconstructed niche we inhabit. Brandom (1994) and Sellars (2007) initiate such a theoretical reversal by emphasizing the philosophical priority of material inference over formal logic. They argue that formal relations allow us to express explicitly what we must already know how to do in our practical grasp of concepts but cannot substitute for that practical ability even in principle. We need to extend this explanatory reversal by recognizing that the material-inferential proprieties that govern our use of words in turn presuppose a rich practical-perceptual grasp of the ongoing discursive practice that constitutes a natural language. Important aspects of this grasp of the normativity of discursive practice are themselves discursively articulated in the form of vocative, recognitive, and other deictic and indexical uses of linguistic expressions that indicate our socially interactive, practical-perceptual immersion in partially shared circumstances (Kukla and Lance 2009). These circumstantially situated expressions and uses are not dispensable additions to or elliptical contractions of impersonal, decontextualized assertions but instead provide indispensable background to any ability to understand such abstracted and impersonal expressions as speaking to us meaningfully. Such expressive articulations of an ongoing practical-perceptual immersion in a discursively articulated ecological and evolutionary niche are nevertheless always only partial. These pragmatic indications depend upon that worldly immersion even as they usher it into a conceptually articulated space of reasons.[37]

37. A more detailed discussion of how discursive articulation is related to a broader practical-perceptual immersion in a shared ecological-evolutionary niche must be reserved for chapter 4. I note now, however, that this claim for the interdependence of conceptually articulated understanding with our practical-perceptual immersion in the world should not be conflated with any of the various proposals for distinguishing conceptual from nonconceptual content. The difference between these approaches and its importance will be developed in chapter 4, but the reader should not assume that what is being proposed here is some form of "nonconceptual content."

Understanding the evolution of language as a form of niche construction thus also encourages a reorientation of how we think philosophically about linguistic understanding. Grasp of a language incorporates a practical and perceptual involvement with a public practice that is a salient feature of our developmental and selective environment rather than consisting primarily of mental representations of syntactic, logical, and semantic structure. Human beings develop in and adapt to a world pervaded by public discursive performances, and our practical-perceptual capabilities are shaped through that development. The resultant capabilities allow us to make situationally competent discursive performances that help reproduce a partially shared discursive environment. Our own discursive performances are part of a flexible responsiveness to a dispersed but salient feature of our normal human environment. This conception of linguistic competence foregrounds the widely known role of early exposure to a natural language in the acquisition of normal linguistic ability. Language is a preeminent example of the intimate entanglement of human bodily skills with specific, concrete features of our niche-constructed human environment, which is a discursively articulated world. Linguistic understanding is not a mental representation or other "internal" structure that *interfaces* with practical-perceptual involvement in our surroundings to produce conceptually articulated performances. Linguistic understanding is a practical-perceptual capacity that is *integral* to the unified phenomenon of skillful bodily responsiveness to an environment pervasively shaped and marked by the cumulative history of that ongoing interaction.

Understanding the evolution of language as a distinctive form of niche construction nevertheless may seem to solve one problem at the expense of creating another more troubling difficulty. Niche construction theory, presented in this way, may seem to make the evolutionary emergence of language intelligible at the cost of making unintelligible its responsiveness to rational, conceptually articulated norms. Understood as a complex form of skillful practical and perceptual coping with an ecological inheritance that includes the ongoing reproduction of discursive practice, language may seem to reduce to a complicated form of robust tracking and flexible responsiveness to the perceived environment. Such an account would then seem to treat utterances in a language as merely noises that, quoting Haugeland's comments about nonhuman animals' functionally adaptive responses to their environment, exhibit an "[ersatz intentionality that] can 'mean' [nothing] other than what normally elicits [them] in normal circumstances" (1998, 310). If language and thought are just a more complex form of practical/perceptual coping with our selective environment, how does a gap ever open between how our utterances or thoughts take the world to be and the actual circumstances to which they are responsive? If we are only talking about practical/perceptual coping with a world that includes human vocalizations, why should it matter for *us* any more or any differently than for flocks of birds or herds of sheep that the eliciting circumstances for our vocalizations often include vocalizations by other organisms? Put in different philosophical terms, treating conceptual understanding as a behavioral form of niche construction may seem to avoid a mysterious invocation of Kantian freedom as responsive to rational norms only by reverting to the objective side of the traditional Kantian dualism, which would seem to leave no space for conceptual normativity. The problem, then, is to understand how such a conception of language as behavioral niche construction can still adequately account for the normativity of conceptual understanding.

The next chapter addresses this issue by setting it in a broader context. My central concern in this book is to develop a more adequate naturalistic sense of conceptual understanding in scientific practice. This first part of the book is concerned with the normativity of conceptual understanding more generally as background to thinking about scientific understanding. In this chapter, I have turned to questions about the evolution of language and have proposed that language emerged through protolinguistic niche construction that drew upon and transformed early hominids' inherited capacities for robust tracking and flexible responsiveness to their perceived circumstances. In the next chapter, I return to the question of the relationship between language and conceptual understanding. Evolutionary naturalists such as Dennett or Millikan have proposed a cognitive continuity between human beings and nonhuman animals that situates human language within a more general account of intentionality and representation that also encompasses the more robust and less "sphexish" forms of animal behavior.[38] Conceptual understanding would then be a much more general phenomenon than language, even though linguistic understanding dramatically expands the scope and articulation of conceptual capacities. My proposal takes a different route. I endorse the evolutionary

38. "Sphexish" refers to the kinds of behavior that are directly and inflexibly cued to environmental circumstances, although they can be linked together in extended chains to produce relatively complex patterns that nevertheless form rigid behavioral sequences. The term reflects Wooldridge's (1963) characterization of the egg-laying behavior of the sphex wasp as exemplary of this kind of behavioral pattern. Sterelny (2003) uses the term 'detection agent' for organisms whose behavioral repertoire is mostly "sphexish" in this sense.

cognitive continuity between humans and other animals in our capacities for robust tracking and flexible responsiveness to relevant features of our selective environments. Such perceptual responsiveness is not yet conceptually articulated. Conceptual normativity emerges with the development of language as a highly articulated and integrated form of behavioral niche construction.[39] The domain of conceptual normativity is not thereby limited to linguistic performances, however, or even to thoughts and actions that can be appropriately understood in terms of tacit linguistic commitments. The availability of language as an integral part of our organismic way of life instead opens a space of normative accountability that extends beyond language in any narrow sense to incorporate many domains of human activity that are not themselves readily expressible linguistically.

39. My account remains agnostic concerning whether language emerged first as relatively autonomous from other conceptual capacities or whether its emergence was coincident or even integrated with other conceptual abilities, such as systematically interconnected uses of equipment with norms for its appropriate use or music, dance, and other expressive behavior.

FOUR

Language, Social Practice, and Conceptual Normativity

I—Setting the Problem

The previous chapter introduced an alternative to familiar accounts of the evolution of language: language is a preeminent example of the evolutionary importance of niche construction. Spoken language is a salient and pervasive feature of the environment in which human beings normally develop into functioning adults, and in most human communities, written language is now similarly pervasive.[1] Such a developmental environment only exists because it is reproduced anew in each generation. The coevolution of human beings with languages has thereby made natural languages a reliable, central component of our biological inheritance from preceding generations. Under those conditions, our species has evolved under selection pressure to facilitate normal development of capacities to recognize

1. Even for people who are illiterate, or illiterate in the predominant local language, written language is a recognizably pervasive and influential feature of their environment. People learn to respond to the discursive significance of written language even when they mostly cannot discern what is being said. Written signs are themselves perceptually salient, and they reorient everyone's practical-perceptual orientation within that environment, even for those who cannot read their content. The perceptual-practical pervasiveness of written language is a sufficiently recent phenomenon, however, that it is unlikely to have generated the kinds of neurological reorganization or genetic fixation that facilitate the rapid and early development of oral linguistic competence.

and produce expressions in natural languages after relatively limited exposure to that language at a sufficiently young age. Languages themselves in turn evolved in response to the changing capacities for language learning because only those features and structures of a language that are readily learnable under current conditions will be reliably reproduced in subsequent generations. Languages change in response to changing human capacities and performances in many ways, including grammaticalization and functional reanalysis (Tomasello 2008), full or partial genetic assimilation of language learning, and the "stretch-assimilate" process in which genetic assimilation or environmental scaffolding of some aspects of language learning enables the development of more complex forms (Dor and Jablonka 2000, 2001, 2010). If language emerged through a coevolutionary process of niche construction, however, then linguistic understanding cannot be primarily a matter of internal representations in the mind or brain, even though the human central nervous system has evolved significantly under selection pressure for language learning and use (Deacon 1997; Jablonka and Lamb 2005; Dor and Jablonka 2010; Bickerton 2014). Language is first and foremost a public practice that we learn to track and respond to perceptually and practically. Davidson once remarked that there is no boundary between "knowing a language and knowing our way around in the world generally" (2005b, 107). That is true in significant part, however, because linguistic expressions and capacities are salient components of the human world we learn to negotiate.

This approach to the evolution of language suggests a reconception of the relation between language use and conceptual understanding. Philosophical analyses of the relationships among intentionality, language, and conceptual understanding differ significantly, as we saw in chapter 2. Understanding language as a form of niche construction and developmental coevolution both clarifies these relationships and introduces new challenges. The clarification results from recognizing the frequent conflation in philosophical discussion of two different forms of "intentional" directedness and their constitutive normativity. There is clearly a kind of directedness in many organisms' perceptual-behavioral responsiveness to their surroundings. Yet in several important respects, perceptual registration and response in nonhuman animals is not conceptually articulated.[2] So the question that must be addressed is the

2. The point of the qualification is not to suggest that human capacities for perception are discontinuous from those of nonhuman animals; on the contrary, I take human perceptual capacities, and the character of human perception per se, to be quite comparable to those of other organisms

relation between an organism's perceptual-practical responsiveness to its surroundings and its capacities for conceptually articulated understanding to the extent that it possesses and deploys them.

Perception typically opens onto the world only in what Akins (1996) calls a "narcissistic" way, such that aspects of the world show up as differentially significant for the organism's life activities rather than as indicating objective properties of environmental events. Moreover, what is registered perceptually is systemically interlinked with an organism's behavior and physiology. This interlinking is partly due to the ways in which perception itself involves movement; perceptual "input" motivates further specific movements of perceptual exploration and indicates how things are accessible to such exploration (Nöe 2004).[3] The tight link between perception and behavior is more fundamentally governed by the "narcissistic" character of perception, however. Organisms register perceptually those environmental features that would motivate different behavioral or physiological responses, and those responses affect how the organism orients and sets itself perceptually. Some organisms function as "detection agents" whose behaviors are directly cued by specific environmental differences (Sterelny 2003). Even those organisms capable of robustly tracking more complex perceptual configurations also still respond to features of their environments, or combinations of features, that motivate different responses. Such robust tracking capacities presumably arose over time in response to environments whose significance for the organism's behavior was multifactorial and informationally translucent (Sterelny 2003).

Such intertwined perceptual and behavioral responsiveness is aspectual in an important sense: the life activities of an organism are affected by and responsive to some aspects of its physical surroundings and not others. The biological environment of an organism comprises the pattern of surrounding conditions to which the organism's physiology, development, and evolution are directly or indirectly responsive. What the organism is and does is part of a larger system that incorporates its bodily responsiveness to an environment configured by the organism's own life activities and bodily processes as relevant to its ongoing way of

in our lineage (and in many respects to be less discriminating). The question is only whether an organism's linguistic or other capacities for conceptually articulated understanding changes the place of perception within the organism's overall behavioral and cognitive economy and consequently changes *how* perception is directed toward and about what is perceived.

3. The scare quotes indicate my dissent from an understanding of perception as providing "input" that then motivates behavioral "output" via intervening "central processing." Perceptual and behavioral responses to circumstances are thoroughly entangled.

life. Organisms nevertheless do not thereby "take" their surroundings *as* meaningfully configured by their life activities. The close connection between perceptual inputs and behavioral responses makes it impossible to specify either the environment of an organism independent of its characteristic life activities or vice versa.[4] The organism's developmental and selective environment is thus specified by the conditions to which it does or would actually respond, and there can be no further standard (apart from the de facto normal range of responsiveness characteristic of organisms of that kind) that could define a norm against which its actual response could be understood as mistaken or otherwise deficient.[5] Such a tight coupling between organism and environment can nevertheless lead to highly flexible and adaptive behavior by some organisms in response to perceived differences within their environment. Such flexibility arises from a capacity to track and respond differentially to combinations of perceptual features, including novel combinations; to the sequential interactions among the organism's perceptual-behavioral responses, which in turn establish strong feedback relations with the perceived behavior of other organisms; and from the subsequent tuning of the organism's responsiveness to further inputs due to its own prior interactions.

This intimate connection between what an organism can register perceptually and how it responds physiologically and behaviorally contributed to Haugeland's (1998, ch. 10, 12, 13) dismissal of even the most complex perceptual/behavioral repertoires of nonhuman animals as merely "ersatz" imitations of intentional (i.e., conceptual) content. For all the flexibly goal-directed appropriateness of much animal behavior, including abilities to generate and sustain novel responses to changing circumstances, such behavior cannot be either correct or mistaken but only normal or abnormal and adaptive or maladaptive. There is no gap that would permit the attribution of any "content" or aim to the

4. The coupling is not close in the sense of being invariant. Expression of the same genes under relevantly similar conditions can differ stochastically in some respects due to developmental "noise" (Lewontin 2000), and behavioral responses to perceptual inputs display similar ranges of variation. The coupling is instead close in the sense that the organism's perceptual capacities have been shaped in evolution and development by what is significant for its behavioral responses, which have been reciprocally shaped by its characteristic perceptual inputs.

5. I argue in chapter 8 that such species-typical patterns of response (which have analogues at higher taxonomic levels) are appropriately understood as law governed. Only an inadequate conception of laws and nomological necessity leads to the common mistake of thinking that where there is evolutionary contingency and variation within populations, the biological functioning of organisms within a relevant taxon cannot be law governed. For more extensive discussion of such laws of functional biology, see Lange (2007, 2000a) and chapter 8 of this book.

behavior distinct from what organisms of that kind normally do in response to behaviorally relevant differences in circumstances.

An important reason for this inability to ascribe intentional content within its normal pattern of responsiveness is that the goal-directedness of an organism's physiology and behavior is holistic. For example, we might be inclined to think that other organisms take some things in their environment *as food* because of the relatively good match between what they eat and what is edible for them. But there are many edible things they do not eat, even when hungry. Sometimes individual organisms make mistakes relative to their normal pattern. But other cases may not fall in their normal pattern of recognition and response, perhaps (from an evolutionary perspective) because the requisite discriminative capacity would be too energetically or cognitively costly. In that case, we would have to say that these animals do not respond to what they do eat "as food," but "as energetically and cognitively accessible food." But of course that category also has exceptions, which must in turn be added to a more complex description of the original response. Similarly, we might be inclined to interpret a lion chasing a springbok as having failed if the springbok eludes it. Lions that always caught animals they chased, however, might well be less successful in the biologically relevant sense than lions whose behavioral repertoire differently balanced the likelihood of catching and eating prey against the energetic and opportunity costs of the chase. Within such a behavioral repertoire, any specific occasion of "failure" to catch its prey would instead exemplify a successful strategy. There is no principled stopping point to that process of qualifications to the supposed as-structure of the organism's behavior, short of its entire normal behavioral pattern in response to its normal environmental range. Conceptually articulated intentional comportments, by contrast, involve both directedness toward identifiable aspects of the world and a discernible content to how they *take* those aspects of the world to be, such that these two components of intentional directedness can diverge.

From a naturalistic standpoint, of course, Haugeland's characterization of nonhuman animal behavior as "ersatz intentionality" can only be a picturesque way to highlight differences between the intimate interlinking of nonhuman animals' perceptual registration with their behavioral response and the dual normativity of conceptually articulated intentional directedness. An organism's perceptual openness to an environment configured as such by its own characteristic forms of activity and exploration stands on its own without comparison to our idiosyncratically constructed developmental niche. In avoiding efforts to characterize other animals' ways of life by their degree of resemblance to or even "aim" toward ours, we circumvent the kind of human exceptionalism that treats the absence of language in nonhuman animals as a deficiency or limitation in their cognitive capacities. Recognizing such differences between flexibly practical-perceptual and conceptually articulated modes of directedness within an environment still has far-reaching consequences for how we understand each mode. As Kathleen Akins noted, "Virtually all naturalistic theorists agree on an important methodological point, namely, that a naturalistic theory [of intentionality] should start with the static perceptual case" (1996, 340), despite more specific differences among them on various issues. If other animals' practical-perceptual responsiveness and our forms of conceptual understanding are "intentional" in fundamentally different ways, then this common starting point is likely to be misleading, unless one explicitly accounts for the differences.

Accounts of language and conceptual understanding as behavioral niche construction must still recognize the underlying perceptual capacities as broadly continuous among humans and many nonhuman animals. The relevant capacities are no longer cases of *static* perception, however; animal perception requires active responsiveness and also reflects the tight coupling of perceptual uptake and behavioral response. More important, a niche constructive account does not thereby take perceptual directedness and responsiveness as a *model* for understanding other modes of directedness toward a system's surroundings. It instead takes perceptual capacities as a *prerequisite* for linguistic and conceptual understanding because the acquisition of language is first and foremost the evolutionary development of a novel perceptual-behavioral capacity through repeated cycles of behavioral niche construction. The eventual result is nevertheless a quite different mode of engagement with and directedness toward the organism's surroundings. Moreover, although language is integral to the emergence of conceptually articulated intentionality, I argue, conceptual articulation is not confined to comportments that are or can be articulated in language. Once a capacity for conceptual understanding emerges, it extends beyond language and linguistically articulated thoughts to inform perception and action more generally. Understanding that mode of conceptually articulated intentionality, and how it is related to the practical-perceptual capacities from which it was forged in human evolution, is the task of this chapter.

The strategy of beginning with perception in accounting for intentionality and conceptually articulated content has obvious appeal for anyone of a broadly naturalistic bent, but it also presents an underlying dilemma. Perception seems an attractive model because the relationship between perceiver and perceived is causal and to that extent scientifically explicable. The aspectual character of intentionality may also seem readily manifest both in the differences among various actual and possible sensory modalities and in the straightforwardly perspectival character of perception. Familiarity with perceptual illusions and suboptimal perceptual standpoints suggests the possibility of a perceptually recognizable gap between how things show up and how they are: illusions and other inadequate presentations can be corrected by further perceptual exploration. The normativity of perceptual presentation seems to be articulable within perception itself, since subsequent perceptually based corrections seem to suggest a notion of how one *should* have perceived the object in contrast to how one actually did perceive it.

The dilemma nevertheless arises because the very features of perception that make it seem initially attractive as a model for conceptually articulated intentionality may also undermine its prospects. Precisely what has classically seemed to make intentionality difficult to understand philosophically or scientifically is the possibility of relations to objects that are not causal relations and that could not be causal in the case of nonexistent intentional objects. Different modes of conceptual presentation of an object do not seem to differ from and relate to one another in the same ways that different sensory modalities differ and coordinate. Moreover, upon further consideration, the seemingly normative character of perceptual presentation may not actually be *perceptually* manifest. The recognition that perceptual presentations are sometimes illusory requires going beyond perceptual relations. Why isn't the familiar appearance of the Müller-Lyer illusion, for example, assimilated as how equal-length lines "should" look in certain contexts rather than as a somehow deficient or incorrect appearance of their lengths? For that matter, why is a subsequent perceptual presentation understood as *correcting* a prior appearance rather than merely changing it? There is a difference between changes in an organism's perceptual-practical orientation toward its surroundings in the course of its ongoing life activity and an understanding of it as having corrected an earlier mistaken appearance. Understanding perception as self-correcting requires showing on other grounds that perception involves *taking* its surroundings *in* some way that might then be recognized as *mis*taken.

My discussion of perception so far, and my suggestion that linguistic understanding is at base a kind of practical-perceptual skill, may seem to heighten these concerns. Instead of taking perception as a model for conceptually articulated intentionality, I have emphasized the differences between them. Perception registers not how the world is in some respect but how salient aspects of the world solicit subsequent behavior on the part of the organism, including further perceptual exploration. Perception is thereby linked not to internal representations that might exemplify more traditional notions of intentional content but instead to the bodily activity that it solicits. An organism's perceptual capacities do have a normative relation to the world but one governed by normal goal-directed functioning rather than correct representation. As we have seen, the result is that an organism's normal perceptual uptake and behavioral response cannot be mistaken about what it perceives: what its perceptual appearances indicate are solicitations of a behavioral response.

In emphasizing the perceptual and performative character of language, this approach may also seem to introduce additional philosophical difficulties. Everyone recognizes that language use involves perceptual and practical skills, but these skills have long been seen as merely instrumental to, rather than constitutive of, the understanding that is enabled by language. The crucial accomplishments of linguistic understanding seem to be symbolic-semantic and syntactical. Language is distinctive because the significance of its token performances is displaced from their immediate, perceptible circumstances of utterance. Language also permits novel expressive combinations, both in the basic form of predication and in the recursive recombinability of syntactic units to produce more complex forms of semantic significance. By contrast, which sound (or other perceivable indication) is conventionally associated with those semantic contents has long seemed completely arbitrary. To focus on the skills of perceptual recognition and performative skill in producing new verbal expressions may seem to place the wrong considerations in the forefront. We would thereby risk accounting for the evolution of language in ways that overlook what matters in its contribution to the development of conceptual understanding.

The only way to respond to these concerns is to show how to account for conceptually articulated content and its normative authority and force while treating perception as closely coupled with behavioral response and language as behavioral niche construction. There are multiple steps to this account. The first step is to recognize that the initial emergence of limited forms of symbolic displacement, as proposed and understood in different ways by Bickerton (2009) and Deacon (1997) among others, is not yet sufficient for language or conceptual understanding. Only the further development and articulation of such protolinguistic capacities, as both interconnected with one another and integral to a whole socially articulated way of life, could overcome the primary orientation of practical-perceptual intentionality toward present circumstances.[6]

Once we understand in broad terms how such development might have taken place, we need to develop three philosophical points that then collectively account for the emergence and further articulation of conceptual understanding in human life. The first point will concern the role of what Kukla and Lance (2009) call the vocative and recognitive dimensions of discursive practice in enabling speakers to call one another to account for their discursive and other performances. The second point is what I call the partially autonomous character of linguistic practice. This feature of linguistic practice is what most directly enables a gap to open between conceptual content and intentional directedness, which cannot occur simply through perceptual and practical responsiveness to nondiscursive environmental circumstances. The third consideration is the distinctive character of the social practices that language both exemplifies and makes possible. The social character of linguistic practice thus turns out to be indispensable to conceptual normativity. In chapter 2, we saw Haugeland argue against the possibility of accounting for intentionality and conceptual normativity in social terms. Haugeland's arguments are indeed decisive against a social account of conceptual normativity if we accept a familiar and widely accepted model of how social practices institute norms. That model misunderstands social practices, however. A more adequate account of social practices enables understanding how the authority and force of conceptual normativity are socially constituted and sustained. The result, however, is to reconceive conceptual normativity as grounded in the temporality of discursive practices as well as in their dependence upon discursive interaction with our biological environment. These three considerations are conjoined in the emergence of language as a partially autonomous, vocative-responsive, social practice, which is an integral component of a behaviorally reconstructed evolutionary niche. Within that context, the conceptually articulated performances and capacities that language enables have a distinctive temporal and modal character that enables us to understand their characteristic normativity. In this chapter, I develop this revised conception of the social character

6. Bickerton (2014, ch. 4–5) explicitly recognizes that symbolic displacement is not yet sufficient for language but only for a kind of "protolanguage." He takes the defining difference to be between expressions that can only be combined serially like beads on a string and expressions that can be merged pairwise to form more complex constructions. In this respect, his view interestingly converges with Davidson's (2005a) insistence on the distinctive role of predication, which differentiates sentences (which can be true or false) from lists of names, which cannot.

of language and discursive practice more generally. Chapter 5 works out its implications for how to think about conceptual normativity and the space of reasons. The second part of the book then shows how and why we should think of scientific understanding in these terms.

II—Language as a Social Phenomenon

Deacon (1997) and Bickerton (2009), among others, have argued that the principal barrier to the evolution of language was overcoming organisms' strong practical-perceptual orientation toward responsiveness to their present circumstances. The difficulty was for organisms to recognize (and produce) vocal or other perceivable expressions as salient features of their surroundings, whose significance was nevertheless symbolically displaced from those surroundings. Once that difficulty has been overcome, and a protolinguistic practice that allowed symbolic displacement became integral to an organism's way of life, it seems much less difficult to see how that practice could lead to more complex, articulated, and flexible forms of discursive practice. Under new selection pressures for ease and reliability of language acquisition and use, such novel expressive capacities could readily arise through phenotypic plasticity and its genetic assimilation, the feedback relations of niche construction, and the resulting coevolution of language and organism.[7] By contrast, the known forms of nonhuman animal communication reinforce rather than displace organisms' involvement in and "narcissistic" orientation toward their surrounding conditions. So the fundamental difficulty is to understand how it was possible to get from there to here and, to the extent that the evolutionary history is accessible, to understand how it happened in the hominid lineage.[8]

7. Bickerton (2014) argues that the neurological reorganization needed to allow for effective production and interpretation of strings of symbols in real time, without encountering combinatorial problems in interpreting their conjoined meaning, marks the transition from protolanguage to language. This issue is undoubtedly important, and Bickerton's suggestion is attractive that universal grammar as conceived in Chomsky's (1995) later minimalist program *is* what results from that neurological reorganization. Recognizing the connection between this problem and the problem of predication (the unity of the proposition) nevertheless suggests that the transition from protolanguage to language had already occurred at the point at which strings of symbols were understood as a complex whole that needed to be disambiguated rather than as a string of expressions to be interpreted individually.

8. Accounts of the origin of a phenomenon need not thereby determine its present form, and so one might worry that an account of the evolutionary emergence of language and conceptual understanding might only describe the prehistory of language, without thereby determining its subsequent evolution and present-day form. This worry is misplaced when directed toward an account

Bickerton and Deacon make a compelling case that the principal issue for understanding the evolution of language is the emergence of symbolic displacement. Each then goes on to argue that the initial steps toward grasping symbolic displacement required changes in hominids' early way of life that would create strong selection pressure for the ability to produce and understand symbolic expressions freed from their immediate circumstantial indications. That pressure would need to be strong enough for long enough to overcome early hominids' established cognitive orientation toward and sensitivity to relevant features of their current circumstances. Virtually all prior selection pressures for perceptual and cognitive adaptation likely favored perceptual sensitivity and appropriate responsiveness to life-relevant environmental conditions. Both Bickerton and Deacon emphasize that the selective difference that most likely directed human evolution toward language was the need to coordinate group action oriented toward spatially distant situations. Yet initially, from the organism's perspective, orientation toward a spatially distant situation is just a special case of what Sterelny calls a "translucent environment." Translucent environments present "a complex relationship between the incoming stimuli that the organism can detect and the features of relevance to it" (Sterelny 2003, 27), in this case, features that are spatially distant. Yet such translucent complexity normally favors *more* robust and flexible capacities to track multiple perceivable direct or indirect guides to behavior, which would enhance the organism's prospects for survival and reproduction (Sterelny 2003). Selection for symbolic understanding thus had to counter the very sensitivity to relevant environmental circumstances that had undoubtedly been entrenched throughout the organism's perceptual and behavioral repertoire.

---

of language and conceptual understanding as forms of niche construction, for several reasons. First, a niche constructive account endorses the recognition that the earliest forms of linguistic or protolinguistic communication need not resemble the phenomenon as it has subsequently evolved. One of the advantages of the approach is that it need not postulate an initial emergence of complex linguistic forms but instead posits the early emergence of a protolanguage that lacks many of the features familiar in languages today and that only gradually evolved into more complex capacities and performance. Second, an account of language as a form of behavioral niche construction is not merely an account of the origin of language but also an account of its ongoing reproduction and transformation through the present day. Once language was in place as a public practice that is reliably reproduced developmentally as an integral part of human beings' developmental and selective environment, it would continue to evolve with the human lineage. Where neo-Darwinist accounts of language evolution would typically characterize genetic changes that were adaptive in early hominid environments and thereby became fixed in the population, niche constructive accounts characterize a continuing process of developmental uptake and partial genetic assimilation. Third, as we shall see below, a niche constructive account brings out the importance of contemporary features of language and conceptual normativity that other accounts do not readily explicate. Thanks to an anonymous reviewer for indicating the need to address this worry at the outset.

Limited evidence about early hominid life leaves room for considerable speculation about just how they encountered such sustained, effective selection pressures for symbolic displacement, but any intelligible account faces evident constraints. First, these selection pressures would have to engage fitness-relevant aspects of hominid life; the predominant focus of animal communication systems on food, sex, and predator-avoidance (Hauser 1996) is strongly suggestive of the likely possibilities. Bickerton highlights three other likely constraints. A protolinguistic species would need to be a social animal with wide-ranging patterns of movement in which organisms are often dependent upon distant conditions, such as widely and irregularly scattered food sources; this peripatetic way of life would have to involve fission-fusion patterns of social life so that persisting groups do not already share information about conditions elsewhere; and, finally, their way of life would have to depend upon effective collective action, which would require recruiting others to act together at distant sites (Bickerton 2009, ch. 6–7, 11). Bickerton proposes a specific hypothesis for how these constraints were satisfied, appealing both to ecological changes confronting early hominids and to extensive paleo-archaeological evidence for the emergence and proliferation of a distinctive kind of hand ax (Bickerton 2009, 143, 213, 220).[9] I find Bickerton's scenario highly plausible, but his proposed constraints upon any viable account seem more reliable than any specific hypothesis intended to meet them, including his own.

Bickerton's proposed constraints upon explanations of how early hominids could have encountered selection pressure for symbolic displacement are important, but they also raise a further crucial concern.

9. Bickerton's specific hypothesis is that selective pressures for linguistic communication arose in the transition to a new predominant food source and associated way of life. On this hypothesis, some early hominids became territorial scavengers of large animal carcasses, using hand tools to both cut open the carcasses before other scavengers could do so and defend the scavenging site against predators and other scavengers. The pressures for effective communication about distant conditions supposedly arose because these hominids would have to explore widely in small groups but rapidly convene larger groups when food sources were found. This hypothesis, discussed later in this chapter, is not without competitors. Deacon (1997), for example, had previously suggested selection pressure for language to maintain parental investment and pair-bonding in the context of fission-fusion foraging or hunting. Both Bickerton's and Deacon's hypotheses implicitly argue that the emergence of collectively cooperative action, with recognized differentiated roles and expectations of shared reward, was integrally part of this route to language. Tomasello (2008) puts the emergence of cooperative activity front and center as quite different from the ways that other great apes coordinate their activity with others but do not cooperate. In this important respect, Tomasello's view reinforces Deacon's and Bickerton's approach, but his account accords no role to public language as a form of niche construction. As I argue below, my account does not depend upon which account, if any, is correct about the selection pressures that led to the emergence of symbolic displacement in language.

Even if Bickerton's own account or one of its competitors were correct in their characterization of the early selection pressures for language, their hypothesized protolinguistic capacities to report life-relevant conditions at a distance are still not yet sufficient to generate conceptual capacities. Utterances that correlate with specific environmental conditions, and thereby motivate collective action in statistically effective response to those conditions, do not yet achieve displacement, even if the relevant conditions are spatially distant. Some animal communication systems already do serve to call attention to and motivate behavior toward perceptually absent conditions, with the warning cries of vervet monkeys for different "kinds" of predators (Cheney and Seyfarth 1990) as perhaps the best-known example. The selective significance of such communications arises precisely because imminent dangers or opportunities are nevertheless often not perceptually accessible to most members of a group.[10] We can thus think of such performances as indirect extensions of an organism's capacities for perceptual discrimination and appropriate responsiveness. Responding to a warning cry is not so different from responding to a characteristic motion in high grass rather than to the perceived approach of a leopard. Either case amounts to a perceptual indication for how to respond to the organism's current circumstances.

The problem is not just that such environmentally responsive verbal expressions do not have articulated content that could distinguish reference under different aspects (such as "danger," "aerial predator," "eagle," its direction or style of attack, or the appropriate avoidance response), although indeed they cannot do so. I placed scare quotes around the word 'kind' in the previous paragraph to indicate the expressive indeterminacy of vervets' or other animals' warning cries. The vervets' cries are directed responses *to* impending predation that might be avoided by timely responses, but they do not involve any understanding, classification, or even representation *of* anything *as* an animal, a predator, a danger, or an indication to flee in a specific way. The vervets' cries and their responses to those cries configure their circumstances and their own way of living in those circumstances as an actual pattern in the world, without ever achieving a symbolic or conceptual articulation of those circumstances and responses. The emergence of such warning cries as guiding differential responsiveness are indeed a form

10. We need not take up for my purposes the controversies over whether such adaptations require some mechanism of group selection or whether kin-selection models are sufficient to understand how such "altruistic" behaviors arose. For extensive discussion of this issue, see Sober and Wilson (1999).

of behavioral niche construction but not one that can reasonably be regarded as even protolinguistic.

The underlying problem is the fundamental difference between indirect perceptual presentation and symbolic conceptualization. The vervets' warning cries do not describe or represent anything even indirectly, at least not without an equivocation upon the concept of a representation (Horst 1996). Such behaviors correlate with distant or otherwise imperceptible conditions to motivate appropriate behavioral responses. They are continuous with many animals' familiar abilities to track and respond to multiple, translucent indications of life-relevant surrounding conditions in subtly different ways. Such communicative capacities expand the range of surrounding circumstances to which an organism can be perceptually and practically responsive but do not break from their focused orientation toward behaviorally relevant features of their current circumstances. Indeed, these capacities would more likely intensify such practical-perceptual orientation toward current circumstances. An organism would need to track both the current vocal or gestural performances of other organisms and the significance of hidden or distant features of their current circumstances, which together have a different practical and perceptual significance than either would by itself. Such tracking of multiple perceptual configurations and their behavioral significance in translucent environments requires *closer* attunement to the current situation rather than displacement from it. Such communicative capacities would only enable the organism to respond more adaptively to the entire complex of behavior-cum-circumstances, and that is not a route to language or conceptual understanding.

This problem will arise for any attempt to begin with the representational/reportorial role of language as the key to its emergence and differentiation from other animals' flexible responsiveness to complex environmental circumstances.[11] In emphasizing the barriers to the evolution of symbolic displacement, Bickerton and Deacon have thus also given good reason to think that symbolic relations between utterances and situations in the world cannot have evolved directly. The obvious alternative is that symbolic displacement arose indirectly through the appropriation of more complex expressive capacities that first arose in a different way. One need not imagine that symbolic representation or articulated judgment

11. In this respect, my account differs from Tomasello's (2008) approach, even though there is considerable common ground on many details. Tomasello seeks to understand the emergence of referential expression directly from pointing and gaze-directing gestures, whereas I am arguing that while these could be complex and effective forms of behavior, they cannot lead directly to symbolic displacement.

arose suddenly or directly. What may instead be needed initially is the emergence of vocal utterances or gestures that are proximally responsive to other expressive behavior (and thus only implicitly and indirectly responsive to their nonvocal circumstances) rather than directly responsive to those circumstances.[12] In that case, what eventually became language would have to emerge as a *social* activity before it could ever be discursive and conceptually articulated. Speech behavior would need to comprise a sufficiently complex pattern of response to other speech behavior. Verbal expression would thereby become partially dissociated from its immediate perceptual environment because it would be responsive to a relatively independent, social, "conversational" context. Yet social relations are not altogether distinct from their environing circumstances. The intertwining of social relations and other circumstances would become especially important if Bickerton is right about the evolutionary importance for early hominids of coordinating collective action at a distance. What I characterize below as the partial autonomy of discursive practice—the responsiveness of vocal expressions primarily to other vocal expressions, yet without complete disconnection from accountability to environmental circumstances—might allow for the possible divergence between what is said and what ought to be said, which has otherwise seemed inexplicable. What is appropriate in response to its proximate social and "conversational" context might nevertheless be mistaken in its broader practical and perceptual situation.

Multiple lines of reasoning convergently suggest the plausibility of such an indirect, social, evolutionary origin of language. The first consideration is the obvious point that humans are primates; we and our common primate ancestors are social organisms with complex, hierarchical relationships among conspecifics that play important roles in reproduction, defense against predation, and access to food. The attitudes, intentions, and resulting behavior of others in its social group are among the most salient and vital features of a primate's environment. Sensitive perceptual attentiveness to what others are doing and its implications for their future behavior, alongside the ability and inclination

12. Tomasello (2008) argues that such capacities likely first arose through communicative gestures, since other great apes do not use vocal expressions in ways that are communicatively directed toward other animals and do use an extensive range of expressive gestures to get or direct the attention of others. I remain agnostic on this point, although Tomasello is surely right that the initial uses of vocal expressions for directed communication were likely conjoined with bodily gestures in ways that facilitated their communicative uses. For reasons given below, however, I take it that the development of vocally expressive and auditory discriminatory capacities were a crucial step in the development of anything like human language.

to engage the perceptual attention of others, are richly refined aspects of our primate heritage (Dunbar 1996; Tomasello 2008). Moreover, if that were not the case, it would be very hard to see how the perceptual and practical skills involved in language could ever have emerged. Language requires focused, discriminating perceptual and practical attention to others, both in listening and in speaking to them and in attracting and sustaining their attention.

Robin Dunbar (1996) puts forward one of the more extensive arguments for locating the evolutionary origins of language in transformations of hominid social life. Dunbar began with the importance of grooming behavior in primate social life as a possibly ancestral trait.[13] Group behavior matters in avoiding predation, and he argues that the time and energy most primates devote to grooming establishes and cements social hierarchies, alliances against predators, and mating relations within groups. Well-known ecological challenges likely placed early hominids under selection pressure to form larger social groups, and he argued that language evolved for the analogous function of securing and assessing alliances within larger, more amorphous social groups.[14] Bickerton (2009, 27–28) argued against Dunbar's hypothesis that it does not account for the needed selection pressures for symbolic displacement. Bickerton is surely right to reject Dunbar's claim that language emerged directly from the exigencies of maintaining the cohesion of hominid group life. Dunbar's argument gets greater traction, however, and also strengthens Bickerton's own line of argument when they are combined in the right way.

Changes in the diet and foraging behavior of the great ape species provided an important background for Dunbar's argument. The transition from a forest environment to more open savannah greatly increased the

13. Tomasello (2008) has argued for the emergence of distinctively human cooperative behavior via the appropriation of common primate capacities for gestural expression, with the shift to a vocally expressive repertoire coming very late in the evolution of human communication. He is surely right that expressive gestures had to be integral to the ways in which protolinguistic primates attended to one another and their shared circumstances. Yet part of his reason for thinking that primate vocal capacities were initially irrelevant here is that vocal expression first directs other primates' attention to the speaker rather than to some relevant circumstances (in contrast to pointing, gaze directing, and other attention-directing gestures), and that it is more expressive of emotion than communicative of information (2008, 226–32). Dunbar's argument relies on those very features of vocal utterance, however, and Tomasello himself also notes that the public character of vocal expression, even when directed vocatively, also enables others to track its relations to other expressions and activities in forming a "reputation" within a group, which also matters to Dunbar's hypothesis.

14. Tomasello (2008) also notes that great apes will differentially cooperate with other apes that have previously been cooperative within the group; the capacity for tracking the social "reputation" of others in a group is clearly ancestral in the primate lineage.

risks of predation (Dunbar 1996, especially ch. 1, 2, 4, 6). Bickerton's argument significantly expands this effort to situate the evolution of language within relevant ecological changes. Bickerton draws upon optimal foraging theory to explicate a transition in early hominid life from catchment scavenging to territory scavenging. This reasoning reinforces Dunbar's primary point, since the assembly, mobilization, and defense of larger social groups (which is central to Dunbar's view of the need to maintain reliability and cohesion within such groups) would become especially salient for territorial scavengers who used primitive tools. Bickerton points out that early hominids clearly used hand axes in large numbers, and these tools would enable opportunistic scavenging of large animal carcasses before other predators could break through their skins. Under those circumstances, hominid bands would not merely be more vulnerable to predation by traveling to more irregular food sources in open country. They would become actively attractive to and competitive with other predators trying to drive them away from scavenging sites they occupied for extended periods of time.

Combining the two hypotheses also makes Bickerton's criticism of Dunbar irrelevant, if what Dunbar's hypothesis explains is the emergence of expressive activity that does not yet involve displacement or conceptual content. Dunbar's hypothesis is more plausible to explain an articulated, expressive repertoire that could then be appropriated for different purposes as opposed to a direct argument for the evolution of symbolic displacement. Articulated vocal expressiveness would allow others to recognize and respond to the emotional states and practical orientations of others in a group. Such sustained attentiveness to others could both express and sustain group allegiance, as does grooming behavior. Moreover, two other important features of language become more intelligible in light of this combined hypothesis that symbolic displacement in language appropriated an expressive repertoire that had initially helped secure social cohesion.

Dunbar recognizes that his social origins hypothesis explains the simultaneous local cohesion and broader diversification of languages. He does not, however, highlight that an increasingly fine-grained articulation of vocal expression would be a straightforward further consequence of this explanation. It is easy to overlook the significance of the human ability to hear and produce subtle differences in vocal articulation, without which any capacity for conceptually articulated expression would confront severe limits in its expressive range and communicative efficacy. If a primary function of verbal articulation was initially to secure intragroup alliances within large, amorphous social groups, that role would explain why verbal expression within a group converges on similar patterns and diverges from other groups' patterns.[15] The challenges of maintaining mutual recognition and alliance within larger fission-fusion groups of hominids would produce pressure for both conformity to local patterns of talk and recognizable divergence from outsiders' expressive patterns. If linguistic expression arose from the sustained devotion of time and attention to other members of a group, to express and secure group commitment, then a likely consequence would be the ability to recognize and produce relatively fine-grained differences between "dialects." What better way to identify one's membership in a group than to display evidence of extended participation in its distinctive vocal exchanges?

Dunbar also does not discuss another feature of language, which his account nevertheless helps accommodate. Kukla and Lance (2009) point out that a public linguistic practice requires the ability to call other agents, and to recognize and respond to such calls. These vocative and recognitive aspects of linguistic practice are often taken for granted in philosophical reflection upon language. Even the most impersonally directed reports nevertheless have an ineliminable vocative and recognitive dimension. Kukla and Lance note, "If it is not part of the structural aim of a speech act to make a claim on someone and demand recognition of this claim, then that speech act fails to have any actual, lived pragmatic force at all; part of what makes a speech act a *claim* is that it seeks normative uptake from agents capable of recognizing normative claims" (2009, 163–64). In our familiar linguistic practices, the vocative and recognitive roles of speech acts serve to focus both speaker and listener upon the content of a claim, and the content in question may be partly independent of the conversational context. Yet in trying to understand how vocal expressions could first acquire an articulated content, vocative and recognitive considerations move to the forefront. The availability of vocal performances through which organisms call upon one another and respond to such calls, and the ability to express recognition or direct others' attention, could be "recruited" to direct attention and action toward distant or otherwise absent circumstances. The ability to call one another and to recognize oneself as called to respond would then precede the articulation of such calls into contentful claims about something else.

15. Tomasello (2008) does recognize and highlight the importance of recognizable similarities in behavior as important for establishing and sustaining group allegiance.

Our problem has been understanding how symbolic displacement in language could evolve despite the barrier presented by most organisms' flexible perceptual and practical attunement to their actual circumstances. How might the emergence of a social-expressive repertoire, including the ability to call others and recognize oneself as called, help overcome this barrier? Bickerton provides an important clue in his criticism of Dunbar's appeal to the social function of gossip to bridge the gap between mere vocal expression and conceptually articulated language: "Novelty is the soul of gossip. But there's no way in which a tiny number of words can be permuted to express a wide range of new items. You'd need at least several dozen, more likely a few hundred words before you could begin to do that. But you'd never get that far unless the first few words already had a substantial payoff" (Bickerton 2009, 28). Bickerton thereby highlights one aspect of the central difficulty for understanding the evolution of language: grasping how even a rudimentary discursive practice could get started so as to overcome the substantial evolutionary barriers to symbolic displacement. Kanzi's limited achievements show that the primate lineage has a latent capacity to acquire linguistic understanding *if* a perceptually and practically accessible linguistic practice were already prominent in their early developmental environments. Yet that initial barrier, in retrospect, was insuperable for all but one primate species, or perhaps even one small subpopulation of that hominid species, if language acquisition initiated its reproductive isolation and eventual speciation.

Any account of language evolution that posits direct selection for representation and information exchange must confront this difficulty head on. Such capacities would only be useful at all after the achievement of extensive representational articulation, cohesion, and precision. Its initial selective grip would be hard to understand. By contrast, the problem does not arise if articulated vocal expressiveness originally served functions other than reportorial/representational. A limited initial expressive repertoire would not be pointless if the initial evolutionary "payoff" reflected needs to recognize, sustain, and coordinate larger and more amorphous social groups. Moreover, if these initial expressive and recognitive capacities were adaptive for both individual organisms and groups,[16] then their further elaboration would benefit from multiple se-

16. Neo-Darwinist arguments (notably kin selection) during the last decades of the twentieth century led to general dismissal of group selection as a significant factor in evolution, but subsequent work has shown that group selection can play a role wherever variation between groups is more significant than variation within groups (Sober and Wilson 1999; Dor and Jablonka 2000, 2001). Behavioral niche construction exemplifies the kind of selective circumstances in which group

lective processes. These processes include straightforward selection of those organisms to which the following applied: protolinguistic facility that led to more effective and central integration within cohesive groups; sexual selection within such groups for vocal articulateness and responsiveness; group selection for more effective coordination of collective action; neural reorganization and some genetic assimilation of initial neural plasticity for more rapid acquisition of the relevant perceptual, practical, and social skills; and a subsequent "stretching" and "ratcheting" of these assimilated capacities, which would expand and intensify discursive practice (Deacon 1997; Avital and Jablonka 2000; Dor and Jablonka 2000, 2001; Tomasello 2008; Bickerton 2014).

A vocative-recognitive expressive capacity that first served social purposes of cohesion, affiliation, and collective orientation would not lack all broadly "semantic" significance even at the outset. Expressions of mood, attitude, orientation, and attention-direction, which were already part of the great apes' gestural and vocal capacities, also correlate with circumstances characteristically relevant to what is expressed (exemplified by the vervet monkey calls that ambiguously indicate distinctive fears, warnings, or evasive actions, as well as differences among predator species and styles of attack). Charles Taylor (1985, ch. 10) long ago highlighted uses of language (e.g., saying "Hot, isn't it?" on a sweltering day) that create or sustain a "public space" of mutual directedness toward common circumstances, seeking to share an attitude or orientation rather than to inform. Taylor envisaged social expressive uses of an already-articulated linguistic ability to establish a shared social orientation, but the appropriation of abilities whose primary function was social and expressive could proceed in the other direction, gradually acquiring semantic significance.[17] What makes such appropriation possible is that an organism is not merely a bounded physical entity but a pattern of goal-directed responsiveness to the environment relevant to its way of life. For an organism with sufficiently sophisticated tracking skills, to be attentive and responsive to other organisms is also to

selection can play a prominent role. Moreover, there may be plausible reason to think that the constriction of the hominid lineage is itself the result of group selection for linguistic or protolinguistic ability. Behavioral niche construction then would play an evolutionary role in speciation comparable to that long accorded to geographic isolation, but the "isolation" it secures is not spatial but behavioral, and it only isolates individual organisms as participants in the group's distinctive behavioral repertoire.

17. Tomasello (2008) proposes that the establishment of something like common space as part of collective intentionality arose more directly and centrally rather than via a vocative, emotionally expressive and responsive repertoire. I see the two as arising together and do not see the need to speculate about their order of emergence.

be responsive to how their changing environment matters to those organisms and to itself. This mutual entanglement of organism and environment takes on new levels of complexity, however, once this socially responsive behavior itself becomes a salient and influential dimension of an organism's developmental and selective environment.

I have been arguing that Bickerton's (2009) arguments for the emergence of symbolically significant language—as a response to problems of socially coordinated action within large, amorphous, fission-fusion groups of primates—do not work on their own. These arguments nevertheless become much more plausible if these social hominids were already vocally articulate as well as gesturally expressive and also perceptually attentive to such articulation as expressively significant. Dunbar's hypothesis accounts for just this possibility.[18] What was needed to get a proto*language* under way was an already extant, interrelated expressive repertoire that could then be utilized to coordinate group behavior more flexibly, extensively, and effectively.[19] The implicit semantic significance of that expressive repertoire could gradually take on more prominent roles and correspondingly refined forms through ongoing cycles of niche construction.

Such an account of language evolution would have at least three important consequences for how to think about linguistic and conceptual normativity in light of its evolutionary origins. The first consequence, one I have been emphasizing throughout the book, is that language and conceptual normativity are products of behavioral niche construction. Philosophical conceptions of language and conceptual understanding would accordingly shift from emphasis upon internal, "mental" representation to outwardly directed skills for recognizing, responding, and contributing to an ongoing public activity that was integral to early hominid life. Our characteristic capacities for linguistic and conceptual understanding are bodily skills for perceptual discrimination and practical expression. These skills were not initially different in kind from other organisms' robust and flexible responsiveness to multiple, discordant

18. Much of the empirical detail of Tomasello's (2008) account of the emergence of human social cooperativeness as essential to the evolution of language can also be constructively assimilated within this conjoined account, without needing to take on his gratuitous appropriation of an account of intentionality (taken from Grice [1989] and Searle [1982, 1995]) that presupposes what we are seeking to explain in an evolutionary context.

19. Although it matters to have an articulated expressive repertoire in place, for which vocal expression is the obvious candidate, one need not think that vocal expression functioned in isolation. Indeed, its integration with gestural expression, and especially with the many ways in which primates demand and direct one another's attention, is part of the argument. Tomasello (2008) is especially instructive on the importance of these latter functions.

indications in translucent environments. These perceptual and practical capacities were nevertheless transformed by the gradual emergence and refinement of multilevel reflexive interrelations. Linguistic understanding involves tracking one's own and others' vocal expressions simultaneously with respect to an assimilated repertoire of expressions, a local "conversational" context, and the place of that conversation within a larger social-biological environment. Moreover, it requires the ability to adjust subsequent performances and recognitions in response to the relations among these multiple contexts.

The second consequence directly follows from this conception of language as practical-perceptual responsiveness to a behaviorally constituted ecological-developmental niche. Language is first and foremost a public practice that only exists in being continuously reproduced and transformed through the consequent coevolution of human cognitive capacities in their inherited linguistic developmental niche. As Ruth Millikan notes about her understanding of language as public, "The phenomenon of public language emerges not as a set of abstract objects, but as a real sort of stuff in the real world, neither abstract nor arbitrarily constructed by the theorist. It consists of actual utterances and scripts, forming crisscrossing lineages" (2005, 38). These lineages are themselves partially constitutive of the practice in the sense that participants make and understand utterances as iterable iterations of ongoing patterns. To use a *word*, for example, as Kaplan (1990), Millikan (2005), Ebbs (2009), and I (Rouse 2002, 2014b) have all variously argued, is to use recognizably the same expression that others have used and can use again *as* they have used it. Such patterns of practice cannot be captured either as a de facto regularity in what people do or as a definite rule that governs its constitutive performances.[20] Patterns of word use are not regularities, for multiple reasons: they encompass erroneous, deviant, and other idiosyncratic uses; there is no practice-independent way to specify the domain of performers or performances for which it would supposedly be regular; practices often depend upon differences as well as similarities among practitioners' performances; and the pattern shifts over time, such that objectively similar performances can have a different status within the practice at different times.[21] The "sameness" of iterative per-

20. For earlier developments and defenses of this claim, see Rouse (2002, ch. 5; 2006; 2007). See also further development of these arguments later in this chapter.

21. There are multiple kinds of normative status involved: whether a performance is part of the practice at all (even if deviant or incorrect); what role it is performing in the larger practice, such that it is subject to normative assessment with respect to that role; and of course, how it is (and ought to be) assessed with respect to that role, for example, as correct or incorrect, appropriate or

formances of a practice is thus reflexively constituted by the ability to perform and track the iterative sequences. The resulting patterns are also not rule governed because the relevant rule is never fixed.[22] These patterns are constituted by and reproduced in part due to the interactions among the performers and performances that contribute to the pattern. Practices are constituted as such by the mutual normative accountability of their performances to "norms" that are always at issue within the practice. To say these norms are "at issue" is to say not merely that the norms change over time. It indicates that change results in significant part from the ongoing effort to sustain a common practice accountable to norms, even though what the norms are is not yet settled. That is part of why it matters to understand language as a *social* practice more fundamentally than it is a semantic practice. Utterances get their semantic significance from the mutual accountability of its practitioners to one another *as* situated in a partially shared context and responsive to one another in that context.

The third consequence, building upon the first two, is what I have been calling the partial autonomy of linguistic practices. Language is partially autonomous in at least three senses. First, as I just noted, linguistic expressions are iterative of and iterated by other linguistic expressions; they are linguistic in being iteratively interrelated as instances of what thereby become linguistic types. Second, the vocative and recognitive aspects of linguistic practice make token expressions proximally responsive to other linguistic tokens, such that utterances typically make sense within a mostly intralinguistic context.[23] Utterances are normally understood in a conversational context; even written texts

inappropriate, novel or familiar, or interesting or banal, with the possibilities for normative assessment themselves open-ended.

22. I argue in the next section and the subsequent chapter that the contestability within a social practice of what norms govern the practice are what distinguishes such biosocial practices from the kinds of biological normativity of even the robust and flexible patterns of perceptual-practical response to their environments characteristic of many nonhuman organisms.

23. Declarative sentences typically have what Kukla and Lance (2009) call agent-neutral inputs and outputs: what is claimed in uttering that sentence is expressible by anyone who has appropriate warrants, and the claim, if warranted, ought to be taken into account by anyone. Focusing upon declarative assertions with this normative structure has encouraged most philosophers to put the pragmatics of utterances into the background of how to think about conceptual articulation and contentfulness. Yet even the most agent-neutral and decontextualized content only actually makes a claim on anyone through having been uttered to them in some specific context. In such contexts, moreover, claims are interpreted and assessed not merely as context-free truth claims but for their relevance to the conversational context: if the contextual relevance of the claim isn't clear, we often find ourselves asking what the speaker meant by interjecting that claim into the conversation (where it would not be sufficient to identify what was meant with the conventional meanings of what was said).

invoke an assumed audience and make assumptions about what other claims can be taken for granted in context (Kukla and Lance 2009). The reason I have emphasized the evolutionary priority of a social-expressive vocal activity as a prior condition for symbolic displacement is the need to generate these first two features of the partial autonomy of linguistic expression.

These first two respects in which linguistic activity has a certain degree of autonomy provide the background for a third form of partial autonomy, which I take to be decisive for understanding the evolution of a capacity for symbolic displacement. Recent work in lexical semantics highlights the limited and distinctive "semantic envelope" of human languages oriented toward specific aspects and features of the world (Dor 2000; Dor and Jablonka 2000, 2001; Levinson 2000). As Dor and Jablonka note,

> A survey of the world's languages reveals a very surprising fact: languages are definitely not all alike, but the semantic categories which are reflected by grammatical complexities in natural languages belong to a very constrained subset of all the categories which we can use to think, feel and conceptualize about the world: some semantic categories turn out to be grammatically marked in language after language, whereas some others consistently do not participate in the grammatical game. Specifically, no language we know grammatically marks the distinction between friend and foe, or between interesting and boring events. The categorical distinctions between animate and inanimate entities, telic and atelic events, factual and hypothesized events are reflected in virtually every language we know, and so are the distinctions between different spatial relations and time configurations. (2000, 39)

I think it likely that the evolution of language is informed by the differences between semantic categories that are embedded in complex grammatical relations retained across languages and semantic categories that are not grammatically significant. Semantic categories that are unmarked grammatically can be important for many aspects of human life, but their importance is only expressed in language rather than by language itself. Those structures embedded in grammar presumably mark differences that played an important role in the coevolution of human beings and languages. There would have been considerable selective pressure for ease, speed, and reliability of acquisition of the structural distinctions that played a central role in whatever emerging system of expressive communication initiated the coevolution of languages and vocally expressive human beings.

Daniel Dor (Dor 1999, 2000; Dor and Jablonka 2000) has highlighted several distinctive semantic differences with these grammatical characteristics. The grammatical behavior of verbs and their arguments, for example, indicate structural differences among four types of basic events—activities, accomplishments, states, and achievements—and within those classifications, various grammatically differentiated subclassifications (e.g., between verbs that distinguish activities by their manner of motion or by their surface contact or between verbs that distinguish accomplishments by their directed motion or a resulting change of state). The differences among verbs belonging to different event types make a difference to which constructions employing them are grammatical or ungrammatical, whereas otherwise important differences between verbs within the same type do not affect the grammar of their sentences. The relations among these event types are also grammatically distinguished, as are additional considerations having to do with the factuality or nonfactuality of embedded components, and among the factual components, the speaker's knowledge or ignorance of those facts.

These kinds of grammatically operative semantic differences are suggestively consistent with Bickerton's (2009) and Deacon's (1997) hypotheses that symbolic displacement emerged as a means of coordinating group action at a distance and Bickerton's (2014) hypothesis that minimalist "universal grammar" marked the neurological accommodation and disambiguation of "protolanguage," but those are questions for further empirical research in multiple fields. The point that matters to my argument is the claim that language emerged with a limited "semantic envelope" that was marked in structural relations within and among linear sequences of vocal expressions. This point matters in two interconnected ways. First, it reinforces my earlier suggestion that language emerged from the appropriation of an extant repertoire of vocal expression that already played a role in hominid social life. On this hypothesis, representational relations between vocal expressions and objects or situations in the world were not what directly enabled the emergence of symbolic displacement. Language instead gradually emerged as *structured* combinations within an already available and salient expressive repertoire, whose use enabled more effective coordination of action among members of larger, amorphous social groups. The second way that the limited semantic envelope of language matters, however, is the significance of the *partial autonomy* of linguistic expression. Linguistic expressions are normally internally related to other linguistic expressions, both in a "conversational" context and as tokens of expressive types. Yet these "conversational" exchanges are also situated within and responsive to other aspects of perceptual and practical response to an environment. In the case of early hominids, surely what mattered most was their relation to fitness-relevant features of the environment.

This multidimensional partial autonomy of protolinguistic expressions is important for understanding the decisive transition from a sophisticated practical-perceptual responsiveness to circumstances to a genuinely symbolic and conceptual understanding. Recall that pervasive selective pressure for close coupling between perception and flexibly appropriate responsiveness to an organism's surroundings is an important evolutionary barrier to symbolic displacement. In most organismic lineages, other organisms' vocalizations are just one among many indicators of effective response to a translucent environment. The partial autonomy of linguistic expression instead situates vocal expression within expressive and conversational contexts that are partially independent of other aspects of the organisms' immediate circumstances. The result is a *dual* practical-perceptual tracking of the environment: tracking vocal expressions in relation to their conversational and expressive contexts (other recent utterances and other uses of the same expressions) and tracking these larger patterns of "intralinguistic" expression in the context of broader perceptual and practical responsiveness to circumstances. This dual tracking provides a basis for distinguishing between appropriate utterance (i.e., appropriate in its "intralinguistic" contexts) and correct utterance (i.e., appropriately responsive to the circumstances of one's broader perceptual/practical immersion in the world).

Language thereby first emerged as a specialized expressive repertoire with a limited, structured "semantic envelope," which also enabled new patterns of expression and uptake. As this repertoire took on increasingly central roles in early hominid life, it thereby became a more integral part of their selective environment. Such effective integration would lead to multiple forms of coevolutionary selection pressure for a mutual coadaptation between languages and human cognitive and expressive/recognitive capacities. Continuing cycles of niche constructive adaptation have led to the extraordinary expansion and diversification of human expressive and articulative capacities within that recognizably persistent semantic-syntactic envelope.[24] From a naturalistic point of view, the consequences for how to think of the resulting conceptual

24. Dor and Jablonka talk solely of the "semantic envelope," but they do so to emphasize that many of the structures that linguists identify as purely syntactic have semantic significance. The

capacities are complex. First, on the account I have been sketching, conceptual understanding is grounded in perceptual and flexibly responsive behavioral capacities that are broadly continuous with those of other animals, especially our evolutionary kin in the primate lineage. That continuity is not diminished by the greatly enhanced and specialized vocal and auditory abilities as speakers and listeners that arose through coevolution with our discursive niche. Second, however, the development of symbolic displacement in language is a genuine evolutionary novelty in the human lineage. Kanzi shows in retrospect that the capacity to acquire such abilities via early development in a discursive environment was already latent in our common primate ancestors, but phenotypic expression of this capacity was blocked by the absence of the requisite developmental and selective niche.

A third straightforward consequence of this novelty is that what emerged is not merely linguistic ability as an isolated trait. The genetic and cognitive assimilation of capacities for language learning does have some autonomy, and languages do have a somewhat specialized semantic-syntactic envelope. The decisive evolutionary novelty was not just language itself, however, but a capacity for symbolic displacement and conceptual understanding that then extends beyond language narrowly construed. Further extension of conceptual capacities occurs in two mutually reinforcing ways once a capacity for symbolic displacement is initially established through linguistic niche construction. First, the key innovation was the ability to recognize expressive activities, including one's own utterances or other performances, as having significance and accountability beyond their surrounding circumstances. Once vocal expressions were understood in this way, and became integral to human social life, other expressive productions could gradually be undertaken and taken up with comparable transcendence of their circumstances. Conceptual significance has thereby accrued to such nonlinguistic forms of expression such as music, dance, drawings/diagrams/maps, bodily adornment, games, and so much more. More strikingly, however, they also include the making and use of *equipment*: not just individual things that can be used instrumentally, as many other organisms do, but interrelated complexes of equipment understood as available and appropriate *for* some tasks and not others and assignable to differentiated

result, however, is that these semantic differences are embedded in the syntactic structure and processing of linguistic expressions, and I therefore prefer "semantic-syntactic envelope."

social roles.[25] Equipmental complexes are among the most tangible and extensive forms of conceptually articulated niche construction; whereas linguistic niche construction initially takes the form of ephemeral behavior that must be continually reproduced, the making and refining of equipmental complexes combines behavioral with physically persistent forms of niche construction. Haugeland's discussion of a telling example illustrates how equipmental complexes are heritable forms of niche construction that are also symbolically/conceptually articulated:

> How much of what a culture has learned about life and its environment is "encoded" in its paraphernalia and practices? Consider, for example, agriculture. . . . Crucial elements of that heritage are embodied in the shapes and strengths of the plow, the yoke, and the harness, as well as the practices for building and using them. The farmer's learned skills are essential too; but these are nonsense apart from the specific tools they involve, and *vice versa*. . . . Hence, they constitute an essential unity—a unity that incorporates overall a considerable expertise about the workability of the earth, the needs of young plants, water retention, weed control, root development, and so on. (1998, 235)

Conceptual understanding was enabled in whole or in part by the development of language but not thereby limited to what can readily be linguistically expressed.

A second and more decisive extension of the capacity for symbolic displacement and conceptual understanding absorbs the entire human perceptual and practical repertoire within the space of conceptual normativity. Language is first and foremost a specialized, partially autonomous practical-perceptual capacity and, to that extent, is continuous with our evolutionary heritage as animals and primates. Once language has been sufficiently articulated and centralized within the human way of life, however, our other perceptual and practical capacities also acquire a broadly conceptual significance. Everything we perceive and do can have further discursive significance as events trackable in relation to their broader discursive/symbolic context. Sellars described that context as "the space of reasons," and I will be arguing over the next several chapters that it is extensionally equivalent to our discursive biological niche.

25. The locus classicus for the recognition of integrated complexes of equipment and social roles as distinctively human forms of intentional directedness is Heidegger ([1927] 1962, div. I, ch. 3–4).

This recognition that conceptual understanding transforms our perceptual and practical capacities returns us to the dispute between McDowell and Dreyfus over the scope of conceptual understanding but with a new basis for properly locating their concerns. Dreyfus was rightly attentive to the many domains of human life that are not readily or constructively articulable within the semantic-syntactic envelope of language, which remains limited even though it has evolved and expanded over time.[26] This broad range of nonlinguistic perceptual and practical skills centrally includes the perceptual and practical capacities for vocal and written expression and auditory and visual recognition that enable linguistic understanding.[27] Dor and Jablonka tellingly characterized the still expanding scope of our capacities for expression and understanding that far exceed easy linguistic articulation:

> The expressive envelopes of different languages are different in interesting and subtle ways, but they all share a common core. Types of messages which fall comfortably within this core are best suited for communication through language. Types of messages which do not comfortably comply with it turn out to be more difficult to communicate through language. Many other types of messages, which do not comply with this scheme at all, turn out to be virtually impossible to communicate through language. Interestingly, many of the messages which turn out to be very difficult to communicate through language seem to be very well suited for communication through other means of communication: we can *mime* and *dance* them, use *facial expressions* and *body language* to express them, *paint* and *draw* them, write and play *music*, prepare *charts* and *tables*, write *mathematical formulae*, screen *movies* and *videos*, and so on. (2000, 40)

Dreyfus was wrong to conclude that these domains of human life constitute a realm of nonconceptual content. Language first enabled symbolic displacement and conceptual normativity, but the resulting conceptual capacities then extend beyond what is readily articulable linguistically. All these other forms of expressive activity are themselves conceptual in the sense of being expressive, significant, and normatively accountable

26. I discuss Dreyfus's criticism of McDowell, and his advocacy of a domain of "nonconceptual content," in chapter 2.

27. Bickerton (2014, 42–45) highlights the fact that the cognitive process of sentence formation and the reverse process of sentence interpretation take place below the threshold of conscious awareness and reflective control: "Humans are no more aware of what their brains are doing while they are speaking or listening to others than spiders are when they are spinning webs or bats when they are hunting insects" (44).

far beyond their responsiveness to and import for their immediate practical-perceptual circumstances. Moreover, the linguistic domain, in the narrow sense that fits within the semantic-syntactic envelope incorporating the grammatical structures of human languages, is not even the relevant characterization of our evolved linguistic capacities. Language in this narrow grammatical sense is still only a *partially* autonomous domain, and transformative relations among linguistic utterances and other performances and recognitions go in both directions. As a result, with language in place, all our perceivings and doings also have a broadly linguistic significance. In the other direction, the expressive capacities available within the narrow semantic envelope of languages have expanded over time, so that a more expansive domain of conceptual content can be explicitly taken up linguistically, through indexical, demonstrative, anaphoric, and recognitive expressions and performances. In these ways, the dependence of conceptual understanding upon language as an evolutionary novelty, with its limited core semantic-syntactic envelope, is fully compatible with John McDowell's (1994) insistence upon "the unboundedness of the conceptual." The conceptual domain in this sense is not narrowly linguistic but incorporates our entire active, expressive, and receptive engagement with our biological niche, which is the Sellarsian space of reasons.

III—The Sociality of Conceptual Normativity

What would be accomplished by a successful development of this approach to the evolutionary origins of conceptual understanding? I argued in chapter 2 that an account of conceptual normativity needs to begin with perceptual and practical involvement with a surrounding environment and then show how such interaction with the world becomes conceptually articulated. Conceptual understanding requires a mode of engagement with some aspect of the world that can distinguish an articulated determination of how an agent, speaker, or thinker *takes* some aspect of the world to be from how it *is*. Such a conception must be accountable to its intended "object" such that its conception of that aspect of the world can be mistaken.[28]

28. "Object" is in scare quotes, both because objects are only identifiable within a broader grasp of a situation and because the object of a discursive engagement is often at issue within that engagement. This latter aspect of the "objectivity" of objects is discussed in chapter 5.

My development of this explanatory strategy has been framed by Haugeland's arguments against the possibility of explicating the normativity of conceptually articulated understanding through either of two familiar, initially plausible approaches. Haugeland argued that neither the teleology of biological maintenance and reproduction nor the institutional authority of social norms within human communities could suffice to explicate the dual normativity of conceptually articulated intentionality as both aspectual and truthful. The biological way of life of an organism determines and is determined by its environment, composed of those parts of the world with which it is developmentally, physiologically, and selectively interdependent. Haugeland argued that the extensional determination of a biological environment by an organism's way of life cannot also take up that environment *as* aspectually understood. The social practices of a community, on the other hand, can institute the myriad articulated distinctions that structure games, rituals, organizations, laws, social proprieties, and so forth, but they cannot make those practices accountable to anything beyond what participants in the community accept or regularly do.

My account responds to Haugeland's arguments in a twofold way. First, I argue that an adequate account of conceptual normativity requires the integration of biological teleology and social practice; neither alone is sufficient. Second, an adequate account of conceptual normativity also requires revisions to our familiar conceptions of biological evolution and of social practices. Chapter 3 and the first sections of this chapter have already worked out key elements of this revised approach to the biological evolution of conceptual understanding. Conceptually articulated understanding arises from the evolution of language and other forms of symbolically articulated activity as forms of behavioral niche construction. These behavioral patterns have evolved along with us as part of the developmental environment in which we normally mature as human beings and to which our biological physiology, development, and reproduction are adapted. Although this account of behavioral niche construction provides a better understanding of human evolution, this revision by itself cannot account for how conceptual *normativity* evolved. As I noted at the end of the previous chapter, language and other symbolic expressions might then only amount to especially convoluted examples of the highly differentiated ways of life that can result from biological evolution. Articulated vocal and other "symbolic" expressions would then stand alongside diverse mating rituals, hunting or foraging strategies, communicative dances or songs, and other evolved behavioral patterns. Their normative teleology would not differ from such anatomical, physiological, or developmental peculiarities as peacocks' tails, ruminants' digestive tracts, or the syncytial development of *Drosophila*. Talking (along with drawing, dancing, praying, game-playing, and so forth) in mutually responsive ways would simply be part of what this organism happens to do in response to, and in partial reconstitution of, its biological environment.

A more adequate conception of these forms of behavioral niche construction as social practices makes a difference for understanding conceptual normativity. Familiar philosophical or social-theoretical accounts of the normativity of social practices appeal to accepted rules or predominant regularities in the social life of a community as setting communal standards for the correctness or appropriateness of performances by individual members of that community. Haugeland's arguments were rightly directed against these familiar accounts. Such conceptions of social normativity could only account for the limited sense in which individual utterances and actions can be correct or incorrect as performances of an extant social practice. They do so at the cost of then being unable to understand the normative accountability of the entire practice in turn. Individual performances could be socially incorrect in the sense that they deviate from what others normally do or from communally accepted norms. The community's regular behavior or accepted norms would then be criterial for correctness, however, and could not in the same way be understood as open to correction or criticism.

Such regularist or regulist conceptions of social life could not account for how social practices were directed toward, and accountable to, anything other than their own continuation. Notably, they provide no resources for understanding the intentional directedness of what is said or done in such communities as accountable to broader patterns of environmental interdependence for their correctness or incorrectness. They also take for granted, but then cannot account for, how and why belonging to that community and acceding to its norms could be authoritative for its members; for who is or should be included as a member of the community; or for how it matters whether and how such practices are continued. That is why regulist or regularist conceptions of social practices and social norms could not contribute to understanding conceptual normativity, even if they were embedded within an account of those practices as also biologically evolved. Conceptual understanding involves both normative accountability beyond its own actual performances and open-ended capacities for critical reflection. The conceptual domain cannot be adequately understood in ways that would block reflective assessment of the social regularities or presupposed norms that are supposedly constitutive of its norms.

An alternative conception of social practices and their normativity can circumvent these limitations. We should not think of the social-normative dimension of practices in terms of either behavioral regularities or accepted norms, rules, or conventions. I have discussed this alternative conception of social practices (which I call a *normative* conception, in contrast to the familiar regulist and regularist alternatives) extensively elsewhere (Rouse 2002, ch. 5; 2006; 2007). Here I will summarize the key elements of that account for a conjoined social-biological understanding of conceptual normativity. As a first consideration, a normative conception of social practices does not identify a practice by any exhibited regularities among its constituent performances or by their accountability to an independently specifiable rule or norm. On this conception, a practice is instead held together by the interactions among its constitutive performances, which constitute their mutual accountability. One performance can respond to another, for example, by trying to correct it, drawing inferences from it, translating it, rewarding or punishing its performer, mimicking it, iterating the "same" performance in different circumstances, circumventing its effects, and so on.

Intimations of this conception of practices, as patterns of responsive interrelations among their constitutive performances, can be retrospectively recognized in other familiar discussions of social life. Robert Brandom suggested that "we can envisage a situation in which *every* social practice of [a] community has as its generating response a performance which must be in accord with another social practice" (1979, 189–90). Such a chain of responses need never terminate in an objectively characterizable regularity.[29] Michel Foucault's conception of power, as "a mode of action which does not act directly and immediately upon others, [but] instead acts upon their actions" (1982, 220), likewise emphasizes patterns of mutual interaction among performances rather than any

29. Brandom's terminology in this passage is somewhat different from mine, but that should not engender confusion. By "social practice," he means something more like what I would call "kinds of performance within a practice": examples of "social practices" in his sense might include patrons presenting tickets at the door of a theatre, ticket takers inspecting the marks on the tickets and ushering their presenters to a seat (or refusing them entry), presenters arguing with the ticket taker, ticket takers calling the police to arrest an interloper, and so on. I am using "practice" in cases like this to refer to the interrelated complex of performances that together compose the practice of putting on and attending theatre productions. In the case of language, Brandom has in mind speech acts and the uses of their component words or phrases as "social practices," which I am treating as constitutive performances within a larger pattern of discursive practice.

supposed similarities or shared presuppositions or norms. Further examples include Donald Davidson's (1986) effort to characterize linguistic interpretation without reference to a shared language[30] and Marcel Mauss's (1979) discussion of distinctively French and American ways of walking developed by imitation of and various responses to how others walk.[31] We can also assimilate to this strategy Wittgenstein's well-known remark that requests for justification of a practice must eventually encounter a stopping point at which one can only say, "This is what we do" (1953, par. 217). Wittgenstein is often read as appealing to a social regularity, but his remark can instead be heard with the inflection with which a parent tells a child, "We don't hit other children, do we?"[32] Such statements or rhetorical questions do not describe regularities in children's actual behavior. On the contrary, parents make such comments precisely because children do hit one another. Parents do so, however, in response to or anticipation of such "deviant" behavior in order to hold it accountable to correction. Children's behavior in turn is only partially accommodating to such correction: sometimes obeying, sometimes challenging or circumventing corrective responses, sometimes disobeying and facing further consequences, and so forth.

This conception of social practices, as a network of mutually interactive performances, is not yet sufficient to account for conceptual normativity, however. The problem is that these mutual interactions cannot, by themselves, explicate how performances in a social practice could be directed toward, and accountable to, anything other than their own continuation. A second crucial feature of practices, normatively conceived, is that these patterns of interaction continue over time with an orientation toward *how* the practice continues in the future and the broader significance of that outcome. Alasdair MacIntyre's conception of a tradition also exemplifies an interactive conception of social practices but does so in a way that would highlight this second consideration: "What

30. Davidson (1986) may have a more expansive understanding of this claim than I endorse. He is clearly denying that understanding a language (in the sense of an abstract structure shared with others that provides a basis for interpreting their utterances) is necessary for interpretation, drawing upon our ability to understand mistakes, jokes, metaphors, and a range of other nonstandard uses. That is the claim I endorse. Davidson also seems to think that one might be in the position of a truly radical interpreter, who starts with a collection of token utterances, which can be individuated as words or other linguistic expressions apart from their place in a larger linguistic practice, merely by their auditory or other form of similarity. Such a collection of utterances would thereby be understandable as comprising an idiolect, independent of its place in any larger pattern of social practice. That claim I reject. For further discussion of this issue in Davidson, see Ebbs (2009, ch. 4–5).

31. Mauss would fit here on my reinterpretation of Mauss (Rouse 2002, ch. 5) in response to Stephen Turner's (1994) criticism of the very idea of a social practice.

32. I adapt this interpretation and the example from Wheeler (2000, ch. 6).

constitutes a tradition is a conflict of interpretations of that tradition, a conflict which itself has a history susceptible of rival interpretations. If I am a Jew, I have to recognize that the tradition of Judaism is partly constituted by a continuous argument over what it means to be a Jew" (1980, 62). Judaism, like any other significant tradition of social practice, cannot be identified by elements shared throughout its history; there typically are no such elements. What it is to be a Jew is instead contested among the performances that constitute the ongoing practice of Judaism, in all their historically interrelated complexity. These performances both iterate and respond to other performances, which thereby are held together as belonging to a practice. I characterize the relations among these iterative responses in terms of what is "at issue" in a practice and what is "at stake" in how it continues. Various performances take up an ongoing practice and continue it in partially conflicting ways. These differences locate and focus what is thereby at issue among these conflicting continuations of prior patterns of performance. What is at stake in the practice is the difference it would make to resolve those issues in one way rather than another. In MacIntyre's example, what is at stake among conflicting interpretations of the practice are what it would then mean to be a Jew and to practice Judaism and how those differences matter. But those differences are not already settled, and there is usually no agreed-upon formulation of what the issues and stakes are. Working out what is at issue in a practice, and how the resolution of the issues matters, is what the practice is "about."

Most philosophical conceptions of normativity nevertheless presume that determinate norms must already govern the performances accountable to them and thereby already determine what is at stake in the practices they "govern." Such conceptions can allow for the practitioners' epistemic uncertainty about these norms but not any metaphysical indeterminacy in the norms themselves. This presumption that social normativity presupposes determinate, authoritative norms is also shared by many naturalist critics of normativity (e.g., Turner 1994, 2010, 2014; Roth 2003, 2006, forthcoming); in denying the existence of any such norms, they conclude that any apparent normative accountability must also be explained away. On a normative conception of practices, however, what is at issue and at stake in practices is not just subject to epistemic uncertainty but also open textured and partially indeterminate in a perspectivally varied way. This open-textured normativity is the third key feature of normatively constituted practices. Brandom (1994) again exemplifies this conception in this respect. He characterizes the normativity and semantic contentfulness of discursive practices in terms of the essentially perspectival objectivity of conceptual norms:[33] "Each perspective is at most *locally* privileged in that it incorporates a structural distinction between objectively correct applications of concepts and applications that are merely subjectively taken to be correct. But none of these perspectives is privileged *in advance* over any other. . . . Sorting out who *should* be counted as correct, whose claims and applications of concepts should be treated as authoritative, is a messy retail business. . . . [T]here is no bird's-eye view above the fray of competing claims from which those that deserve to prevail can be identified" (Brandom 1994, 600, 601, my italics). The participants are each committed to their accountability to norms that are up to not just them but with no way to determine the norms except through further ongoing interaction.

The normativity of practices, on such accounts,[34] is expressed by the mutual accountability of their constitutive performances rather than by a determinate norm to which those performances are each somehow already accountable. What they are mutually accountable *for* is what is at issue and at stake in whether and how the practice continues. How the practice will continue is not already settled but always remains prospective. The continuation of a practice is constrained by past performances—subsequent performances are accountable to them for their intelligibility as continuations of the "same" practice. Otherwise, they would replace the practice rather than continue it. Those past performances are nevertheless also partly reinterpreted by the subsequent development of the practices to which they belong. Indeed, that is the point of introducing the phrases "at issue" and "at stake," which refer anaphorically to the contested directedness of the mutually interrelated performances of a social practice. Performances of a practice are directed toward and accountable to "something" (an issue and what is at stake in the possible resolutions of that issue) that outruns any particular expression of what it is.

People often do make explicit judgments about what is at issue and at stake in the practices in which they participate. Such judgments, how-

33. In *How Scientific Practices Matter* (Rouse 2002, 247–54), I argue that the metaphor of visual perspective is not the best way of thinking about how our performances are situated within larger patterns of practice, but that argument does not affect the central point of the passage quoted here.

34. Brandom is not alone in this conception of normative accountability without a determinate norm toward which performances are and should be accountable. In a different tradition and idiom, Foucault likewise rejects any "sovereign" standpoint "above the fray" from which competing political or epistemic claims can be definitively assessed, colorfully expressed by the claim that "in political thought and analysis, we still have not cut off the head of the king" (1978, 88–89). For a more extensive discussion of a parallel sense of epistemic normativity without epistemic sovereignty, both in Foucault and more generally, see Rouse (1996a, 2003).

ever, are typically efforts to express what is *already* at issue and at stake in the practice. Moreover, part of the point of making such judgments is to contest alternative, partially conflicting conceptions of the same issues or stakes. We might imagine trying to stand "outside" of an ongoing practice (to view it from "sideways on" in McDowell's terms), in order to identify definitively the norms that really do, or should, govern its performances. Any such efforts are instead assimilated within the practice, however, as one more contribution to shaping what it will become, and how that future would matter to present performance. Arthur Fine nicely summarized this inability to interpret practices from "sideways on" in the case of scientific practice: "If science is a performance, then it is one where the audience and crew play as well. Directions for interpretation are also part of the act. If there are questions and conjectures about the meaning of this or that, or its purpose, then there is room for those in the production too. The script, moreover, is never finished, and no past dialogue can fix future action. Such a performance . . . picks out its own interpretations, locally, as it goes along" (1986a, 148). Language and other symbolically expressive and conceptually articulated practices share this open-endedness. Their normative accountability is an essentially temporal phenomenon, a mutually interactive accountability toward an unsettled future continuation. That future would nevertheless encompass its past and present performances as iteratively interrelated and reinterpret them in terms of their place in this reconfigured pattern of practice.

I call this way of understanding social practices and the mutual accountability of their constitutive performances a "normative" conception of practices, because it does not reduce normative considerations to nonnormative ones or eliminate them altogether. This feature of the account is what worries naturalist critics of normativity, such as Turner (1994, 2010) or Roth (2003, forthcoming). They insist that invocations of normative authority call for explanation and that an explanation in normative terms would be question begging. The criticism is misplaced, however; it would indeed be question begging to appeal to any determinate, authoritative norms to account for how normative considerations ever acquire authority or determinacy, but my account does not do so. I am arguing instead that normative authority and its open-textured contentfulness arise from holistic interrelations among the performances that thereby come to make up a social practice. The inability to characterize those performances or their interrelatedness in nonnormative terms does not make them naturalistically inexplicable, for we are also not taking for granted any characterization of them in normative terms.

An illuminating parallel to Donald Davidson's account of meaning can help us to see why it is not question begging to understand conceptual normativity by appealing in this way to holistic interrelations among the performances that comprise a social practice and between these performances and the larger patterns of practice they help compose. Jonathan Bennett once made a parallel objection to Davidson's approach to understanding meaning: "It is part of a philosopher's task to take warm, familiar aspects of the human condition and look at them coldly and with the eye of a stranger. . . . Davidson is not at [a proper analytical] distance. He stands in the thick of the human situation, helping himself to things that he finds within reach—things like the concept of language, [or] sentence" (Bennett 1985, 619). Contra Bennett, I think that a naturalistic account of conceptual normativity *must* proceed from "in the thick of the human situation" in this way. It is one thing to look at particular social practices or conceptual relations with "the cold eye of a stranger"; strangers inhabit different social practices and conceptual relations and draw upon them in explicating what they find unfamiliar in what others say and do. It is another thing altogether to try doing so for social practices and conceptual normativity generally. That would be an effort to view conceptual relations to the world from "sideways on" (McDowell 1984, 1994), as if we were not already in the midst of language and conceptually articulated norms. If conceptual normativity structures our environmental niche and our socially interactive way of life within it, as I have been arguing, then there is no alternative to explicating it from within. The result need not be question begging, however. We can recognize and understand the holistic interconnectedness of social practices and their mutual normative accountability, and the inability to explicate this interconnectedness from sideways on, without positing or taking for granted any prior specifications of norms as authoritative.

Language and other conceptually articulated practices do not merely involve holistic interrelations among their constitutive performances, however. The normativity of conceptual practices must also be accountable to the biological environment to which they belong in ways that allow a characteristic two-dimensional normativity. We must understand how performances of conceptual practices articulate distinct aspects of their environment by *taking* them in some definite way. That taking-as is distinct from the determination of what those performances are actually dealing with or directed toward within a broader physiological and behavioral way of life, such that the taking-as is accountable to what is environmentally taken-up for its correctness or truth.[35] The concept of objectivity has often been invoked to express how various performances or practices are accountable to the world for their content and/or their correctness. The next chapter takes up this question of how our biologically evolved, niche-constructive linguistic and other conceptually articulated social practices can display such a two-dimensional normative accountability and whether and how it should be expressed as a form of objective accountability.

35. I distinguish "taking-up" aspects of one's physical surroundings within a biological environment, from "taking-as" in some definite way that is accountable to what is taken-up, in chapter 2. This distinction of two aspects of intentional comportments parallels Cummins's (1996) distinction between the targets and contents of representations.

FIVE

Two Concepts of Objectivity

The two previous chapters argued that language and other conceptually articulated practices emerged in the human lineage through behavioral niche construction and that languages themselves then coevolved with human capacities to understand and use linguistic and other conceptually significant expressions. The neurological basis for rudimentary forms of conceptual understanding was already present in the primate lineage, but the realization of those capacities confronted serious developmental and evolutionary barriers. Most organisms with a sufficiently complex and flexible behavioral responsiveness to their surroundings achieve that flexibility through closely attentive responsiveness to multiple, conflicting perceptual indications. Other organisms' close attunement to their selectively relevant environment blocks uptake of one another's vocal or gestural expressions as disconnected from the immediate behavioral relevance of those expressions. Understanding how an organism's expressive repertoire could become detached from its immediate behavioral and physiological significance has thus become widely recognized as the central problem in accounting for the evolution of developed capacities for displaced, articulated conceptual understanding. This reformulation of the issue nevertheless simplifies another explanatory problem. Once a rudimentary discursive practice somehow becomes integrated within a social organism's way of life in fitness-relevant ways, it is easier to understand the evolution of more complex and extensive forms of discursive performance. With conceptually articulated, displaced forms of communication as a fitness-relevant feature of those organisms' normal developmental environment, selection could readily favor more rapid learning and more complex deployment of the relevant discriminative and expressive capacities.

This approach to a naturalistic account of conceptual normativity as intelligible within a scientific understanding of nature has two primary consequences. The first consequence builds upon recent challenges to the predominant philosophical, linguistic, and psychological approaches to understanding language and other conceptual capacities. Conceptually articulated understanding on this approach is a practical, socially mediated skill in tracking and producing conceptually significant discursive performances in their linguistic, conversational, and broader practical contexts. Rather than embodying self-contained representational states that only then inform perception and action, the resulting capacities are perceptually and practically responsive to a saliently discursive environment. The exercise of these capacities also differentially reproduces them as salient and selectively significant features of the developmental environment of subsequent human generations. While human neural organization has, without a doubt, been significantly transformed by selection pressures for the acquisition and extension of linguistic capacities, neural organization has to be understood as part of a larger functional system that includes bodily capacities for perception and vocal expression, along with the public discursive practices to which they are responsive.[1]

The second consequence of this approach is that the most distinctive feature of *conceptually* articulated practices is a characteristically two-dimensional normativity. We have seen that living organisms are goal-directed processes whose normativity is one-dimensional. As Okrent reminds us,

> There is a central respect in which Darwin was the greatest Aristotelian of the nineteenth century. Darwin agrees with Aristotle—and disagrees with Christianity—on the central issue of whether individuals are evaluable in a non-arbitrary fashion even if they were not made by some rational creator. Darwin even agrees with Aristotle in his judgment concerning which things are so evaluable: living things. For Darwin and

1. The "enactive approach" developed in Nöe (2004, 2009) is exemplary of efforts to see the relevant functional system for human experience and understanding as composed of active bodily skills and the biological environment to which they respond rather than the brain by itself, the brain coupled with the peripheral nervous system, or even the body apart from its environment.

Darwinians, *living* organisms are those individuals that carry the principle of nonarbitrary normative evaluability in themselves. Nonarbitrary standards for evaluating goal-directed events are borrowed from non-arbitrary standards for evaluating the entities in which they occur.[2] (2007, 68)

The crucial Aristotelian insight is that living entities are goal-directed processes whose constitutive goal is the continuation of that very process. As Richard Dawkins succinctly put the underlying point, "The minimum requirement for us to recognize an object as an animal or plant is that it should succeed in making a living of some sort. . . . You may throw cells together at random, over and over again for a billion years, and not once will you get a conglomeration that flies or swims or burrows or runs, or does anything, even badly, that could be remotely construed as working to keep itself alive. . . . Staving off death is a thing you have to work at" (Dawkins 1986, 9, quoted in Okrent 2007, 69). The life processes—or better, the developmental life cycles (Griffiths and Gray 1994)—of organismic lineages are sustained over time through ongoing intra-action with what is thereby coconstituted as their developmental and selective environments. Organisms utilize various capacities and affordances of their surroundings and are vulnerable to environmentally mediated disruption of those constitutive abilities. These life processes can then succeed or fail at the goal of sustaining themselves. Whether failure manifests an "internal" organismic malfunction or the unsuccessful adjustment of its normal functioning to the available environmental affordances, the only nonarbitrary normative standards in play are those defined by the goal of self-maintenance.[3] Failure amounts to the diminution or disappearance of the organismic lineage.

The partial autonomy of linguistic and other conceptually articulated practices allows for a second dimension of nonarbitrary normative assessment. These practices, like the organismic ways of life to which they

2. Where Okrent characterizes organisms as "individuals," I would introduce two qualifications. First, organisms are only bounded as individuals as components of a larger pattern of intra-action with their developmental and selective environment. To that extent, organisms are patterns that constitute what Barad (2007) calls "phenomena" rather than individual entities. Second, drawing upon recent work on the ubiquity of microbial symbiosis as playing indispensable functional roles in the life patterns of eukaryotic organisms, I would follow Gilbert, Sapp, and Tauber (2012) in treating such organisms not as eukaryotic individuals but as symbiotic "holobionts." Okrent himself calls attention to the problem of determining which levels of biological organization are goals toward which selection can be directed (2007, 99–103) but then develops his argument with the presumption that familiar eukaryotic macrobes are the primary levels of biological goal-directedness.

3. The "self" in question is not simply the organism as a body but the phenomenon that incorporates its selective and developmental environment: the boundary between body and environment is constituted within this larger intra-active process.

contribute, are also maintained through the ongoing, goal-directed reproduction of the practices themselves. Natural languages, equipmental complexes such as carpentry or agriculture, or expressive practices such as dance or drawing only exist through their ongoing reenactment.[4] The proximate responsiveness and accountability of their performances to one another then constitute norms of appropriate performance that are partially independent of their contribution to organismic success.[5] In the case of linguistically articulated performances, such two-dimensionality enables the differentiation of what the performance "*says*" from what it is "about" or directed toward. The former concerns whether a performance is appropriately responsive to other elements of the practice itself; the latter concerns whether and how those uses, and the proximate norms that govern them, are to be assessed within the overall behavioral economy of an organismic way of life. Conceptual normativity is two-dimensional rather than merely comprising two distinct forms of normative accountability, but not merely because the same performances are accountable to different standards of assessment. The relevant standards are also holistically interconnected in ways that partially transform the character of the standards themselves.

In this chapter, I begin to explore and explicate the two-dimensionality of conceptual normativity. The strategy of this explication is neither to "bake a [normative] cake out of [nonnormative] ingredients" (Dretske 1981, xi) nor to take conceptual normativity or rationality as sui generis (McDowell 1994). The aim is instead to begin with the goal-directed normativity of biological lineages and understand how to account for conceptual normativity as a biologically explicable extension of our organismic way of life.[6] McDowell has rightly criticized the kind of "philosophical revisionism" that "takes its stand on one side of a [dualistic]

4. The difference between reenactment of an extant practice and the production of a similar performance that is not part of a pattern of practice (or the initiation of a new practice that replaces or diverges from its predecessor) will be discussed below as marking the temporal constitution of the normative authority and force of conceptually articulated practices. For present purposes, I simply note that there are such practices within human ways of life and call attention to how their normativity seems to diverge from that of the organismic goal-directedness within which they are situated.

5. As we shall see below, these "norms" are never fully determinate and are only specifiable anaphorically. In a strict sense, therefore, I could speak of the normativity or normative accountability of these performances without referring to norms. The account developed in this chapter could then be characterized as showing how there could be normativity without norms, at least on most interpretations of "norms" familiar from philosophy and social theory. In my preferred vocabulary, I talk about what is "at issue" and "at stake" in social practices instead of its norms, where issues and stakes are only specifiable anaphorically and interactively.

6. Millikan (1984, 2005) shares this broad strategic approach, but she draws upon different biological resources and tries to explicate intentionality and conceptual normativity in response to different philosophical approaches to these phenomena.

gulf it aims to bridge, accepting without question the way its target dualism conceives the chosen side [then] constructs something as close as possible to the conception of the other side that figured in the problems, out of materials that are unproblematically available where it has taken its stand" (1994, 94). I avoid such unsatisfactory revisionist strategies by starting with accounts of conceptual normativity as sui generis and showing how to reconstruct their relevant features as biological phenomena. I do so initially by considering two different ways of thinking about discursive practices as "objectively" accountable.

The concept of objectivity has a surprisingly short but complex history given its pervasive role in thinking about conceptual normativity. In the first section of the chapter, I briefly consider the familiar sense of objectivity as an epistemic concept applicable to the assessment of judgments as knowledge claims. This first section sets the stage for introducing a different way of thinking about objectivity. The second section explores this alternative conception of objectivity as a norm for conceptually articulated understanding that is a prerequisite to epistemic assessment. This alternative has emerged explicitly in recent philosophical work by Donald Davidson (1984, 2001), John McDowell (1994), Robert Brandom (1994), and John Haugeland (1998). Their approach nevertheless is often not recognized as advancing an alternative conception of *objectivity*, even though both conceptions can be traced to central themes in Kant's (1998) *Critique of Pure Reason*.

There are two reasons for regarding the work of Davidson and his successors as advancing an alternative conception of objectivity rather than as changing the topic. One reason to understand their work in these terms is that the two conceptions are competitors. If Davidson and these left-Sellarsians are correct, their alternative approach would supplant epistemic objectivity as a conception of how thought and action are accountable to the world. Their accounts aim to show why epistemic conceptions of objectivity are both unattainable and superfluous. In thus dispensing with the more familiar epistemic conceptions of objectivity, in my view, they rightly recognize the constitutive two-dimensionality of conceptual understanding. From this vantage point, familiar difficulties confronting epistemic conceptions of objectivity arise in part because these conceptions effectively collapse the two dimensions of conceptual normativity into one. Epistemic assessment is indeed important, but it cannot be adequately understood in isolation from the discursive context in which the knowledge claims to be assessed acquire conceptual content.

Davidson's, McDowell's, Brandom's, and Haugeland's ways of thinking about *conceptual* objectivity nevertheless turn out to reproduce some of the problems confronting the epistemic conceptions. Recognizing how developments of this second conception of objectivity retain problematic vestiges of their predecessors provides a second reason for identifying these two conceptions of objectivity as alternative treatments of the same issue. Moreover, we can thereby more readily grasp how to circumvent these challenges to understanding how thoughts and actions are engaged with and accountable to the world. With this background, the third section of the chapter develops a constructive, naturalistic account of how conceptual understanding is normatively accountable. I thereby address the worry, raised at the end of chapter 3, that reconceiving language and other conceptually articulated practices as forms of discursive niche construction would account for the evolution of language in an unsatisfactory way that would not allow for its intentional directedness and consequent normative accountability. In showing how the Sellarsian space of reasons *is* our continually reconstructed biological niche, this alternative approach to the two-dimensionality of conceptually articulated understanding thereby offers a more adequately naturalistic account of our conceptual capacities.

I—Epistemic Objectivity

Objectivity is most familiar as an epistemic concept expressing a norm for objective *knowledge*. Lorraine Daston and Peter Galison argue that this concept emerged in the nineteenth century as mechanical objectivity: "the insistent drive to repress the willful intervention of the artist-author, and to put in its stead a set of procedures that would, as it were, move nature to the page through a strict protocol, if not automatically" (Daston and Galison 2007, 121). Its meanings have since proliferated. "Objectivity" has been attributed to various aspects of inquiry supposedly conducive to knowledge: disinterestedness, emotional detachment, rule-governed procedures, quantitative methods, openness to criticism, responsiveness to evidence, or accountability to a mind-independent reality, among others. Their advocates have also accorded different epistemic roles to these marks of objective inquiry or objective knowledge, ranging from methodological advice on how best to conduct inquiry to standards proposed as criteria for knowledge. The historical emergence of objectivity as a normative standard for inquiry or its products has often been explained as a response to the geographic and social expansion of inquiry; it compensates for the loss of direct personal assessment of scientific credibility with increased social or geographic distance. Commentators from Nietzsche (1998), to Daston and Galison, to Porter (1995) also connect rhetorical recourse to epistemic objectivity with institutional or political weakness: those who cannot effectively assert their authority instead tout their objectivity, advancing their claims while deferring responsibility for them.

Despite its relatively recent historical emergence with proliferating interpretations, the concept of objectivity has now become a prominent and perhaps even the primary term expressing how human thought and agency is responsible and accountable to something not subject to our own will or authority. For that very reason, various conceptions of objectivity, and even the very idea of objectivity, have also been the target of widespread criticism throughout the social sciences and humanities. These criticisms take on different significance once we recognize the multivalence of the concept, however. My aim in this section is to sketch some of the most salient critical strategies and responses and their significance for understanding conceptual normativity more generally.

Many critics of epistemic conceptions of objectivity only target some of its multiple meanings, typically in order to advocate a revised version of epistemic objectivity. Much of the feminist-philosophical literature on objectivity takes an explicitly revisionist critical stance, for example, in arguing for more expansive or inclusive conceptions of objective methods of inquiry or standards for the assessment of the objectivity of knowledge claims (Hankinson-Nelson 1990; Longino 1990; Harding 1991; Lloyd 1996). What I call nostalgic criticisms of the concept of objectivity come from an opposing direction. Nostalgic critics insist that objectivity in one or more of its guises is an unattainable, perhaps even undesirable, epistemic ideal. Yet they also insist that the fulfillment of this ideal would be necessary for knowledge claims to have the authority or universality often attributed to them. Nostalgic critics thus ironically retain the authority of whichever conception of objective knowledge they criticize. The *ideal* must be sustained in order to maintain the significance of denying that this ideal could ever be attained or approached.[7]

7. Readings of epistemic theorists as nostalgic critics of objectivity or its philosophical surrogates are inevitably controversial since almost no one deliberately aspires to nostalgic criticism. I nevertheless read some key contributions to the early social constructivist literature in the sociology

Revisionist and nostalgic critics of various accounts of objectivity still work within the conceptual space of objectivity understood as an *epistemic* norm: their questions concern what it would mean to ascribe objective knowledge and what would be an appropriate basis for doing so. Other recent criticisms of epistemic conceptions of objectivity cut more deeply in questioning whether a concept of objectivity appropriately expresses how knowledge claims are accountable to the world. In this section, I will focus on three broad lines of criticism that raise deeper concerns about the very idea of objectivity as an epistemic norm. These considerations are initially important for my purposes by preparing the ground for understanding an alternative way of thinking about objectivity. Later in the chapter, we will also see how they let us recognize some residual limitations in this alternative approach.

The first of these critical challenges to epistemic conceptions of objectivity calls attention to their interdependence with the paired concept of subjectivity. Daston and Galison emphasize that mechanical objectivity was understood as aiming to avoid or overcome the intrusion of subjectivity into scientific inquiry. Any influence of the epistemic subject was to be removed or minimized. Yet the resulting expressions of the ideal then typically take the form of alternative subject positions. Emotional detachment, disinterestedness, strict proceduralism, undogmatic open-mindedness, attentiveness to evidence, and many other suggested antidotes to subjective distortions of knowledge are put forward as better ways to be an epistemic subject. Revisionist critics of epistemic objectivity, for example, most commonly work within this conceptual space, arguing that various ways of positioning the inquirer are more or less conducive to objective inquiry or objective knowledge. Such conceptions of objectivity, originally advanced as ways to let the object speak for itself without intervention or imposition by inquirers, instead direct

of scientific knowledge as nostalgic in this sense. Nostalgic contrasts are built in to well-known summary claims such as Steven Shapin and Simon Schaffer's concluding assertion that "it is ourselves and not reality that is responsible for what we know" (1985, 344) or Andrew Pickering's early claim that "[although] many people do expect more of science than the production of a world congenial to social understanding and future practice, . . . the history of High-Energy Physics suggests that they are mistaken. . . . There is no obligation upon anyone framing a view of the world to take account of what twentieth-century science has to say" (1984, 413). Steve Woolgar's (1982) identification of the function of irony in the sociology of scientific knowledge called attention to how the significance of many sociological accounts seemed to depend upon what I am calling a nostalgia for objectivity, but Woolgar's (1988) own subsequent invocations of reflexivity seem nostalgic in the same way. Nostalgic criticism of epistemic objectivity is also widespread in some strands of the continental philosophical tradition, notably in some readings of Derrida's (1967a, 1967b) criticisms of the "metaphysics of presence."

sustained attention back toward the knowing subject, the subject's positioning in inquiry, and the critical assessment of that positioning.

This critical approach to conceptions of objectivity as the proper positioning of epistemic subjects has guided critics in different directions. One response to this line of criticism has been to understand the sciences and other forms of conceptual understanding as practices or discourses rather than as relations between knowing subjects and transcendent objects. This response provides one route to the alternative to an epistemic conception of objectivity, which is introduced in the next section. That response also points toward conceptions of scientific understanding developed in much of the interdisciplinary field of science studies and in part 2 of this book. Such projects do not treat sciences and other forms of thought and understanding primarily as efforts to represent the world (or objects within it) within a language, theory, or research program. They instead understand scientific practitioners and other knowers as interactively caught up within and responsive to the world around them. "Practices" in this sense are not just the sayings and doings of practitioners (as analogues to the subject-positioning of knowers in relation to "external" objects); practices incorporate the things "practiced" on, with or amid, and the discursive articulation of the practitioners' situations as a field of intelligible possibilities. I postpone further discussion of this strategy until the next section.

An alternative response to the criticism of objectivity as a form of subject-positioning relocates a recognizably epistemic conception of objectivity. Mechanical objectivity, aperspectival objectivity, disinterestedness, and related forms of subject-positioning implicitly identify objectivity as a "ground-level" norm applicable within inquiry or the assessment of knowledge claims. Discussions of objectivity in this sense concern how one ought to conduct inquiry or its assessment as an aspiring knower. In philosophical reflection upon scientific knowledge, however, objectivity more often functions at a metalevel. Asking whether a claim, a method, or a stance is objective usually involves what I call "epistemic ascent," paralleling Quine's (1960, 271–76) more familiar notion of semantic ascent. Semantic ascent is a shift from talk about things to talk about talk about things. Epistemic ascent is a similar shift in how we think about reasoning and justification. At ground level, we find scientists' or other knowers' reasons for choosing research projects, using or eschewing methods, accepting some claims, entertaining others, rejecting those that do not stand up to assessment, and never even considering those that do not stand out as serious possibilities. Epistemic ascent moves to a metalevel, asking whether these reasons, or the *class* of reasons to which they belong, are genuinely good reasons.[8]

Much recent philosophy and sociology of science is committed to epistemic ascent. Many postempiricist accounts of the objectivity or rationality of scientific inquiry have rightly been characterized as "metamethodological": methodological considerations function at the ground level, often guided by theoretical understanding of the domain of inquiry, whereas philosophically articulable norms supposedly govern the rational adjudication of competing research programs as holistic programs encompassing both theory and method.[9] Metamethodologists' sociological critics in the tradition of the "sociology of scientific knowledge" (SSK) work at the same metalevel, denying that *any* first-order reasons can transcend their contingent local circumstances. The successor debates over scientific realism also involve epistemic ascent in a comparable way. Within debates over realism, epistemic objectivity first emerges on the antirealist side. If we had direct access to objects themselves, we could assess our representations of them by direct comparison. Without such access, we must assess our forms of inquiry and systems of belief from within. Objectivity then becomes a metalevel surrogate for truth-as-correspondence. If our methods of inquiry or reasons for belief are objective, then we can be reassured that we are on the right path of inquiry even if we can never reach its end. But scientific realists play the same game. Their abductive arguments for realism as the best explanation of scientific successes (Boyd 1980) are indirect, nonconstructive metalevel arguments. The conclusion is supposedly that "mature" scientific theories are referentially successful and *approximately* true. But "approximate truth" is just another form of reassurance that we *really* are on the right path even though its end still lies ahead.

Arthur Fine (1986a, 1986b) exemplifies a second critical strategy that challenges the efficacy of any attempted move to a metalevel to escape the difficulties of ground-level conceptions of epistemic objectivity. Placing the passage I quoted at the end of the preceding chapter in its dialectical context, we see Fine repudiating the epistemic ascent that is attempted by realists and antirealists alike: "The realisms and antirealisms seem to

8. Epistemic ascent is an alternative way to characterize the positions and approaches that my earlier work (Rouse 1996b, esp. ch. 1–2) interprets as undertaking "the legitimation project."

9. Lakatos (1978), Laudan (1977), or Longino (1990) are good examples of metamethodological projects in the philosophy of science. Longino's work is distinctive in proposing a metamethodological program for the assessment of inquiry as a social activity rather than for the assessment of scientific reasoning directly.

treat science as a sort of grand performance, a play or opera whose production requires interpretation and direction. . . . [But] if science is a performance, then it is one where the audience and crew play as well. Directions for interpretation are also part of the act. . . . The script, moreover, is never finished, and no past dialogue can fix future action. Such a performance . . . picks out its own interpretations, locally, as it goes along" (1986a, 148). Epistemic ascent is an attempt to step outside of our situated scientific reasoning and view scientific understanding of the world "sideways on" in search of reassurance that our methods and theoretical commitments have not altogether lost touch with the world, experience, or rational methods of inquiry. Fine argues that the reassurance sought through such a "sideways," or metalevel, view of scientific understanding is impossible. The supposed moves to a metalevel assessment can only provide additional ground-level reasoning that would need to be secured in turn. Such unattainable epistemic security would also be superfluous, however. Science is a risky game, and once we have checked and double-checked and critically assessed our judgments from within the overall practice of inquiry, no further or higher reassurance is available. None is needed, however, since any specific reasons for doubt are open to further reflective assessment.

Some feminist critics develop similar arguments. Donna Haraway (1991, ch. 9), for example, dismisses both epistemic objectivity and epistemological relativism as different versions of what she calls the "god-trick." Situating knowledge claims and their critical assessment in specific historical settings marked by differences in social power and epistemic access does not thereby block objective accountability: "So I think my problem, and 'our' problem, is how to have *simultaneously* an account of radical historical contingency for all knowledge claims and knowing subjects, a critical practice for recognizing our own 'semiotic technologies' for making meanings, *and* a no-nonsense commitment to faithful accounts of a 'real' world, one that can be partially shared and friendly to earthwide projects of finite freedom, adequate material abundance, modest meaning in suffering, and limited happiness" (1991, 187). Like Fine, Haraway asks that we acknowledge and take responsibility for our own partiality and finitude, and our accountability in and for partially shared circumstances, without the illusion of epistemological transcendence that could secure claims to knowledge once and for all.

Responses to Fine and Haraway are nevertheless instructive. Both are widely interpreted as making yet another move within the game of epistemic ascent instead of opting out of the entire project. Fine has been read alternatively as a moderate scientific realist (Musgrave 1989) or more commonly as defending a new form of antirealism. Others (e.g., Zammito 2004) similarly read Haraway as either a revisionist defender of epistemic objectivity or a nostalgic "postmodernist" skeptic for whom objectivity is unattainable. Moreover, even those readers of Fine or Haraway who understand their intentions to opt out of that philosophical game often resist, taking their criticisms instead as invitations to play the same old game better. The "stake-in-the-heart" move remains elusive.[10]

A third critical strategy challenges a metaphysical conception of knowers' relation to the world that is implicit in either conception of epistemic objectivity: as ground-level subject-positioning or as meta-level ascent. Both conceptions implicitly seek to assess knowers' relation to the world from "sideways on" but inevitably fail to do so. Epistemic objectivity as an ideal presumes a gap between us as knowers and the world to be known. An objective method, stance, attitude, or disposition is put forward to bridge that gap. But any such proposal as a form of subject-positioning finds itself firmly placed on our side of the gap between us as knowers and the world as "beyond" our representations of it. The objection is that the gap between knowers and the world is thereby conceived in advance in a way that renders it unbridgeable. Moreover, this conception can itself be challenged as a dogmatic presupposition that we should reject. Hegel famously characterized such epistemological preconceptions that make the recurrence of skepticism or relativism inevitable as a "fear of error [that] reveals itself rather as fear of truth" (1977, 47). This self-defeating fear of error repeatedly calls forth efforts to refute skepticism or relativism yet again. As Heidegger ([1927] 1962, sec. 43) once suggested, the problem is not that the refutation of skepticism has yet to be accomplished once and for all but that it continues to be attempted again and again out of a dogged commitment to an underlying conception of a gap between knower and world to be known.

II—Conceptual Objectivity

I regard these three linked strategies for criticism of epistemic conceptions of objectivity as compelling. I will not try defending that judgment here, because I think the residual appeal of epistemic conceptions of objectivity depends upon the lack of an apparent alternative more than upon doubts about the appropriateness of these critical responses.

10. Susan Oyama (2001) introduced this term for parallel efforts to opt out of "nature/nurture" debates in ways that would resist being resituated as a move within the same debates.

To short-circuit the recurrence of self-defeating conceptions of epistemic objectivity, we would need a more adequate alternative understanding of our relation to the world as inquirers and agents that does not reproduce these problems. Both revisionist and nostalgic conceptions of epistemic objectivity do reproduce the problems. They still function within the conceptual space of knowers who need to transcend their own concepts and representations to encounter the world itself, as if those concepts and knowledge claims were somehow meaningful apart from their ongoing use in the midst of a larger pattern of worldly interaction. I therefore take up a different response to these criticisms that moves the concept of objectivity away from epistemology. Like much else in philosophy, this response traces back to Kant (1998), but I will not consider its Kantian roots. I will instead address its emergence in work by Davidson, McDowell, Brandom, and Haugeland with two aims. First, I will explore how and why they should be understood as offering a reconception of *objectivity*. Second, despite my sympathy with and indebtedness to their work, I will indicate some limitations of that conception and some ways around them.

Davidson and these "left-Sellarsians" relocate the question of objectivity. Instead of asking how knowledge could be objective, they ask how knowledge claims could even purport to be objective—that is, they ask what it is for our performances to be *about* objects and *accountable* to them at all. The issue then concerns objective conceptual content rather than objective knowledge.[11] Epistemic conceptions of objectivity only come into play once some claim to knowledge has been formulated and recognized as a claim. The epistemic question is then whether that claim is true, or objectively justified, and objectivity is invoked to settle this question, whether at the ground level or through epistemic ascent. Davidsonians and Sellarsians think that understanding how our performances are meaningfully accountable to the world renders epistemic objectivity superfluous. With Fine and Haraway, they think epistemic questions can only be settled within ongoing inquiry, which answers holistically to norms of conceptual objectivity.

I begin with Davidson (1984), partly because he is among the first to speak of *objectivity* in this way and partly because later criticisms of

11. Haraway's (1989, 1991, 1997) work shares with Brandom, Haugeland, and McDowell the insistence that understanding meaning or conceptual content is a more basic and important issue than is assessing the adequacy of specific claims whose content is taken as already determinate. She makes this point in the context of ongoing critical engagement with specific practices of conceptual articulation and deployment in and around the sciences rather than through a philosophical account of how conceptual articulation is objectively accountable.

Davidson are instructive. Davidson notes that any assessment of truth or error presupposes an interpretation of the meaning of the sentence or thought to be assessed. Yet interpretations of meaning and assessments of truth are accountable to the same evidence, drawn from the circumstances in which a sentence is uttered and those in which other sentences are uttered using the same words in recombination with others. With two variables to solve for—truth and meaning—and only one body of evidence to constrain the solution, problems of interpretation would be intractable without a principled way to fix one of the variables. Davidson argues that the only defensible way to do that is to maximize truth to solve for meaning. Otherwise, any interpretation attributing errors to a speaker would invite the retort that the error is in the interpretation. Error can be attributed selectively but solely on the grounds that other interpretations would require attributing still greater error. Davidson famously summarized his understanding of interpretation and conceptual understanding in terms of objectivity: "In giving up dependence on the concept of an uninterpreted reality, something outside all schemes and science, we do not relinquish the notion of objective truth—quite the contrary. . . . Truth of sentences remains relative to a language, but that is as objective as can be. In giving up the dualism of scheme and world, we do not give up the world, but re-establish unmediated touch with the familiar objects whose antics make our sentences and opinions true or false" (1984, 198).

Objectivity in Davidson's sense is no longer epistemic objectivity, however. What is supposedly settled objectively in a holistic interpretation of a speaker that maximizes truth is not whether the speaker has true beliefs. The truth of the bulk of a speaker's beliefs is a criterion of interpretation, not its outcome. Against the background of our causal involvement in the world, conceptually articulated through our linguistic abilities, all speakers having mostly true beliefs is a routine consequence (despite some indeterminacy in just which beliefs these are). The consequence for epistemology is that one need not bother to refute skepticism or relativism but can instead "tell the skeptic to get lost" (2001, 154). Of course, Davidson's or similar views cannot reassure us that any particular claim is true. Along with Fine and Haraway, they take the fate of particular claims to be resolved through ongoing engagement with one another and our shared surroundings. Brandom nicely summarizes the resulting relocation of objectivity: "The objectivity of conceptual norms . . . consists in a kind of perspectival *form* rather than [any] cross-perspectival *content*. . . . Sorting out who should be counted as correct . . . is a messy retail business of assessing the comparative authority of competing evidential and inferential claims. . . . That issue is adjudicated differently from different points of view, and although these are not of equal worth, there is no bird's eye view above the fray of competing claims from which those that deserve to prevail can be identified" (Brandom 1994, 600–601). No privileged standpoint of objectivity achieved through epistemic ascent, no bird's- or god's-eye view from above the fray, is needed, however. The objective accountability of the entire practice is secured when we understand it as conceptually articulated at all.

I share this strategy for thinking about conceptual understanding as both precluding and obviating epistemic conceptions of objectivity. Subsequent criticisms nevertheless suggest that no one has yet adequately carried out this strategy. Davidson's own attempt to circumvent epistemic ascent was prominently challenged by McDowell (1994). McDowell shares Davidson's aversion to efforts to refute skepticism or relativism, but he also argued that Davidson himself was not entitled to that dismissal. Davidson conceived of causal or experiential relations to objects as "outside" the semantic space of meaning and justification, notoriously concluding that "nothing can count as a reason for holding a belief except another belief" (Davidson 1986, 310). McDowell argues that Davidson thereby reconstitutes, against his own intentions, a hopeless conception of language and thought as a self-contained game disconnected from the world, a "frictionless spinning in a void" in his picturesque phrase. McDowell was not thereby proposing a new skeptical riddle, now concerning meaning rather than truth (in contrast to Kripke 1982). He endorsed Davidson's intended shift from treating objectivity as an epistemic concept to recognizing that any conceptually articulated understanding is thereby already objective. McDowell's conclusion is instead that accomplishing that shift requires a more expansive account of intentionality and conceptual understanding than Davidson himself allows. Experiential or causal relations to objects must themselves be brought within the space of reasons and conceptual spontaneity.

McDowell, Brandom, and Haugeland each in his own way then attempts to show how conceptual understanding really does reach out to be accountable to and constrained by objects themselves. McDowell (1994) appeals to the passivity of conceptually articulated perceptual receptivity to provide the needed "friction"; Brandom (1994) claims that the game of giving and asking for reasons incorporates our causal relations with objects in perception and action; Haugeland (1998, ch. 13) argues that only an "existential commitment" to preserving an "excluded zone" of conceivable but impossible occurrences can allow objects themselves to govern what we say and do. I have already argued in *How Scientific Practices Matter* (Rouse 2002) that each of these accounts of the objectivity of conceptual understanding fails. I will not reconstruct these arguments in detail. I instead highlight their common and all-too-familiar form, for which McDowell's criticism of Davidson is the prototype. Each view develops its own model of conceptual understanding as a Sellarsian "space of reasons": Davidsonian radical interpretation, McDowell's second-nature acculturation as rational animals, Brandom's game of giving and asking for reasons, or Haugeland's account of constitutive skills, standards, and commitments. Each then tries to show how performances within this space of reasons are genuinely constrained externally, by objects, experience, or the world. Their critics, myself included, respond that only the semblance of constraint has been demonstrated: we are left with a "frictionless spinning in a void," a second nature disconnected from any explicable relation to first nature, a self-contained game of intralinguistic moves in which perception and action always remain "external," or a self-binding commitment with no greater normative authority and force than New Year's resolutions.

This argument pattern should also sound familiar. It takes analogous form to the objections we reviewed earlier, against the metaphysical presuppositions of epistemic objectivity understood as a ground- or meta-level subject-position. Nor is the parallel merely coincidental. Epistemic objectivity was conceived as an *epistemically preferred* subject-position. Davidson and the left-Sellarsians instead construe conceptual objectivity as a *constitutive* subject-position for rational agents aspiring to knowledge of objects. Each then offers a self-defeating conception of subjects seeking to transcend the very limitations that define them as epistemic or intentional subjects. Their efforts to relocate such transcendence of "subjectivity," from epistemic justification to conceptually articulated understanding, nevertheless still reproduce the problematic pattern that made epistemic conceptions of objectivity self-defeating.

III—Conceptual Normativity as Evolutionary Niche Construction

These two conceptions of objectivity differ in the locus for which they ascribe objective accountability to the world. Epistemic accounts of objectivity locate such accountability in the correctness or incorrectness of judgments or claims; conceptual accounts locate it in the holistic articulation of the contentfulness of judgments or claims, which opens them to assessment by reasons and evidence. Common to both versions is an understanding of us as thinking and knowing subjects (whether as individuals or as discursive communities) who "have" conceptions of things in the form of mental representations or intralinguistic discursive commitments. "Objects" (*Gegenstände*) stand "against" these conceptions as external normative constraints upon what we (should) think, say, and do, via their experiential or causal impingements upon us from "outside." In each case, their externality to the conceptual or epistemic domain (ascribed in order to provide the needed constraint or "friction") blocks any effective engagement with epistemic justification or conceptual understanding.

My account begins differently. We are not subjects confronting external objects but organisms living in active interchange with an environment. An organism is not a self-contained entity but a dynamic pattern of interaction with its surroundings (which include other conspecific organisms). The boundary that separates the organism proper from its surrounding environment is not the border of an entity but a component of a larger pattern of interaction that is the organism/environment complex.[12] In the absence of appropriate interaction with a suitable environment, there is no organism because the organism dies. Death *is* the cessation of the constitutive ongoing pattern of interaction that is an organism *making* a living in its environment. After the organism's death, and especially after the extinction of its lineage, there is also no environment. An "environment" is the "belonging together" of various aspects of the organism's surroundings *as* collectively enabling/sustaining life.[13] This pattern is teleological and hence normative: it has a goal,

12. As Karen Barad (2007) points out, for that reason, "interaction" is a misleading notion here in suggesting that two self-contained entities then interact. But in the case of organism and environment, the two are only differentiated from one another by the maintenance or reproduction of the larger pattern that contains both of them. She thus introduces the term 'intra-action' for patterns in the world that constitute significant boundaries that only exist as components of the larger pattern. I endorse Barad's point but continue to use the more familiar word 'interaction' with an expanded sense that incorporates intra-active interactions.

13. For these reasons, some philosophers of biology take the relevant focus of evolutionary biology to be not the individual organism but developmental systems that incorporate the environmental resources needed to maintain the recurrent process of development that constitutes a lineage. Thus Griffiths and Gray argue that "an evolutionary individual is one cycle of a complete developmental process—a life cycle. . . . Developmental systems include much that is outside the skin of the traditional phenotype. . . . This raises the question of where one developmental system and one life cycle ends and the next begins. . . . [Our current proposal] converges on our older idea that an individual is a life cycle whose components cannot reconstruct themselves when decoupled

and it can succeed or fail in attaining that goal.[14] The goal, however, is not something external to the goal-directed process but is instead the continuation of the process itself: organisms in environments are what Aristotle (1941, bk. IX) called *energeia* ("actualities"), goal-directed processes whose goal or end is present in the process itself.

Strictly speaking, the life process is not confined within an individual living organism in its local environment but rather in the organism's lineage with its corresponding environmental lineage.[15] The death of individual organisms within a lineage is an integral part of the ongoing process of its lineage sustaining a living. The lineage thereby also maintains an enabling or affording environment (the "agent" here that brings about this "maintenance" is not the individual organism, or even the collective set of organisms within the lineage, but the self-sustaining process that incorporates organisms and environments). Individual organisms and their local environments are constitutive subpatterns of the larger pattern of the lineage. That an organism's environment cannot simply be identified with its physical surroundings, identifiable independently of the ongoing life process, becomes especially clear when we consider "weed" species. Weeds (including parasitic organisms that cause fatal infectious disease or immunity in their hosts) are organisms whose life processes make their current surroundings uninhabitable for them. As a consequence, they are essentially mobile lineages: their way of life is to "colonize" disturbed settings and then move on, drawing upon whatever environmental resources enable that mobility. Their environment is not any particular disturbed setting that provides an opportunity to maintain their lineage but the pattern of shifting from one to another that incorporates the dynamic interaction of developing/reproducing/dying organisms with their changing/shifting environmental circumstances.

from the larger cycle" (2001, 209, 213). There are further complications to these relations among organisms, life cycles, and lineages, which Wilson (2005) helpfully surveys.

14. Organisms can have goals (and even *be* goal-directed in their activities) without having any explicit awareness of their own goal-directedness, in the sense that the goal *explains* what they do (in this sense, of course, goal-directed activity includes physiological functioning such as breathing or blood circulation, as well as both tightly cued and flexibly responsive behavior). For a careful analysis of the goal-directedness of organisms, see Okrent (2007, ch. 2–4).

15. Biologists have not, to my knowledge, spoken of environmental lineages corresponding to organismic lineages. The concept nevertheless straightforwardly follows from the conjoined recognition that the biological environment of an organism is only definable in relation to the organism's way of life, that organismic lineages change over time, and that niche construction plays an important role in that evolutionary process (here "niche construction" explicitly incorporates migration that changes the selection pressures on a population).

In this respect, a lineage of organisms and environments has an ontological character that Haugeland (1998, ch. 10) also discerned in chess games. Chess pieces must be physically realized in some form or other, but they are not identical with any of their physical realizations. Not only can the same game be continued in radically different and discontinuous physical media, but the same pieces and positions are preserved through such changes within a single game. One cannot castle with a rook that has previously moved, even if the move was made with the ivory set on the patio and the castling was attempted after the game was moved to the plastic set in the den or was continued by e-mail correspondence. Haugeland concluded, "Chess games are a kind of *pattern*, and chess phenomena can only occur within this pattern, as *subpatterns* of it. The point about different media and different games is that these subpatterns would not be what they are, and hence could not be recognized, except as subpatterns of a superordinate pattern with the specific structure of chess—not a pattern of shape or color, therefore, but a pattern at what we might call '*the chess level*'" (1998, 248). A living lineage is likewise a kind of pattern whose constituent organisms-in-environments are subpatterns of their superordinate pattern at the life level. There needs to be *some* material continuity throughout that pattern—living organisms belong to lineages rather than types—but not necessarily substantial continuity.

This pattern of goal-directed self-reproduction is thus continuous but not stable over time. Life processes evolve through their *differential* reproduction, where "reproduction" is understood more expansively as also incorporating development and metabolism, through which the life pattern continually reproduces itself differentially (as Richard Dawkins noted in a passage I cited earlier: "staving off death is a thing you have to work at" [1986, 9]). Even in the short run, organisms with a flexible behavioral repertoire that is responsive to multiple, possibly countervailing aspects of their environments exhibit a more complex pattern of differential behavioral and physiological response. Niche construction, including the cycles of recurrent niche-destruction characteristic of weeds, then introduces a new level of nonlinearity to the evolutionary process, as the selective environments to which evolution is responsive do not merely change over time but coevolve with the organisms. Behavioral niche construction, especially the niche-constructive emergence of discursive practices, nevertheless introduces a fundamentally different kind of normativity into the human lineage.

For other organisms, the goal of their physiology and behavior is the ongoing maintenance and reproduction of the life pattern that is their lineage. That goal is irreducibly deictic, since the goal is to sustain *this* temporally extended and changing pattern, *this* way of making a living in this environment, even if as a result, the present configuration of that way of life and its selectively relevant environment changes significantly (genetic/genomic, developmental, and physiological/metabolic capacities and actual environmental circumstances of course impose very stringent constraints on how the way of life of an organismal lineage might change over time to maintain itself in response to various selection pressures).[16] What discursive niche construction and conceptually articulated understanding then add to this teleological dimension of any living lineage is a second level of goal-directedness. Human behavior is directed not merely toward the goal *that* its life pattern continues but also toward *what* that life pattern will be.

To see how and why this is so, consider first how discursive niche construction works. Human organisms began to evolve capacities for conceptual understanding when they developed a partially autonomous expressive/responsive repertoire that eventually became recognizable as language.[17] There was an autonomous linguistic practice to the extent that linguistic expression was proximally responsive to its local conversational and broader intralinguistic contexts. This autonomy was only partial to the extent that whole extended chains and patterns of linguistic exchange were also held accountable amid broader practical-perceptual engagement with other aspects of the speakers' environment. The ability to track both the intralinguistic and the broader practical-perceptual significance of linguistic utterances opened the possibility of a gap between how one takes things to be and how they are.

The partial autonomy of conceptually articulated performances is then writ large by the emergence of differentiated but interconnected

16. Mark Okrent (2013) identifies this point as Aristotelian: "For Aristotle, the answer to the question 'What is it?' when asked of an organism is supplied by appealing to the organism's essence, and that essence coincides with the type of organism the individual is. This type prescribes a certain pattern of organic activity and a certain way of making a living. What Sammie, my pet Sheltie, *is*, is a dog, and being a dog involves him in surviving as a dog by living in a doggie way, structurally, metabolically, and behaviorally. No doggieness, no Sammie." (2013, 145). I think this claim is right (about both Aristotle and animals), but only if one follows Witt (1989) in recognizing that Aristotelian essences are not universal types but instead are the causes of the unity of individuals. Where Okrent speaks of "the type of organism the individual is," I would therefore substitute "the place of the individual within a lineage," where the lineage is understood to be a pattern of organism/environment interaction and not just of organisms.

17. My account remains officially agnostic concerning whether language emerged together with more complex forms of interrelated equipment, skills, and social roles or whether a rudimentary language was their enabling precursor. When I speak of "language" in this context, that should therefore be understood as a shorthand expression for whatever complex of conceptually articulated capacities and performances were initially conjoined with early hominids' protolinguistic abilities.

domains of social practice. Recall from the preceding chapter that practices should not be understood as social regularities: they do not consist of various agents performing in similar ways or sharing background beliefs or presuppositions. Practices instead are composed of performances that are mutually interactive in and with partially shared circumstances. The intelligibility of various performances within a practice normally depends upon the anticipation and achievement of appropriate alignment with others' performances and their circumstances toward some "end." Ends in this sense, however, are not something external to a practice for which the performances of the practice are merely instrumental. Practices, like the biological lineage to which they belong, are instead Aristotelian *energeia*, "ends present in the [practice]" itself (1941, bk. 9).

Although practices are constituted as Aristotelian ends through the ongoing mutual alignment of various performances and circumstances, the performers and circumstances are usually only partly accommodating. One person's performances only make sense if others act appropriately and the equipment, materials, and circumstances cooperate. In response to various misalignments within ongoing practices, human agents adjust what they do, sometimes by changing their own performances, sometimes by trying to affect what others do or rearrange the circumstances, and most commonly doing some of each. These patterns of mutual responsiveness and recalcitrance typically focus a practice on specific issues. Issues arise wherever some adjustment of performances or circumstances seems called for to allow the practice to proceed intelligibly; what is *at* issue is what adjustments are called for in order to sustain the practice intelligibly. Moreover, as discursively articulate beings, we may respond to those issues in part by trying to *say* what the issues are and what inferential and practical consequences arise from this explication. These efforts to talk through what is at issue in a practice, including responding to divergent interpretations of the issues, are themselves further performances within the practice, however. Through these ongoing interactions—or intra-actions, to use Barad's (2007) more perspicuous term—practices evolve and articulate themselves.

The temporal extension of these patterns of recalcitrance and mutual responsiveness plays a crucial role in constituting normative authority and force within practices. To understand why this is so, we must first recognize what it means for some phenomenon to be normative.[18]

18. What follows is a characterization of the two-dimensional normativity of conceptually articulated understanding. For discussion of the relation between the content, authority, and normative force of such capacities, and the goal-directed normativity of organisms, see section I of chapter 11.

Normative phenomena involve interplay among their content, authority, and a distinctive kind of force. The binding "force" of meaning, justification, law, rights, and so on is not merely causal force nor is it equivalent to coercion, even if coercion has a role to play. As Rebecca Kukla notes, "Something is authoritative only if it is binding, and makes a claim on the subject of its authority. Furthermore, for it to genuinely bind or make a claim, its authority must be legitimate. There can be no such thing as real yet illegitimate authority, since such 'authority' would not in fact bind us; the closest there could be to such a thing would be coercive force which makes no normative claims upon us" (2000, 165). Normativity also involves at least the capacity to *recognize* normative authority and to be bound to it in part *through* recognition of that authority. To mean something by an utterance is in part to be able to recognize and respond to the appropriateness or inappropriateness of those words. To speak *about* an object or an issue is in part to be able to hold one's performances accountable to it. And so forth. Of course, one need not correctly recognize what, if anything, authoritatively binds one's performances; if someone's utterances are senseless or her actions unjust, it may be because she is mistaken in understanding what can be sensibly said or what claims justice can make upon her. But if she were constitutively incapable of recognizing or responding to justice or sense, then these normative concerns could have neither authority nor binding force for her.

Kukla emphasized this constitutive capacity to recognize normative authority in order to call attention to its retrospective temporality. If normative authority and force are constituted by their legitimacy, and legitimacy depends upon agents' capacity to recognize that legitimacy as binding upon their performances, then any normative authority must always *already* be in place. Moreover, one must always already be a norm-responsive agent in order to be bound by normative authority. If not, semantic, epistemic, or ethical/political norms could have no authoritative force to their claims. Unless we were already in normative space, no appeal to meanings, reasons, evidence, rights, or goods could bind us with authority. Recognizing this feature of normative authority might seem to undermine its very possibility. Kukla instead argues that normative authority is *performatively* constituted and reconstituted by a distinctive kind of *misrecognition*. In *taking* myself to be subject to normative authority, I thereby retroactively bind myself within the space of reasons.

Althusser (1971) offers what has become a classic example: when a policeman shouts, "Hey, you!," and I turn my head, I misrecognize myself as one legitimately called to respond. I nevertheless thereby performatively constitute or reconstitute myself as bound to the authority of the state and not just compelled by its coercive force. Althusser presented such misrecognition as the classic form of ideology. Kukla argues that misrecognition can instead exemplify the essentially *mythical* legitimation of normative authority. From social contract theories to Sellars (1997) on the mythical constitution of the epistemic authority of sense experience, philosophers have proposed stories of how people came to be bound to normative authority within the space of reasons. These stories are not and cannot be literally true narratives of a past transition that somehow brought us into the space of reasons. The enacting, telling, and retelling of the story are instead what do the work. We call ourselves to self-recognition in our stories and in answering the call retroactively make ourselves into agents already bound by the requisite authority. We also *live* such stories of misrecognition, for example, in language learning.[19] We adults take children's babbling and vocal imitation as if they were utterances in a language and respond accordingly. In picking up on the practice, they and we retroactively constitute them as speakers, capable of novel thoughts and expressions precisely by being already bound to norms of meaning and rationality.

Kukla's account is instructive as far as it goes, but I have argued that she is only accounting for one side of this temporally extended process (Rouse 2002, 352–58). Kukla addresses the retroactive constitution of the authority of the space of reasons. In taking retroactive constitution to be sufficient, she implicitly presumes that once normative authority binds us, its content is already determinate. In this respect, her account is deeply Kantian: for Kant, reason already tells us what any moral law could demand; his critical task is to show how we can be free and bound to answer its call. We may not yet or ever know which claims are justified or true, or what justice demands of us, but we are already normatively bound by those claims through a capacity to discern and recognize them.

What Kukla overlooks is that her account of the retroactive constitution of normative authority has a parallel in the prospective constitution of its content. To understand why, we need to return to my earlier account of what I call a normative conception of the practices we participate in as human organisms, such as scientific, political, economic, or more generally, discursive practices (see chapter 4). Practices only exist

19. For discussion of how stories can be lived as well as told, see Carr (1974) and Rouse (1996b, ch. 6).

in continuing to be reproduced. If people stopped producing, exchanging, and consuming goods and services, there would be no economy; if no one ever again uttered sentences in English, the language would die; and if no one taught or undertook courses of study or research, there would be no university. In taking up these practices, however, we constitute ourselves as bound by their normative authority in the ways Kukla indicates. The problem is to understand what *are* the norms to which practitioners thereby performatively bind themselves.

Remember that we cannot appeal to social regularities or collectively presupposed norms within a practice: there are no such things, I have argued, but more important, if there were they would not thereby legitimately bind us. Any regularities in what practitioners have previously done does not thereby have any authority to bind subsequent performances to the same regularities.[20] The familiar Wittgensteinian paradoxes about rule following similarly block any institution of norms merely by invocation of a rule, since no rule can specify its correct application to future instances (Wittgenstein 1953). Practices should instead be understood as comprising performances that are mutually interactive in partially shared circumstances. The intelligibility of performances within a practice then depends upon the anticipation and partial achievement of appropriate alignment with others' performances and their circumstances, toward what I described above as their "end," as Aristotelian *energeia*. Through discursive niche construction, human beings have built up patterns of mutually responsive activity. These patterns make possible newly intelligible ways of living and understanding ourselves within this discursively articulated "niche."

The normative "force" that binds us to one another in patterns of practice, and makes us responsive to these issues, comes from their ends, the very possibilities they provide for intelligible ways to understand and enact ourselves in the world. Robert Brandom long ago described this binding potential as "expressive freedom": "Expressive freedom consists in the generation of new possibilities of performance which did not and could not exist outside the framework of norms inherent in social practices. . . . Expressive freedom, as the capacity to produce an indefinite number of novel appropriate performances in accord with a set of social practices one has mastered, is an ability which must be exercised to be maintained" (1979, 194). What is at stake in a practice, and

20. Even the life patterns of organisms-in-environments are not regularities in this sense. They reproduce themselves differentially in development and evolution. They are goal-directed toward maintaining their own life pattern, but what the pattern is does not remain fixed throughout.

in the issues that divide its practitioners, are the very possibilities for who and what we might be through involvement in and submission to the practice—that is, their character as ends or *energeia*. Brandom's talk of "norms" is then misleading: norms are not already determinate standards to which performances are accountable but are instead temporally extended patterns that encompass how we have already been living this part of our lives as well as the possibilities open for its continuation. Just what this pattern of practice is—what we are up to, and who we are in our involvement in it—is always partly ahead of us, as that toward which the various performances of a practice are mutually, but not always fully compatibly, directed. The temporal open-endedness of our biological niche construction and that of social practices are two ways of describing the same phenomena. Despite Haugeland's objection to biological or social conceptions of intentionality, there is a possible "gap" between what various performances "mean" and what they actually or normally do. Such gaps are sustained by these performances through their mutual dependence and temporal directedness toward themselves as temporally extended ends—that is, as *energeia* whose character is at issue and at stake within the practice.[21]

21. Astute readers will undoubtedly have noticed a divergence between the sense in which conceptually articulated practices are "normative," which I have just explicated, and the sense in which organisms are goal-directed and thereby "normative" as succeeding or failing in their goal. Although organisms differentiate themselves from what thereby becomes their environment, by acting for the sake of maintaining that differentiation under changing circumstances, they do not also take themselves *as* goal-directed, nor do they articulate *what* the goal of their behavior is. In Brandom's (1994) terms, goal-directedness is "implicit" within the ongoing life patterns that compose an organismic lineage in an environment, but most organisms do not make "explicit" either their goal-directedness or the goal toward which they direct themselves. It is only the emergence of a two-dimensional, conceptually articulated way of life that retroactively distinguishes the goal-directedness of a way of life from the determinacy of its goal as *what* it is directed toward. That is why Okrent insists that the kind of pragmatist account of goal-directedness and intentionality that we each endorse assigns a constitutive role to the *explanation* of behavior: "The pragmatic approach agrees with functionalism in holding that it is the explanatory role of thoughts that is central to an understanding of their intentional status. . . . [A] key distinguishing feature of the pragmatic position that I develop [is] the suggestion that actions have their goals originally . . . only in virtue of the roles they play in the rational explanation of a species of goal-directed behavior, rational action" (2007, 27). Only in the context of a conceptually articulated explanation do organisms manifest themselves *as* goal-directed. Once that context is in place, however, it displays the goal-directedness and rationally explicable normativity of the organism as having been already implicit in the organism's behavior, such that the explanatory account "of what it is for some behavior to have a goal . . . does not appeal to that behavior's being caused in the right way by the intentional states of an agent" (Okrent 2007, 27). The existence of goal-directed organismic ways of life does not depend upon the articulation, or even the articulability of such explanations within a conceptually articulated way of life, but their intelligibility *as* goal-directed is so dependent. There is nothing mysterious about this retrospective intelligibility; it exemplifies the retrospective temporality of all normativity that was just discussed. The goal-directed life patterns of organisms are then "normative" in a sense that is legitimately derived from the kind of normative authority, and response to

This understanding of conceptually articulated practices as subpatterns within the human lineage belongs to the Davidsonian-Sellarsian tradition that emphasizes the "objectivity" of conceptual understanding. Yet the "objects" to which our performances must be held accountable are not something outside discursive practice itself. Discursive practice cannot be understood as an intralinguistic structure or activity that then somehow "reaches out" to incorporate or accord to objects. The relevant "objects" are the ends at issue and at stake within the practice itself. "The practice itself," however, already incorporates the material circumstances in and through which it is enacted. Practices are forms of discursive and practical niche construction in which organism and environment are formed and reformed together through an ongoing, mutually intra-active reconfiguration. People always do have some at least implicit conception of what is at issue in their various performances and what is at stake in the resolution of those issues. They understand their situation in a particular way that takes the form of an ability to live a life within it, a practical grasp of what it makes sense to do, of how to do that, and of what would amount to success or failure in those terms. Such an understanding governs all efforts to work out that understanding by living our lives in particular ways. Yet it is a fundamental mistake to conflate what is at stake in our situation with any particular conception of those stakes.

The reason this conflation is mistaken is that it denies our dependence upon circumstances and the supportive performances of others. It is not up to any of us what is at stake in our situation, precisely because our life possibilities are situated within its extension ahead of us. Realizing any particular conception of who we are, how we should live, and why it matters to live in that way rather than some other way depends upon how other entities respond to one's involvement or interaction with them. Our ongoing activities are vulnerable both to other agents who may not fulfill their roles in a particular conception of their shared situation and to the possible unavailability or unsuitability of other entities for the tasks implicitly assigned to them. Agents often conceive

---

that authority as legitimate, that characterizes the two-dimensional normativity of conceptually articulated understanding ("explanation" *is* the explicit articulation of such understanding). It is only in articulating the two-dimensionality of conceptual understanding, which distinguishes the goal-directedness of its own organismic behavior from the determination of *what* that goal is, that the one-dimensional goal-directedness of other organisms becomes intelligible *as* having already been explicable as succeeding or failing in its goal. Thanks to an anonymous referee for the Press for calling attention to the need to explicate the sense in which organismic behavior is normative, even though it does not ascribe or acknowledge any authority to its own ongoing self-maintenance as its goal.

differently from one another what they are up to in their various interactions, and entities often show themselves differently from how those agents' understand them (although, of course, they show up as recalcitrant only through the effort to understand and deal with them in those ways). Confronting such discoveries, we must respond, typically by revising or repairing our understanding of the situation and of ourselves.[22] Often such revision and repair requires explicit interpretation. That interpretation might involve further articulating just what we were doing and why it matters or engaging critically with others in a process of mutual adjustment of performances and skills. Such adjustments are needed to accommodate or diminish divergences between one's understanding of the situation as a field of intelligible possibilities and how things showed themselves in response to that understanding.

Revisions and repairs of performances and skills are not the only possible response to the resistance of others and the recalcitrance of things, however. In one direction, we sometimes try to go on in the face of resistance and recalcitrance, as if nothing untoward had happened. Perhaps we subtly adjust our performances to avoid encountering those circumstances in which issues arise, learn to ignore the divergences, avoid confronting others, or just live with various forms of incongruity. In the other direction lies a concern for the possibility that the entire practice is suspect, that no revision or repair of ours or others' performances will suffice, and that in this domain of our lives, or our entire life, there is no "there" there. It is not only biological lineages that go extinct: entire social practices, or specific roles within those practices, can die out, including languages, cultures, sciences, games, occupations/skills, arts, and much more.[23] Or rather, these are forms of biological extinction in an extended sense: not the end of a whole biological lineage but of subpatterns of life within the human lineage that are no longer lived and no longer livable in the absence of supporting roles and materials that enable their intelligibility.

22. I take the terms "revise" and "repair" in the specific senses in which John Haugeland used them to talk about responses to various "impossibilities" in one's understanding of a constituted domain: "In the face of a challenge, either a particular exercise of a skill, or that skill itself, can come into question; that is, it could be either that the performance was somehow erroneous (in a sense that is so far neutral between impropriety of performance and incorrectness of result), and should be revised (rectified), or else that the skill itself is somehow defective or inadequate, and should be repaired (modified and improved)" (Haugeland 1998, 334).

23. Haugeland (2013) argues that an understanding of the possible collapse of the intelligibility of practices or ways of life is what Heidegger ([1927] 1962) means by "existential death" and that comportments that undertake responsibility for sustaining the intelligibility of the way of life within which they are situated is what Heidegger meant by "owned resoluteness" (more commonly translated as "authentic resoluteness").

There remains an important difference between the extinction of biological lineages and the "extinction" of conceptually articulated social practices. While both are patterns in the world that only exist so long as they continue to be differentially reproduced, and whose differential reproduction depends upon appropriate alignment with their material surroundings, social practices have a different kind of normative authority and force. Organisms are patterns of activity-in-circumstances that have the goal of maintaining that very pattern (deictically) as the circumstances change. They can succeed or fail at that goal, but they cannot be mistaken about the things they interact with because they have no articulated way of taking-*as* that could open a gap between how they are taken to be and how they are. They respond to their surroundings as an actual setting that more or less flexibly solicits specific responses but not as a space of possibilities whose constitutive end is contested in ongoing performances.

Social practices are ways of life that might or might not continue and within which an individual organism might or might not continue to participate. Consequently, in working out various issues that arise in their ongoing reproduction, what is at issue in the alignments and misalignments of performances within a practice is not only *whether* those forms of interaction will continue but *what* they will be and what place they have in the larger patterns of life activity of the organism and the lineage. Social practices (including languages, sciences, and other forms of conceptually articulated understanding and responsiveness) are integral parts of a human being's niche-constructed environment, the "space" within which we develop into organisms with specific capacities and possibilities. Yet what, or better who, we are thereby becoming is precisely what is at issue and at stake in practices. Moreover, these issues cannot be localized into the ends preferred or chosen by particular individuals. One cannot live a conceptually articulated way of life unless other agents and one's material surroundings accommodate it. How we can live and who we can be depends upon the mutually interactive configuration of a space of intelligible possibilities.

We should not then think of this "intelligibility" or its "rationality" as overarching, ahistorical ideals that are constitutive of conceptual understanding as it has emerged within human evolution (or as norms that govern whether practices or performances are conceptually articulated at all, in the sense in which Davidson speaks of mental events as "governed by the constitutive ideal of rationality" [1980, 223]). What defines a *conceptually* articulated space is its modal character, such that an organism's life activities are directed in response to not only its actual setting but also *possible* ways it might be and consequently toward how things *ought* to be in accord with those possibilities.[24] Whether things are as they ought or need to be for a practice to continue in a particular way and whether others will take up the enabling roles that would sustain the working out of that practice are part of what is at issue in its ongoing development. What is at stake in a practice is whether and how those issues are to be resolved, and thus whether and how the practice can continue as a possible way for human beings to live and to understand themselves and their surroundings. New issues then arise within the practice in response to previous reconfigurations and its involvements with other practices and its broader environmental situation. In the course of such ongoing remaking of ourselves and the world, the normative considerations that "govern" the working out of those issues are themselves part of what is at issue in the ongoing development of various practices and ways of life. Yet the fact that the norms with respect to which practices and their constitutive performances and manifestations are assessed are themselves open to assessment and adjustment does not make them arbitrary or capricious. Human beings develop in a conceptually articulated discursive niche, as a space of normative possibility and intelligibility. The constitutive authority and force of that configuration of possibilities, as mattering for how we can live within them, is temporally constituted. Who we are is a matter of how we have developed in response to prior situations, which configure the prospects for subsequent responses and the difference they would make to our ongoing way of life. Conceptual normativity is grounded in the exigencies of human life within a specific material and historical situation and the significance of the resulting possibilities for how to live a human life in a discursively articulated environment.

24. For more extensive discussion of how to understand the modal dimension of practices, see chapters 8–9.

PART TWO

Conceptual Articulation in Scientific Practice

SIX

Scientific Practice and the Scientific Image

The first part of the book explored how to understand conceptual normativity and human conceptual capacities as intelligibly part of the natural world as scientifically understood. The emphasis was on an evolutionary understanding of conceptual capacities as forms of material and behavioral niche construction that reproduce and transform our developmental and selective biological environment. This part of the book in turn takes up how to situate scientific understanding within that naturalistic, evolutionary account of our conceptual capacities. Naturalism in philosophy is only viable if scientific understanding is a natural phenomenon that intelligibly belongs within a scientific conception of the world.

I—The Scientific Image

What is a scientific conception of the world? Naturalists endorse and rely upon such a conception as the horizon for philosophical understanding. As naturalists, we nevertheless cannot presume an uncontested grasp of what a scientific conception of the world is. Many influential scientists, from Newton to Kelvin, have taken scientific understanding to be consistent with or even subordinate to a theological conception of the world. The sciences have developed in ways that now show more clearly the possibility of disentangling natural science from natural theology.[1] The task of identifying scientific understanding as a ground for and a check upon philosophical reasoning is nevertheless complicated by the continuing entanglement of the sciences with philosophical reflection upon science. Naturalists turn to the sciences themselves for guidance in understanding what a scientific conception of the world is rather than starting from a philosophical commitment to what science must or should be. "Science itself" has always involved reflection upon its own activities, methods, and achievements, however, and such reflection invariably engages contemporary philosophical understanding.

This intertwining of scientific and philosophical reflection on the sciences' aims and methods is no objection to the possibility of a naturalistic orientation in philosophy that is deferential to scientific practice. Naturalists recognize the fallibility of the best current science and perhaps even more so the fallibility of prevailing conceptions of science. The history of the sciences is replete with tensions and conflicts between widely accepted methodological or philosophical prescriptions and the ongoing development of empirical inquiry. The aspiration to remain open to novel theoretical and methodological developments emerging from scientific research is an important motivation for philosophical naturalism. The danger of confusing deference to scientific inquiry with philosophically imposed conceptions of what science ought to be nevertheless requires continuing vigilance.

The Vienna Circle manifesto of 1929 (Neurath 1973) is an instructive example. Otto Neurath, Rudolf Carnap and Hans Hahn published the manifesto to advance a "scientific world-conception" both within the sciences at different stages of maturation and in philosophy, politics, and culture more generally. The Vienna Circle's understanding of a scientific conception of the world was resolutely methodological. The manifesto proclaimed that "the scientific world-conception is characterized not so much by theses of its own, but rather by its basic attitude, its points of view, and direction of research" (Neurath 1973, 306). This attitude was antimetaphysical ("dark distances and unfathomable depths are rejected"), empiricist ("something is 'real' through being incorporated into the total structure of experience"), and formalist (seeking "a neutral system of formulae, a symbolism freed from the slag of historical languages"; Neurath [1973, 306, 308]). The Manifesto sought

1. Rubenstein's (2014) discussion of multiverse cosmologies and the anthropic principle in contemporary physics strongly suggests that this disentanglement is not complete.

guidance from the sciences themselves: Einstein's supposed vindication of Mach's earlier criticisms of absolute space and time as superfluously metaphysical, and the manifesto's own appeal to foundational debates in mathematics among logicists, formalists, and intuitionists as bearing upon the structure of a "scientific world-conception," exemplified such deference. The Vienna Circle's vision did not triumph over its antithetical "metaphysical and theologizing leanings" in central Europe in the mid-twentieth century, but their views did become highly influential in philosophy and the human sciences in the United States and Great Britain. The Vienna Circle exiles also helped shape broader cultural conceptions of science amid the postwar expansion of scientific research and universities (Hollinger 1995).

Wilfrid Sellars (2007, ch. 14), the Vienna exiles' erstwhile ally in the reconfiguration of American philosophy, nevertheless put forward the most striking contemporary challenge to the Vienna Circle's vision of a scientific conception of the world. Sellars did not merely advocate an opposing conception of "the scientific image of man-in-the-world." Juxtaposing Sellars's "scientific image" with the Vienna Circle's "scientific world-conception" shows that each conception of scientific understanding exemplified the other's sense of what was opposed to science. Sellars presented the scientific image as gradually emerging from a perennial intellectual and cultural tradition with its own alternative, comprehensive image. The problem was the apparent conflict between this traditional "manifest image" and the scientific image that it generated. Sellars presented the scientific image as a composite drawn from prevailing scientific theories and their theoretical posits: "man as he appears to the theoretical physicist is a swirl of physical particles, forces and fields" to be integrated with "man as he appears to the biochemist, to the physiologist, to the behaviourist, to the social scientist" (Sellars 2007, 388). The opposing, humanist "manifest image" took as its starting point a conception of human beings as persons—that is, rational, sentient agents accountable to norms: "to think is to be able to measure one's thoughts by standards of correctness, of relevance, of evidence" (Sellars 2007, 374). In this setting, Vienna Circle empiricism shows up as a sophisticated version of Sellars's "manifest image" of ourselves as rational, sentient concept users.[2] For the Vienna Circle, by contrast, the

2. This oppositional view of the relation between Sellars and the Vienna Circle exiles may seem surprising given their close personal and professional relations as well as their alliance in the postwar reconstruction of American philosophy as "analytic" philosophy closely allied with the sciences

metaphysical realism of Sellars's "scientific image" appeared to be an unscientific metaphysics, despite having been drawn from scientific theory. A "scientific metaphysics" could no more constitute a genuinely scientific world-conception for the Vienna Circle than could a theological understanding of God's creation that described it in terms taken from current scientific theories.

Sellars's implicit assignment of the Vienna Circle's "scientific world-conception" to the manifest image foreshadowed later criticisms that logical empiricist norms for scientific concepts, judgments, and explanations were at odds with scientific practice and could claim no philosophical authority over the sciences. Logical empiricists had always acknowledged some divergence between scientific work and its philosophical reconstruction for epistemological purposes, but postempiricist philosophy of science expanded that divergence to the breaking point. Postempiricist challenges to the relevance of logical empiricist philosophical norms to the sciences joined Sellars's account of the scientific image in contributing to the revival of naturalism, replacing empiricism as the dominant science-centered stance within contemporary philosophy (Giere 1985). Sellars's role in advancing naturalism does not imply that naturalists can simply accept Sellars's scientific image or its descendants as exemplary expressions of scientific understanding, however. Not all Sellars's critics share van Fraassen's (1980) commitment to rejecting naturalism. We must instead consider the possibility that Sellars's scientific image, like Vienna Circle empiricism, might be a philosophical preconception of scientific understanding that masks itself as having been drawn from the sciences themselves.

Despite the earlier prominence of van Fraassen's attempt to reclaim the scientific image for empiricism, the most widespread and fundamental current challenge to Sellarsian accounts of the scientific image denies that the sciences even aim produce a single, unified conception of the world. These disunified alternatives to "the scientific image" present scientific understanding as a patchwork that need not even aspire to the ideal of the "Perfect Model Model" (Teller 2001) suggested by

and with a secularized reconception of the mission of universities as research institutions (Hollinger 1995). The extent of their divergence is much clearer in retrospect in light of the shift from empiricism to naturalism as the dominant science-philic philosophical orientation within the United States. Sellars and Quine are two of the American philosophers most closely associated with the influence of the Vienna Circle, but they are also the two most influential figures in the move away from the Vienna Circle's conception of philosophy, philosophical analysis, the rational reconstruction of science, the sense and significance of the a priori, and the respectability of metaphysics.

Sellars's vision. Scientific disunity has itself been conceived in disparate ways: as theoretical understanding distributed among diverse models whose cross-classifications are useful and informative for some purposes and not others (Giere 1988, 2006; Teller 2001); as laws of limited scope whose gerrymandered domains are circumscribed by where their models are even approximately accurate (Cartwright 1999); as the mutual adjustment of highly specialized theory and instrumentation (such that "the several systematic and topical theories that we retain . . . are true to different phenomena and different data domains" [Hacking 1992, 57]); in the collective inferential stability of different sets of laws across different ranges of counterfactual perturbation, responding to the different interests governing different scientific domains (Lange 2000a); or as a recognition of the metaphysical "disorder" of things (Dupre 1993), perhaps most radically expressed in Hacking's speculative vision of scientific understanding as like "a Borgesian library, each book of which is as brief as possible, yet each book of which is inconsistent with every other [such that] for every book, there is some humanly accessible bit of Nature such that that book, and no other, makes possible the comprehension, prediction and influencing of what is going on" (Hacking 1983, 219).

Sellars himself was attentive to the plurality of scientific theories, disciplines, and domains and thought it could be accommodated within his account of the scientific image. He understood the scientific image as an idealization drawn from a more complex and messy practice: "Diversity of this kind is compatible with intrinsic 'identity' of the theoretical entities themselves, that is, with saying [for example] that biochemical compounds are 'identical' with patterns of sub-atomic particles. For to make this 'identification' is simply to say that *two* theoretical structures, each with its own connection to the perceptible world, could be replaced by *one* theoretical framework connected at *two levels of complexity* via different instruments and procedures to the world as perceived" (2007, 389). From my perspective, however, it is moot whether Sellars can indeed successfully accommodate the disunifiers' lessons within such an idealized unity of scientific understanding. Sellars shares with the disunifiers a conception of scientific understanding as representing the world, whether or not these various representations can be unified into a single, idealized, systematic "image." Scientific understanding is taken to be embodied in scientific knowledge. Whether that knowledge primarily takes propositional form or is substantially realized through mathematical, material, visual, or computational models, scientific understanding is mediated in whole or part by a representational simulacrum of the world it seeks to understand.[3] This representationalist vision of scientific understanding has been especially influential among naturalists in the philosophy of mind and language who have often taken mental or linguistic representation as the key to accommodating scientific understanding itself within a scientific conception of the world (Fodor 1979, 1981; Dretske 1981; Millikan 1984; Dennett 1987). If scientific understanding is representational, then a naturalistic account of mental or linguistic representation might suffice at a single stroke to incorporate scientific understanding of the world within the world so understood. I think this aspiration is misplaced, and if so, we need to look elsewhere to grasp what a scientific understanding of the world could amount to. A central part of the difficulty is the aspiration to identify the shape or form of scientific knowledge as a whole, whether conceived as a systematic, unified "image" or as a less systematically integrated set of partial representations for different purposes. Indeed, I shall argue, a more adequate account of scientific understanding must do justice both to its disunified practices and achievements and to the ways in which those divergences nevertheless remain mutually accountable within an interconnected discursive practice.

II—Reconfiguring the Space of Reasons

Sellars himself provides a key formulation for my naturalistic alternative to representationalist conceptions of scientific understanding. In a justly famous passage from *Empiricism and the Philosophy of Mind*, he argued that "in characterizing an episode or a state as that of knowing, we are not giving an empirical description of that episode or state; we are placing it in the logical space of reasons, of justifying and being able to justify what one says" (Sellars 1997, 76). Representationalist conceptions identify scientific understanding with some position or set of positions *within* the space of reasons—that is, as a body of knowledge. I instead locate scientific understanding in the ongoing reconfiguration

3. In talking about the Sellarsian scientific image as representational, I do not mean Sellars's more specific notion of representational "picturing" but the more general notion of scientific understanding embedded in its metaphorical characterization as an "image." Sellars partially disavows those connotations but still holds onto a conception of scientific understanding as embedded in the epistemic product of inquiry as a body of knowledge. In a similar vein, he explicitly characterized the ideal limit of philosophical understanding: "To press the metaphor to its limit, the completion of the philosophical enterprise would be a single model . . . which would reproduce the full complexity of the [conceptual] framework in which we were once unreflectively at home" (Sellars 1985, 296).

of the entire space. The sciences continually revise the terms and inferential relations through which we understand the world, which aspects of the world are salient and significant within that understanding, and how those aspects of the world matter to our overall understanding. Scientific research also enables the expansion of the space of reasons by articulating aspects of the world conceptually. Phenomena within these newly articulated domains can then be discussed, understood, recognized, and responded to in ways open to reasoned assessment.

The sciences thereby sustain and expand a characteristic feature of human life. We are organisms whose way of life configures our surroundings as an environment with which we interact perceptually and practically. Yet our environment is the outcome of intensive and extensive niche construction, including linguistic and other conceptually articulative performances as salient features of our inherited environment.[4] Language only functions within a broader pattern of material and social interaction, in which our way of life and the environment it discloses are at issue for us. Conceptually contentful engagement with the world emerges in this context. Specifically linguistic performances thus play a central role in conceptual articulation but as integral components of a larger field of practical involvement.

The sciences only emerge historically within an already well-established pattern of discursive and material niche construction and need to be understood as extensions of that pattern. Scientific inquiry takes place within the pervasive setting of human life as a conceptually articulated understanding of and engagement with the world. It is a commonplace that the sciences contribute to the attainment and growth of knowledge, which also discredits false belief and undercuts the acceptance of mysterious or incomprehensible powers. That commonplace lies behind a Sellarsian conception of the scientific image, and suitably qualified, it is surely correct. A more fundamental achievement nevertheless underlies and enables the acquisition and refinement of knowledge through scientific inquiry. Scientific knowledge is possible only because of ongoing practical work within the sciences and a broadly scientific culture,[5] which enables relevant aspects of the world to show up intelligibly at all. Through scientific inquiry, human beings

4. For a more extensive discussion of niche construction and its role in human evolution, see chapters 3–4.

5. In chapter 10, I discuss how scientific practice draws upon the conceptual and practical resources of its broader cultural setting. In the present context, I would note only that the establishment of controlled experimental systems as the locus for the articulation of conceptual domains draws upon the much broader effort to establish and maintain standard units of measurement and

are able to talk about, act upon, recognize, and reason about aspects of the world that had previously been inaccessible. The sciences also bring out intelligible interrelations among previously disconnected aspects of the world. Scientific transformations of the space of reasons are not only expansive, of course. These efforts can also close off or render dubious other conceptual relations or whole domains of inquiry that once seemed accessible and intelligible. The critical dimension of science not only falsifies claims but also reconfigures the conceptual space within which such claims once seemed intelligible.

The ability to say what others cannot and to talk about things not within their ken is not just a matter of learning new words; it requires being able to *tell* what you are talking about with those words.[6] As Haugeland once noted, "Telling [in the sense of telling what something is, telling things apart, or telling the differences between them] can often be expressed in words, but is not in itself essentially verbal. . . . People can tell things for which they have no words, including things that are hard to tell" (1998, 313). The sciences allow us to talk about an extraordinary range of things by enabling us to tell about them and tell them apart. To pick a range of exemplary cases, people can now tell and talk about what is subvisibly small (from subatomic particles to cellular organelles), large or opaque (tectonic plates), in deep time (the pre-Cambrian Era), diverse in function or process (retroviruses or syncytial development), astronomically distant (spiral galaxies), or fast and short-lived (chemical kinetics and its intermediate steps). Not so long ago, there was silence rather than error on these and so many more scientific topics.

Despite widespread assumptions to the contrary, the conceptual work that enables such discursive achievements is not merely intralinguistic. The space of conceptually articulated understanding is not confined to a logical space of intralinguistic inferences, which only engages the world at occasional observational and practical interfaces.[7] Practical skills, perceptual discriminations, material transformations of the world, and a socially articulated way of life (including but not limited to scientific

to extend beyond the laboratory setting the isolations, purifications, shielding, and so forth that enable experimental phenomena to manifest clear and intelligible patterns in the world.

6. Strictly speaking, as Putnam (1975) prominently called to philosophical attention, the division of linguistic labor allows people to talk intelligibly about all sorts of things that they are not themselves capable of telling about in this sense or telling apart. Yet someone must be able to tell what is being talked about in some domain of discourse if such talk is not to become a "frictionless spinning in a void" (McDowell 1994, 66). Enabling and sustaining such conceptual engagement with the world is a central part of what the sciences accomplish.

7. For a useful discussion of what is at issue in understanding a domain as a system of distinct components interacting at well-defined interfaces, see Haugeland (1998, ch. 9) and Simon (1981).

life) are integral to opening and sustaining possibilities for conceptual understanding. In this respect, scientific understanding resembles the larger process of discursive niche construction to which it belongs: new patterns of talk express changed forms of life. In many ways, however, scientific conceptualization is *more* intensively interdependent with practical skills, equipment, and the creation of new material arrangements in specially prepared work sites. Even though scientific concepts and theories aim to provide understanding of the world as we find it, their proximate application is usually to the world as we make it in the specialized setting of laboratories or field practices.

The philosophical importance of experimental and other empirical practices goes far beyond their role in assessing the accuracy of scientific claims. Experimental work in the sciences constructs, maintains, and revises whole "experimental systems." Simplified, purified, or constructed elements are brought together in regimented settings that enable their interactions to manifest clear patterns that can help articulate a domain of objects and events. Experimental systems constitute a kind of "microworld" (Rouse 1987, ch. 4), within which the relevant concepts acquire exemplary uses and normative governance. The making of laboratories, experimental systems, and conceptually articulated domains of scientific research thereby go hand in hand in ways that expand and reconfigure the available possibilities for conceptual understanding. Experimental research seeks to make salient and comprehensible many aspects of the world that would otherwise be hidden and inaccessible to us perceptually, practically, or discursively. While inferential relations among concepts and judgments are indispensable to that process of conceptual transformation, the sciences situate those inferences amid more extensive connections to newly accessible or salient features of the world.

III—Understanding the Sciences as Social and Material Practices

Understanding conceptual articulation in the sciences thus requires developing a more inclusive conception of the sciences as social and material practices than has been common in most philosophy of science. Such an account begins with the recognition that sciences are first and foremost research enterprises. "Scientists" in the primary sense are those people who are engaged in empirical inquiry, whether as principal investigators or as active members of a larger research team. The research enterprise, as a distinctive form of niche construction, nevertheless extends far beyond the scientists whose work constitutes its primary focus. Scientific research depends upon an extensive institutional structure, including disciplinary and other professional associations, journals and other publishers, and extending to the universities, hospitals, institutes, government agencies, or corporations in which research activities are situated.[8] Scientific research has become extraordinarily expensive, and its changing sources of financial and other material support are also integral to the research enterprise, especially where they help shape the priorities and direction of research itself. A significant part of that expense is devoted to material resources. The equipment, materials, research sites, and other physical infrastructure of the sciences have become complex and sophisticated but also sufficiently widely used to support their own supply networks.

Part of that wider use of scientific material and equipment reflects the extension of scientific concepts, instrumentation, materials, and practices beyond the research laboratory, as scientific understanding has become materially as well as conceptually embedded throughout modern industrial societies. Laboratories and their component apparatus and skills are not solely research facilities. Nor can the research enterprise be limited to its suppliers, consumers, and institutional context. Scientific research is of limited import unless its achievements are disseminated more broadly and the skills, understanding, and research orientation of its participants are differentially reproduced in succeeding generations.[9] This pedagogical and disseminative role has grown to dwarf the primary research component; science teachers, writers, and administrators, not to mention engineers or technicians, now far outnumber scientists. Scientific education not only produces more scientists but also embeds scientific concepts and scientific understanding throughout a broad swath of society and culture. Even though the active contributors to research are

8. Recognition of the ways in which the development of scientific institutions has been integral to our conception of scientific understanding affects the historiography of science. Work in the history of science (Shapin and Schaffer [1985] and Biagioli [1993] are especially influential examples) highlights how some very important scientific accomplishments arose in contexts with a quite different sense of the aims and significance of scientific research, which were also contested at the time. Such work does not reject our contemporary sense of Galileo's or Boyle's accomplishments as anachronistic but instead highlights the ways in which what it is to do science and to understand the natural world shifts in the course of ongoing inquiry.

9. Emphasis upon the *differential* reproduction of scientific skills and understanding is important. Although scientific practices and understanding are sufficiently continuous over time to talk about their reproduction, we should recognize that concepts, skills, practices, and materials are continually being reconfigured and reconceived in the course of their transmission.

comparatively few in number and fairly readily identifiable, the research enterprise that sustains them is thoroughly entangled with government, the economy, culture, and a wide range of professions and activities not themselves overtly scientific.

I emphasize the expansive scope and pervasive social entanglement of the scientific research enterprise as a reminder of the complexity of scientific understanding. Philosophers of science typically compartmentalize and narrow the phenomena we consider, both because of our parochial disciplinary interests in the sciences as intellectual achievements and in order to make the subject matter tractable. The result has been a highly idealized, disembodied, and largely retrospective conception of scientific understanding and its conceptual content. Philosophers have typically identified science first and foremost with a systematic body of knowledge claims that is nowhere assembled and expressed in that form. It is in any case extracted from the diverse, complex institutional and material settings within which those verbal and mathematical formulations are articulated, understood, and deployed. We have largely neglected the factors involved in determining which research questions are important, which ones are actually pursued, and what forms the pursuit takes.[10] We have likewise neglected how the resulting achievements are understood, deployed, and implemented. Such an idealized and disembodied philosophical conception of scientific understanding is especially unsuited for naturalists. These more concrete, extended developments of scientific understanding have substantially transformed the world we live in and the ways we talk about and understand that world. At the very least, naturalists who endorse a more traditional conception of the scientific image owe an account of how their idealized image is discernible amid the more diverse, messy, and complex practices in which scientific understanding is embedded. Even that might not be enough, however, if we take seriously an account of the evolution of conceptual understanding as a form of niche construction.[11] If scientific understanding is a highly developed aspect of our material and behavioral niche construction, then to abstract away from a more expansive conception of science as a research enterprise would be to give up on a naturalistic account of scientific understanding.

10. Some recent philosophical work has begun to address questions of scientific significance, often in the context of science policy or the politics of science (e.g., Longino 1990, 2002; Kitcher 2001; Douglas 2009).

11. For a more detailed discussion of niche construction, and its importance for understanding language and conceptual normativity, see chapters 3–5.

Understanding the sciences as research enterprises also requires a different conception of their temporality. The predominant philosophical accounts of scientific understanding are retrospective, looking back at the structure and content of what has already been understood and codified as scientific knowledge. That retrospective orientation often persists even in thinking about possible future scientific developments. These possibilities are typically addressed in the future perfect tense by looking forward speculatively to the further development of science, whether or not that speculation is carried all the way to the supposedly regulative ideal of a completed science. Philosophical conceptions of science still typically look forward to looking back from the vantage point of a scientific knowledge not yet actually achieved.

This retrospective philosophical orientation sharply contrasts to the understanding driving the practice of scientific research. Research workers take a more prospective view of their field as oriented toward outstanding problems and opportunities. While they certainly rely upon what has already been achieved, their understanding of the content and significance of those achievements is transformed by their concern to move beyond them. What were once research topics in their own right are now often regarded not so much as achieved propositional knowledge but as reliable effects and procedures usable as tools to explore and articulate new possibilities (Rheinberger 1997). The concepts employed are understood as open textured in ways that both permit and encourage further articulation or correction of previous patterns of use. What scientific claims say about the world is thereby always open to further transformation. Empirical inquiry articulates conceptual understanding, in significant part through the refined skills, practices, and material reconfigurations of the world that allow those concepts to be both meaningful and informative about the world. Above all, the significance of various topics, claims, tools, and issues is organized around their place in a configuration of available and promising research opportunities rather than their role in the systematic reconstruction of current knowledge.

The divergence between scientific understanding of a research field and philosophical conceptions of the scientific image as an idealized retrospective reconstruction arises not only through the orientation of research toward outstanding issues and opportunities. The retrospective assessment of what has already been accomplished in a scientific research field is also restructured by a research orientation. Researchers do, after all, engage in retrospective assessment and produce compilations in specific forms that range from review articles in journals to textbooks and handbooks. These summations do not exemplify anything like philosophers' conception of the scientific image, individually or collectively. Each compilation is intended for specific audiences, and its content is selected and organized for its prospective significance for research. Review articles assess the significance of recent work in the field for subsequent research. Textbooks focus upon the skills and knowledge likely to be needed by the next generation of scientists. Handbooks and atlases are similarly selective and prospective.[12] One cannot instead take the primary journal literature as a distributed repository of the scientific image since it includes conflicting reports as well as preliminary findings understood as both open to and oriented toward further development and correction. Moreover, in many fields, research develops rapidly enough that researchers' grasp of the field typically forges well ahead of the published literature. Actual compilations of scientific knowledge are thus always partial and oriented toward what is significant for specific projects, and the notion of an all-purpose or no-specific-purpose compilation is an oxymoron. I suspect that the conception of a unified "scientific image," a systematic, idealized compilation of scientific knowledge as a whole apart from any specific uses of it, serves specifically first-philosophical projects in epistemology and metaphysics. Naturalists should be worried about this aspiration.

The unusual occasions for a sustained effort to identify scientific consensus on the current state of knowledge provide perhaps the clearest indication of the divergence between researchers' understanding and anything like a Sellarsian scientific image. The reports of the Intergovernmental Panel on Climate Change (IPCC; 1990, 1995, 2001, 2007) have recently undertaken such efforts with extraordinary care and thoroughness in the multidisciplinary domain of climate science.[13] These reports have carefully and comprehensively vetted the research literature and the opinions of research scientists with a thorough review process aiming to correct errors and accommodate critical assessment

12. Daston and Galison's (2007) important study of changing conceptions of objectivity expressed in scientific atlases and handbooks usefully highlights how epistemologically potent concepts such as 'objectivity' not only are terms for philosophical reflection upon science but also function within scientific practice to shape the direction, form, and content of scientific work. Yet Daston and Galison do not explicitly call attention to the prospective orientation of such reference works, as preparation for encountering and making sense of novel cases, even though the importance of this orientation shows up throughout their discussion.

13. Medicine is another area in which efforts are regularly made to articulate consensus, although such efforts are most commonly directed toward recommended clinical practices of diagnosis and treatment. For informative philosophical discussion of medical consensus conferences, see Solomon (2007, 2011).

of preliminary drafts. Disagreement has been recognized and accommodated by incorporating estimated degrees of confidence within the reported results and predictions. The inherent conservatism of the process of consensus formation and its accommodations to vigilant critics and skeptics among the governments responsible for the review process nevertheless strongly suggest that most researchers' own understanding of climate science diverges from the IPCC conclusions, even when they endorse the process and its outcome as expressing scientific consensus. The disagreements do not indicate flaws in the IPCC process or reports but instead highlight the possible divergence between scientific understanding and even the most diligent and thorough determination of a scientific consensus. The ideas of both a scientific consensus and the "scientific image" as a composite characterization of scientific understanding are idealized composites. They are just not the same idealized composite. The issue is whether scientific understanding is adequately expressed as a collective consensus or composite representational "image," even if the "consensus" judgments are qualified by estimates of confidence or reliability.

IV—Looking Ahead

Conceptions of the scientific image as a comprehensive representation of the world and the space of reasons as an intralinguistic domain nevertheless retain the virtues of familiarity and sophisticated philosophical articulation. Philosophers have worked out a rich vocabulary for talking about knowledge and inference in these terms, with a good understanding of how to apply that vocabulary to the methods and achievements of the sciences. What alternative can I offer to familiar accounts of the scientific image as an empirically justified, systematic representation of the world?[14] Without a serious alternative conception of scientific understanding, philosophical naturalists will continue to fall back upon

14. I use 'justification' here in a broad sense that incorporates reliabilist accounts of the authority of scientific knowledge. I conjoin internalist accounts of reasoning and justification with externalist accounts of reliable methods and strategies of inquiry in order to highlight the focus of both upon a mostly intralinguistic domain of statements or propositions ("internalist" and "externalist" are here used in the epistemological sense, rather than according to the use of these terms in the historiography of science). Reliabilists are usually as much concerned with the reliability of scientific knowledge *claims* (even if that reliability is grounded in scientific methods and the perceptual or instrumental detection of entities and their properties) as internalists are with the more narrowly conceived justification of knowledge claims by other statements or propositions.

the familiar presumption that science aspires to a systematic representation of the world, justified within an intralinguistic space of reasoning.[15]

The remaining chapters of this book develop an alternative way of thinking about a scientific conception of the world and the naturalistic philosophical stance that it helps constitute. Familiar accounts of the scientific image and the space of reasons within which it is expressed and justified assume that these are relatively self-contained linguistic or mathematical expressions. I argue instead that conceptual understanding in the sciences involves material, social, and discursive transformations of the human environment taken together. These transformations amount to extensive forms of niche construction (Odling-Smee, Laland, and Feldman 2003). An environmental niche is not something specifiable apart from the way of life of an organism, which in turn cannot be understood except in its specific patterns of interdependence with its environment. A niche is thus a configuration of the world itself as relevant to an ongoing pattern of activity. Yet organismic activities affect their environment in ways that bear upon the subsequent development and evolution of the organism and its way of life. The emergence of discursive practice, conceptual normativity, and ultimately scientific understanding within our evolutionary lineage take the nonlinearity of niche construction and reconstruction a step further. Our way of life and its ecological/developmental environment do not merely affect one another reciprocally. That reciprocal effect is itself part of what we respond to in our environment, such that our own way of life is explicitly at issue in its own reproduction and development.

Thomas Kuhn was widely criticized for claiming that "after discovering oxygen, Lavoisier worked in a different world" (1970, 118), but in a quite straightforward sense, even ordinary "normal" scientific research is world transforming. Scientific practices rearrange our surroundings so that novel aspects of the world show themselves and familiar features are manifest in new ways and new guises. They develop and pass on new behaviors and skills (including new patterns of talk), which also require changes in prior patterns of talk, perception, and action to accommodate these novel possibilities. These developments thereby introduce new ways to understand ourselves and live our lives while reconfiguring or even closing off previously familiar possibilities. Overall, they reconfigure the world we live in

15. There is a crucial role for reasoning and justification even in reliabilist accounts of knowledge since the reliability of various methods or procedures is always indexed to a reference class within which they are or are not reliable, and the determination of the relevant reference class cannot itself be understood in reliabilist terms (Brandom 1994, ch. 4; 2000, ch. 3).

as a normative space, a field of meaningful and significant possibilities for living a life and understanding ourselves and the world.

The sciences thereby articulate the world conceptually. We often say that the world itself does not change but only our social relations and patterns of talk and thought about it. The world is thereby conceived as already articulated into entities and properties, which may or may not be discernible to us; nothing we say or do can change that, except by adding new kinds of human-made artifact. Such dismissive claims presume that changes in how we talk, think, and relate to one another and things around us are *not* already changes in the world. They surely change our practical, perceptual, and socially interactive environment. More important, these changes wrought by scientific work also let the world show itself in new patterns, with newly discriminable elements and new significance. Moreover, these patterns are not themselves intelligible as patterns, except in relation to the correlative forms and norms of pattern recognition embedded in scientific practice (Dennett 1991; Haugeland 1998, ch. 12).

The idea that scientific understanding and other conceptual transformations are also world transforming has sometimes been dismissed as a kind of fuzzy-minded, unscientific, idealist metaphysics (Scheffler 1967; Kordig 1971). How could changes in our thoughts or utterances change the world they describe and not just our descriptions of it? I instead take the world-transforming character of scientific inquiry to be a straightforward commonplace for any naturalistic understanding of ourselves and the world. Naturalists should reject the notion that conceptual understanding is merely a matter of thoughts or utterances in isolation. Conceptual content and authority incorporate patterns of material interaction within an environment. Claiming that changes in conceptual understanding do not change the world implicitly presupposes that changes in conceptual content can take place and be recognized intralinguistically. Changes in language would not then need to involve changes in the world as talked *about* because changes in language would be contained within language. If we instead understand language and conceptual understanding as integral to larger patterns of interaction with the world that constitute our environment and our biological way of life, then the notion that the development and ongoing revision of language change the world should be unsurprising.

In this part of the book, I show how scientific understanding is mediated by experimental practice and theoretical modeling. The sciences transform the world around us and the capacities through which we encounter and live in it and only thereby allow it to be intelligible. The result is a conception of scientific practice as an ongoing reconfiguration of our socially, discursively, and materially articulated environmental niche and thus as comprehensible within a naturalistic account of conceptual understanding. Bringing these two lines of argument together addresses a central coherence condition for any philosophical naturalism: a naturalism that could not account for scientific understanding as part of nature as scientifically understood is fundamentally incoherent. Meeting this condition is not merely an obligation that must be met to sustain a viable philosophical naturalism, however. We gain a richer and more detailed grasp of scientific understanding and scientific practice by recognizing it to be an ongoing process of niche construction. Scientific niche construction involves coordinated shifts that create new material phenomena, new patterns of talk and skillful performance, the opening of new domains of inquiry and understanding, and transformations in what is at issue and at stake in how we live our lives and understand ourselves. The sciences thereby transform the world we live in and our place and possibilities within it. In doing so, they articulate the world as conceptually intelligible. Neither merely "made up" by us nor found to have been already there, conceptual articulation is the outcome of new ways of interacting with our surroundings that mutually reconstitute us as organisms and the world around us as our biological environment.[16]

In the chapters to follow, several aspects of scientific practice receive heightened attention and reinterpretation for their contributions to the conceptual articulation of our niche. Scientific research often requires creating novel phenomena (Hacking 1983, 2009), prototypically in laboratories, but in my expansion of Hacking's sense of the term, the field

16. Strictly speaking, a naturalistic position understood in this way cannot talk about the world as a whole, which is then differently configured by the ways of life of various organisms as selective and developmental environments. Only these interlocking environments allow meaningfully configured manifestations of a world. Yet there is a different way to vindicate talk of the world as it is apart from its significance for us and for other organisms. The environment of an organism is not an enclosed domain with an outside to it. What does not matter to an organism's way of life is not "outside" its environment but is only relatively opaque to its life activities. These forms of opacity and transparency are also mediated by other organisms, since what does not affect an organism directly may nevertheless figure prominently in the developmental and selective environments of other organisms that do matter to it. Moreover, these significance relations are open to change since environments are dynamic. An organism like us, whose way of life is discursively and thus conceptually articulated, thematizes this possibility of disclosing aspects of the world previously hidden from it. Thus from within our way of life, we can understand and comport ourselves toward possibilities of the world being more or other than it appears within the immediate confines of that way of life. The notion of an objective world, a world as it is apart from its meaningful configuration within any specific organism's way of life, thus should refer not to an already determinate configuration of entities but to an *issue* we confront within our ongoing conceptually articulated way of life. This shift in the normativity of conceptual understanding was argued for directly in chapter 5.

and observational sciences also bring new phenomena into play. These phenomena introduce new patterns into the world, which make different aspects of the world salient within our overall way of life. My approach reconceives the empirical face of the sciences in a shift of attention from what we can observe to what various phenomena can show us. Scientifically significant phenomena are structured events that allow patterns or relations to stand out as salient and significant. Observation may seem to involve private, experiential events that contribute to a public practice, but phenomena in this sense are public, mutually accessible features of the world.[17]

In addition to arranging or uncovering new phenomena, scientists also build models of many kinds: analytical-mathematical, computational, physical, pictorial, diagrammatic, verbal, and more. Models are internally structured systems, often ones that produce reliable, trackable responses to operations performed upon them. Juxtaposing these model systems to laboratory phenomena and more complex events also changes what is salient within the modeled events. Phenomena and models highlight the ways in which scientific conceptualization is a public, material process in which meaning arises from patterned interactions within the world rather than from the internal, inferential relations among mental or linguistic representations. The point is not to deny the role of explicit judgments and inferences in scientific understanding but instead to assimilate discursive articulation to the kinds of worldly patterns manifest in natural and experimental phenomena and various kinds of model systems. Inferentially interconnected judgments and their constituent concepts are themselves powerful model systems in just this way. Chapter 7 shows how experimental work and theoretical modeling work together in the development of conceptual understanding. We cannot understand the conceptual "space of reasons" as an ethereal, disembodied play of mental or intralinguistic representations that only then encounter the world as an external constraint determining their correctness or incorrectness. Conceptual understanding is already engaged with the world in ways that are simultaneously concept articulative, empirically accountable, and world transforming.

The modal character of conceptual understanding and scientific practice is the focus of chapter 8. Recent work on scientific practice has often

17. In fact, observation is never merely a private, experiential event but always situates perceptual and practical responsiveness within a larger pattern of material-discursive practice, including how we call others to share in our observational discoveries (Kukla and Lance 2009). See chapters 4–5.

turned away from conceptions of laws and nomological necessity (Cartwright 1989, 1999, 2003; Beatty 1995; Giere 1999; Teller 2001; Kitcher 2002, ch. 6; Bechtel 2006). This shift away from conceiving scientific understanding in terms of laws had multiple sources: recognition that theoretical understanding is primarily worked out in multiple, partial models rather than unifying principles; renewed attention to biology and other sciences that study historical contingencies rather than physical necessities; waning Humean skepticism about the intelligibility of causal relations unmediated by lawlike regularities; and the decline of the deductive-nomological theory of explanation with an accompanying partial eclipse of aspirations to a general philosophical theory of explanation. These concerns challenge some specific conceptions of natural laws and their necessity but not the importance of laws and alethic modalities. Chapter 8 works out a more adequate understanding of laws that highlights their roles in scientific practice for measurement, inductive reasoning, and conceptual articulation. Laws in this revisionist sense are pervasive even in biology, geology, or psychology, sciences often thought to lack laws of their own.

Along with the alethic modalities, this account also emphasizes the normativity of scientific understanding. It is now a commonplace that the European emergence of modern science replaced a normative conception of the world inherited from Greek philosophy with conceptions of nature in terms of causes, mechanisms, laws, or symmetries that leave no obvious place for normativity. In philosophical domains from logic to ethics, politics, and aesthetics, normative considerations have thus been construed as originating with us as rational agents, social beings, affective perceivers, or meaning makers. For naturalists who situate human life within scientifically understood nature, however, such relocations of normativity as instituted within our way of life only postpone the problem. If human beings are natural entities understandable physically, biologically, or psychologically, then any role ascribed to us as sources of normative authority and force must also be situated within a broadly scientific understanding of nature.

Situating conceptual understanding within a biological context closes any supposed abyss between nature and normativity. The way of life of an organism as a configuration of an environment is normatively significant for the maintenance and reproduction of that way of life. In our case, what that way of life is, and thus how things show up as significant within and for that way of life, is itself at issue for us rather than fixed as a relatively stable pattern. The result is the characteristically two-dimensional normativity of conceptual understanding.

Chapters 9 and 10 explore the scientific manifestation of what we have already seen to be the temporally extended constitution of normative authority, content, and force. Our conceptually articulated way of life allows aspects of the world to show up as significant in novel ways, with other seemingly intelligible possibilities closed off or reconfigured. The sciences are especially powerful examples of such conceptually articulated niche construction, most notably in opening whole new domains of inquiry. Scientific research thereby discloses new aspects of the world, new interrelations among familiar aspects, and new possibilities for our self-understanding and way of life. Chapter 9 shows how the newly articulated conceptual relations that constitute such possibilities can be disclosed as having already been authoritative.

Scientific understanding is also highly selective in light of the normative considerations central to scientific practice. Which aspects of the world matter scientifically, which phenomena are worth exploring and understanding, and thus which inquiries are scientifically significant are integral to scientific understanding. Chapter 10 begins with how scientific significance arises within the temporal open-endedness of research and the conceptual field it sustains. Scientific significance emerges from two directions, which illustrate the modal, domain-constitutive character of scientific understanding previously laid out in chapters 8 and 9. Conceptual domains for scientific research are open textured and directed toward further "homonomic" articulation of the domain.[18] Such domains are not entirely self-contained, however. They matter as more than just scholastic exercises because they also have heteronomic inferential or practical bearing upon other aspects of our lives and work.

The book concludes by returning to the question of what it means to be a naturalist in philosophy in light of these reflections upon the normativity of conceptual understanding within scientific practice. Philosophical responses to the selective focus of scientific understanding have mostly worked within a canonically modern conception that situates norm-instituting human activity within anormative nature. Such approaches take scientific significance to be determined either objectively by anormative nature (e.g., determined by the generality of explanatory laws or the specificity of causal relations) or humanistically by our practices and interests. My reconception of scientific inquiry and the world

18. The terms 'homonomic' and 'heteronomic' are taken from Davidson (1980, essay 10) but used in a somewhat different sense. I characterize further conceptual articulation within a law-governed domain as homonomic, whereas connections established between domains are heteronomic. For further explication, see chapter 10.

it discloses instead shows scientific understanding to be normatively constituted in ways irreducible either to objective features of the world or to human imposition or institution. Neither objective nor social-explanatory conceptions of scientific significance provide the resources needed to understand the dynamic, self-transformative orientation of scientific niche construction.

The sciences thus do not merely produce new knowledge and challenge entrenched beliefs. They reconfigure the space of intelligible belief and the concerns that drive further inquiry. Often it is the material practice of scientific inquiry, the careful, methodological construction of experimental systems, and the disclosure of new domains of intelligibility within the world that most thoroughly transform our sense of who we are and what is at issue and at stake in our lives, including how scientific inquiry matters. This second part of the book also shows why recognition of that contingency and situatedness is not a challenge to the authority and importance of scientific understanding within human life. That recognition instead turns us toward the source of that importance and its normative force in the sciences' ongoing partial reconstruction of the world we live in and in the consequent transformative disclosure of who we can be and how we might live.

SEVEN

Experimental Practice and Conceptual Understanding

The preceding chapter introduced this part of the book by arguing that a "scientific conception of the world" or "scientific image" should not be identified with a body of knowledge understood as a composite or idealized scientific description of the world. Scientific understanding instead encompasses the ongoing reconfiguration of the "space of reasons" within which descriptions are intelligible and open to reasoned assessment and development. The sciences enable things to show up for us intelligibly—that is, conceptually—and only within the context of that primary achievement can we understand them as allowing the world to be known in specific respects. So conceived, scientific understanding in practice differs from familiar philosophical accounts of a scientific conception of the world in several key respects. Scientific understanding is selective (most truths about the world do not matter scientifically), contested (albeit within a space of intelligible possibilities), and directed ahead of itself toward possibilities and opportunities not yet fully articulated.

This chapter takes up the role of experimental practice in the articulation of conceptual understanding in the sciences. "Experimental practice" is used very broadly to incorporate a wide variety of empirically oriented research. Observational sciences such as astronomy or field ecology, clinical sciences, or comparative sciences such as paleontology or systematics are also "experimental" in this broad sense. I choose "experimental" as the umbrella term to highlight the fact that even seemingly descriptive or observational sciences typically must undertake material work—with instruments, sample collection and preparation, shielding from extraneous interference, observational protocols, clinical regimens, and more—to allow objects in their domains to show themselves in conceptually significant ways. All empirical sciences materially intervene in the world, transforming aspects of it to allow them to be intelligible and knowable.

Despite renewed philosophical attention to experiment and material practice over the past several decades (Hacking 1983; Rouse 1987; Franklin 1989; Radder 1996; Galison 1997; Rheinberger 1997; Baird 2004; Chang 2004), conceptual articulation in the sciences is still primarily understood to consist in theory construction.[1] In this respect, Quine's (1953) famous image of scientific theory, as a self-enclosed fabric or field that only encounters experience at its periphery, is instructive. Not only does Quine neglect experimental and other empirical activity in favor of a passive conception of perceptual receptivity (as "surface irritations"), but his image of conceptual development in the sciences involves a clear division of labor. Experience and experiment impinge upon us from "outside" our theory or conceptual scheme in ways that can only provide *occasions* for conceptual development. The resulting *work* of conceptual articulation is nevertheless a linguistic or mathematical activity of developing and regulating "internal" inferential relations among sentences or equations to reconstruct the fabric of theory and the web of belief.

Insistence upon the constitutive role of theory and theoretical language within scientific understanding arose from now-familiar criticisms of empiricist accounts of conceptual content. Even many prominent empiricists in twentieth-century philosophy of science are now widely recognized as taking linguistic frameworks to be constitutive for how empirical data could have conceptual significance (Friedman 1999; Richardson 1998). Their "postpositivist" successors have gone further in this direction and not merely by asserting the theory-ladenness of observation. Most recent philosophical discussions of how theories or theoretical models relate to the world begin where phenomena have already been articulated conceptually. As one influential example, James Bogen

1. Experimental practice has received more extensive attention from historians, sociologists, and anthropologists of science but often with even less concern for how experimental work and theoretical understanding engage one another.

and James Woodward (1988) argued that: "Well-developed scientific theories predict and explain facts about phenomena. Phenomena are detected through the use of data, but in most cases are not observable in any interesting sense of the term. . . . Examples of phenomena, for which the above data might provide evidence, include weak neutral currents, the decay of the proton, and chunking and recency effects in human memory" (1988, 306). Similarly, while Nancy Cartwright (1983, essay 7) recognizes that one cannot apply theories or models to events in the world without preparing a description of them in the proper terms for the theory's application, she characterizes this "first stage of theory entry" as an operation on the "unprepared description [which] may well use the language and the concepts of the theory, but is not constrained by any of the mathematical needs of the theory" (1983, 133). Even her first stage of theory entry thus starts from what is already a description. Michael Friedman's (1975) work on explanatory unification made this tendency to take conceptual articulation for granted especially clear by arguing that the "phenomena" scientific theories seek to explain are best understood to be *laws* characterizing regular patterns in the world rather than specific events. If we think about the relation between theory and the world in terms of the explanation of lawlike patterns or descriptions of those patterns, then that relation comes into philosophical purview only after the theory's subject matter has already been grasped conceptually.

Outside philosophy of science, however, questions about how conceptual understanding is accountable to the world have gained renewed prominence in work by John McDowell, Robert Brandom, John Haugeland, and others. To be sure, no return to empiricist accounts of concept formation is in question here. These left-Sellarsians wholeheartedly reject any resurrection of the Myth of the Given, in which conceptual content or normativity could be anchored in the mere occurrence of an experiential or causal event. They locate empiricist accounts of conceptual content as one side of a dilemma, which McDowell suggestively depicts with the image of a treacherous philosophical passage. The rocks of Scylla loom on one side, where attempts to ground conceptual content on merely "Given" causal or experiential impacts run aground. On the other beckons the whirlpool Charybdis, in which the intralinguistic coherence of purported conceptual judgments would instead become a mere "frictionless spinning in a void."[2] In McDowell's

2. McDowell (1984) actually invokes the figures of Scylla and Charybdis in the related context of what it is to follow a rule. There, Scylla is the notion that we can only follow rules by having an

terms, postpositivist philosophy of science has mostly sought to avoid the well-understood perils of conceptual empiricism by steering toward Charybdis, giving primacy to theory and intralinguistic inference in developing and expressing conceptual understanding and setting standards for its own empirical accountability. That division of labor, between conceptual development that is "internal" to scientific theory and its "external" empirical accountability to the products of observation or experiment, nevertheless blocks successful passage between Scylla and Charybdis. Or so I argue.

I—Experimentation and the Double Mediation of Theoretical Understanding

Philosophical work on scientific theories nowadays often emphasizes the role of models in working out their content. Mary Morgan and Margaret Morrison (1999) influentially describe models as partially autonomous *mediators* between theories and the world. Theories do not confront the world directly but instead apply to models as relatively independent, abstract representations. Discussions of models as mediators have nevertheless followed the broader trend within postpositivist philosophy of science by attending more to relations between theories and models than to those between models and the world. Since my aim is in part to understand how the sciences allow aspects of the world to show up within the space of reasons at all, I cannot settle for this starting point. I, too, nevertheless also seek to avoid resorting to any form of the Myth of the Given. We should not think of data or any other observational intermediaries as "Given" manifestations of the world.

My proposed path between Scylla and Charybis begins from Ian Hacking's conception of "phenomena," which is quite different from Bogen and Woodward's use of the term for events-under-a-description. Phenomena in Hacking's sense are publicly accessible events in the world rather than perceptual experiences or events-under-a-description. His account also more subtly shifts the emphasis within accounts of the

explicit interpretation of what the rule directs, and Charybdis is regularism, the notion that rule following is just a blind, habitual regularity in what we do. I adapt the analogy to McDowell's (1994) later argument with different parallels to Scylla and Charybdis, both because the form of the argument is similar and because the figures of Scylla and Charybdis are especially apt there. It is crucial not to confuse the two contexts, however, since when McDowell (1984) talks about a "pattern," he means a pattern of behavior supposedly in accord with a rule, whereas when I talk about "patterns" below, I mean the salient pattern of events in the world in a natural or experimental phenomenon.

empirical dimension of science from what is observed or recognized to what is salient and noteworthy. Phenomena *show* us something important about the world rather than our merely *finding* something there. Hacking's term also has a normative dimension and thus cannot refer to something merely "Given." Most events in nature or the laboratory are not phenomena in his sense, for they show little or nothing with clarity. Creating a phenomenon is an achievement. The focus of that achievement, I suggest, is the salience and clarity of a pattern against a background. Hacking suggested, for example, that "old science on every continent [began] with the stars, because only the skies afford some phenomena on display, with many more that can be obtained by careful observation and collation. Only the planets, and more distant bodies, have the right combination of complex regularity against a background of chaos" (1983, 227).[3] Some natural events have the requisite salience and clarity, but Hacking argued that most do not. Other phenomena must therefore be created in laboratories or other sites of scientific intervention.

I use Hacking's concept to build upon Morgan and Morrison by claiming that theoretical understanding is *doubly* mediated. "Phenomena" in Hacking's sense mediate in turn between models and the world in ways that enable conceptual understanding. Hacking himself may share this thought. He once remarked that "in nature there is just complexity, which we are remarkably able to analyze. We do so by distinguishing, in the mind, numerous different laws. We also do so, by presenting, in the laboratory, pure, isolated phenomena" (1983, 226). I take the "analysis"

3. The term 'regularity' was probably an infelicitous choice in this context. The prominence of this term in philosophy of science and metaphysics stems from Hume's attack on more robust conceptions of causality. Humeans locate scientific understanding in the recognition of regularities, and our habitual expectation of their continuing recurrence, precisely because they think single events have no salient intelligibility. I think Hacking had a very different conception in mind, whose challenge to Humeans was nicely expressed by Nancy Cartwright: "Once a genuine effect is achieved, that is enough. The [scientist] need not go on running the experiment again and again to lay bare a regularity before our eyes. A single case, if it is the right case, will do" (Cartwright 1989, 92). There are, admittedly, some phenomena for which 'regularity' seems more appropriate, such as the recurrent patterns of the fixed stars or the robustness of normal morphological development. The phenomenon in such cases is not the striking pattern of any of the constellations or the specific genetic, epigenetic, and morphological sequences through which tetrapod limbs develop; it is instead the robust regularity of their recurrence. In these cases, however, the regularity itself *is* the phenomenon rather than a repetition of it. Humeans presume that a conjunction of events must occur repeatedly to be intelligible to us. Hacking and Cartwright, by contrast, treat some regularities as themselves temporally extended single occurrences. Phenomena in this sense are indeed *repeatable* under the right circumstances. Their contribution to scientific understanding, however, does not depend upon their actual repetition.

in question here as making the world's complexity intelligible by articulating it conceptually. To take Hacking's suggestion seriously, however, we need to understand how creating or recognizing phenomena could be a distinct mode of scientific "analysis" complementary to nomological representation. Catherine Elgin (1991) makes an instructive distinction in this respect between the properties an event merely instantiates and those it exemplifies. Turning a flashlight in different directions instantiates the constant velocity of light in different inertial reference frames, but the Michelson-Morley experiment exemplifies it. Similarly, homeotic mutants exemplify a modularity of development that normal limb or eye development merely instantiates.

Consider Elgin's example of the Michelson-Morley experiment. The interferometer apparatus allows a light beam tangential to the earth's motion to *show* any difference between its velocity and that of a beam perpendicular to its trajectory as leading to an interference pattern. No difference is manifest when the experiment is properly performed, but this manifestation of the constant velocity of light is contextual in a way that belies the comparative abstraction of its verbal representation. We can talk about the constant velocity of coincident light beams traveling in different directions relative to the earth's motion without mentioning the beam splitters, mirrors, compensation plates, or detectors needed to produce the Michelson-Morley phenomenon. We often represent phenomena in this way, abstracting from the requisite apparatus, shielding, and other surrounding circumstances. Moreover, such decontextualized descriptions express what Bogen and Woodward or Friedman meant by "phenomena." Hacking argued that such decontextualizing talk can be importantly misleading, however. Referring to a different example, he claimed that "the Hall effect does not exist outside of certain kinds of apparatus. . . . That sounds paradoxical. Does not a current passing through a conductor, at right angles to a magnetic field, produce a potential, anywhere in nature? Yes and no. If anywhere in nature there is such an arrangement, with no intervening causes, then the Hall effect occurs. But nowhere outside the laboratory is there such a pure arrangement" (1983, 226). The apparatus that produces and sustains such events in isolation is an integral component of the phenomenon and is as much a *part* of the Hall effect as the conductor, the current, and the magnetic field. The consequent material contextuality of phenomena might then give us pause. Surely conceptual understanding must transcend such particularity in order to capture conceptual generality. Perhaps we do need to *say* what a phenomenon shows, and not merely show it, when we further articulate the world conceptually. I argue instead that the phenomena themselves, and not merely their verbal characterization, can have conceptual significance that points beyond the event or its repeatability.

I begin my reasoning for this claim with critical attention to two important but flawed attempts to ascribe conceptual significance to phenomena themselves as events in the world rather than to those phenomena as already under a description. I think both are moving in the right direction despite failing to pass between Scylla and Charybdis; their failure is consequently instructive. Consider first Hacking's (1992) own account of relations between models and phenomena in terms of "self-vindication." Hacking uses this term both to highlight the codevelopment of experimental phenomena with theoretical models and also to account for how scientific knowledge often achieves a high level of stability. "Self-vindication" occurs in laboratory sciences, because "the . . . systematic and topical theories that we retain . . . are true to different phenomena and different data domains. Theories are not checked by comparison with a passive world. . . . We [instead] invent devices that produce data and isolate or create phenomena, and a network of different levels of theory is true to these phenomena. . . . Thus there evolves a curious tailor-made fit between our ideas, our apparatus, and our observations. A coherence theory of truth? No, a coherence theory of thought, action, materials, and marks" (1992, 57–58). Hacking rightly emphasizes the mutual coadaptation of models and phenomena. We nevertheless cannot understand the empirically grounded conceptual content of models in terms of a self-vindicating coadaptation with their data domain. The reason is that Hacking's proposal would steer directly into McDowell's Charybdis, rendering the conceptual articulation of that data domain empty in its splendidly coherent isolation. Hacking envisaged stable, coherent topical domains, within which the scientific claims in question gradually become nearly irrefutable, by having only limited application to well-defined phenomena they have already been shown to fit. That exemplar of conceptual stability, geometrical optics, tellingly illustrates his claim: "Geometrical optics takes no cognizance of the fact that all shadows have blurred edges. The fine structure of shadows requires an instrumentarium quite different from that of lenses and mirrors, together with a new systematic theory and topical hypotheses. Geometrical optics is true only to the phenomena of rectilinear propagation of light. Better: it is true of certain models of rectilinear propagation" (Hacking 1992, 55).

The problem with Hacking's proposal is that, in supposedly securing the correctness of such theories within their domains, he renders them empty. He thereby helps himself to a presumption that his own account undermines—namely, the sense in which geometrical models of rectilinear trajectories amount to a theory *about optics*. Indeed, Hacking's concluding sentence instructively points to the difficulty: on his account, geometrical optics could only be a model of rectilinear propagation rather than a model of light as rectilinearly propagated. The analysis of light not only is provided by the mathematical model but also incorporates its mutual adjustment with an experimental system, together comprising a "self-vindicating" package. The domain of phenomena to be accounted for then cannot be identified apart from these mutually adjusted mathematical and experimental models. This package of model systems would thereby cease to be *about* anything other than itself. The illusion of empirical constraint is sustained only because we implicitly take the experimental system as a stand-in for the larger domain of events that it models, but that connection is explicitly cut by Hacking's conception of self-vindication. It is one thing to say that geometrical optics only accurately describes some aspects of the phenomena in its domain and thus has limited effective range or only approximate accuracy. It is another thing altogether to confine its domain to those phenomena for which it seems to work. The fine structure of shadows *is* directly relevant to geometrical optics and thereby displays the theory's empirical limitations. Only through its openness to such empirical challenge does the theory even purport to be *about* the propagation of light (rather than just rectilinear propagation). McDowell's criticism of Davidson thus also applies to Hacking's view: he "manages to be comfortable with his coherentism, which dispenses with rational constraint upon [conceptual thought] from outside it, because he does not see that emptiness is the threat" (1994, 68).[4]

Nancy Cartwright similarly tries to implicate material phenomena in the articulation of conceptual content by limiting the scope of concepts

4. An anonymous referee expressed the doubt that McDowell's Kantian worries about conceptual contentfulness only apply globally, whereas Hacking's more local conception situates self-vindicating models within a broader array of conceptual abilities and practices, in which the empirical limitation of the models' scope provides the needed empirical constraint. The problem is that Hacking's conception of self-vindication severs the connection to that wider setting. The mutual adjustment of theoretical and experimental models makes a self-contained analytical package whose concepts apply only internally, without import for or accountability to anything "outside" that could undermine their mutual adjustment.

to the circumstances that they fit with some reasonable degree of accuracy. She characterizes this limitation in terms of the "abstract" character of the concepts in question. The concept of force, for example, is abstract in the sense that it needs more concrete "fitting out" in order to be contentful. Just as I am not *working* unless I am also doing something more concrete like writing a paper, teaching a class, or thinking about the curriculum, so there is no *force* among the causes of a motion unless there is an approximately accurate force function such as $F = -kx$ or $F = mg$ for that causal contribution to motion (1999, 24–28, 37–46). Experimentation plays a role here, she argues, because such functions typically apply accurately mostly to specially arranged "nomological machines" rather than to messier events. Once again, we have a relatively close fit between what the theory is about and what the theory says about it. Cartwright's proposal does somewhat better than Hacking's account of self-vindication by allowing a limited open-endedness to these conceptual domains. She allows that the concept of force extends beyond the models for $F = ma$ actually in hand to apply wherever reasonably accurate models *could* be successfully developed. This extension is still not sufficient, however. First of all, the domain of mechanics then becomes highly gerrymandered. Apparently similar situations, such as various objects in free fall in the earth's atmosphere, fall on different sides of the borders of this domain.[5] Second, this gerrymandered domain also empties the concept of force of conceptual significance and hence of content.[6]

5. Cartwright's account also requires further specification to understand which models count as successful extensions of the theory. Wilson (2006), for example, argues that many of the extensions of classical mechanics beyond its core applications to rigid bodies involve extensive "property-dragging," "representational lifts," and more or less ad hoc "physics avoidance." I suspect Cartwright might take many of these cases to exemplify "the claim that to get a good representative model whose targeted claims are true (or true enough) we very often have to produce models that are not models of the theory" (2008, 40).

6. My objection to Cartwright's view is subtly but importantly different from one offered both by Kitcher (1999) and in a review by Winsberg et al. (2000). They each claim that her account of the scope of laws is vacuous, allegedly reducing to something akin to "laws apply only where they do." Their objections turn upon a tacit commitment to a Humean conception of law in denying that she can specify the domain of mechanics without reference to Newton's laws. Because Cartwright allows for the intelligibility of singular causes, however, she can identify the domain of mechanics with those *causes* of motion that can be successfully modeled by differential equations for a force function. My objection below raises a different problem that arises even if one can identify causes of motion without reference to any laws governing those motions and thus could specify, in terms of causes, where the domain of the laws of mechanics is supposed to reside on her account. My objection concerns how the concepts (e.g., force) applied within that domain acquire content; because the concepts are defined in terms of their success conditions, Cartwright has no resources for understanding what this success amounts to.

The reason that Cartwright's gerrymandered interpretation of concepts renders them empty is that her account conflates two distinct dimensions of conceptual normativity. A concept expresses a norm of classification, with respect to which we may then succeed or fail to show how various circumstances accord with the norm.[7] Typically, we understand not only how and why it matters to apply *this* concept but also how and why it matters to group together these instances *under* that concept, instead of or in addition to others. The difference it makes in each case shows what is at stake in our success or failure to grasp various events in that domain in accord with that concept (e.g., by finding an appropriate force function for them). Both dimensions of conceptual normativity are required: we need to specify the concept's domain (what it is a concept *of*) and we need to understand the difference between correct and incorrect application of the concept within that domain. Otherwise, there is no significance to the concept that groups the "correct" applications together. By defining what is at stake in applying a concept like force *in terms of* criteria for its successful empirical application, Cartwright removes any meaningful stakes in that success. The concept then just *is* the classificatory grouping specified extensionally.[8] This problem parallels the difficulty encountered by Hacking's account of self-vindication: what was supposed to be a theory of the rectilinear propagation of light became merely a model of rectilinearity. In removing a concept's accountability to an independently specifiable domain, Hacking and Cartwright undermine both dimensions of conceptual normativity, since, as Wittgenstein (1953) famously argued, where there is no room to talk about error, there is also no room to talk about correctness.

7. Failures to bring a concept to bear upon various circumstances within its domain have a potentially doubled-edged significance. Initially, if the concept is taken prima facie to have relevant applicability, then the failure to articulate *how* it applies in these circumstances marks a failure of understanding on the part of those who attempt the application. Sustained failure, or the reinforcement of that failure by inferences from other conceptual norms, may shift the significance of failure from a failure of understanding or application by concept users to a failure of intelligibility on the side of the concept itself.

8. One can see the point in another way by recognizing that the scope-limited conception of "force" that Cartwright advances *would* have conceptual content if there were some further significant difference demarcated by the difference between those systems with and those without a well-defined force function. Otherwise, the domain of "force" on her account would characterize something like "mathematically analyzable trajectories" in much the same way that Hacking reduces geometrical optics to models of rectilinear propagation (rather than of the rectilinear propagation of light). As noted in part 1, this two-dimensional normativity to conceptual articulation parallels Cummins's (1996) insistence on the need to provide independent specifications of the target and content of a representation.

II—Salient Patterns and Conceptual Normativity

To move beyond these efforts to *limit* conceptual content to a domain specified in terms of the relevant concepts themselves, we need to think further about phenomena in Hacking's sense. Their defining feature is the manifestation of an apparently significant pattern in the world. Such patterns acquire that significance by standing out against a background. This "standing out" need not be anything like a perceptual gestalt. A few astronomical phenomena are visible to anyone who looks but most require rather more effort. Recognizing the elliptical pattern of planetary orbits required careful observation and extensive analysis of the observed data: the elliptical pattern is not itself directly perceptible. Experimental phenomena require actually arranging things to manifest a significant pattern, even if that pattern is subtle, elusive, or complex. As Karen Barad noted about a prominent recent example of an experimental phenomenon, "It is not trivial to detect the extant quantum behavior in quantum eraser experiments. . . . In the quantum eraser experiment the interference pattern was not evident if one only tracked the single detector [that was originally sufficient to manifest a superposition in a two-slit apparatus]. . . . What was required to make the interference pattern evident upon the erasure of which-path information was the tracking of two detectors simultaneously" (2007, 348–49).

That a pattern stands out, constituting an experimental phenomenon, is thus crucially linked to scientific capacities and skills for pattern recognition. As Daniel Dennett once noted, "The self-contradictory air of 'indiscernible pattern' should be taken seriously. . . . In the root case, a pattern is 'by definition' a candidate for pattern recognition" (1991, 32). This link between "real patterns" and their recognition should not be misunderstood as conferring any special privilege upon our capacities for discernment. Perhaps the pattern in question shows up through the use of complex instruments whose patterned output is discernible only through sophisticated computer analysis of the data. What *is* critical to the notion of recognition, however, is its normativity. To speak of recognition is to allow for the possibility of error. And so the patterns that show up in phenomena must not merely indicate a psychological or cultural propensity for responsiveness to them. Our responsiveness to them, our *taking* them as significant, must be open to assessment. What were once taken to be informative patterns in the world have often been later rejected as misleading, artifactual, or coincidental. The challenge is to understand how and why those initially salient patterns lost their apparent significance, and especially why that loss corrects an earlier error rather than merely changing our de facto responses.

What makes experimental phenomena conceptually significant is that the pattern they embody is informative beyond its own occurrence. To this extent, the salience of natural or experimental phenomena is broadly inductive.[9] Consider the Morgan group's work at Columbia that initiated classical genetics. Their experiments correlated differences in crossover frequencies of mutant traits with visible differences in chromosomal cytology. If these correlations were peculiar to *Drosophila melanogaster*, or worse, to these particular flies, they would have had no scientific significance. Their salience instead expressed the sense of a more general pattern in the cross-generational transmission of traits and the chromosomal location of "genes" as discrete causal factors.

The philosophical issue in such experiments is not how to reason inductively from a telling instance of a concept to its wider applicability. We need to think about reflective judgment in the Kantian sense rather than the inductive-inferential acceptance of determinate judgments—that is, the question concerns how to articulate and understand relevant conceptual content rather than how to justify specific judgments that employ concepts with already-determinate content. The issue is nevertheless still a normative concern for how to articulate the phenomena understandingly rather than a merely psychological consideration of how we arrive at one concept rather than another. In this respect, the issue is a descendent of Nelson Goodman's (1954) "grue" problem. Goodman's concern was not to understand why we actually project the concept 'green' rather than 'grue,' for which various evolutionary and other considerations provide straightforward answers. His concern was why it is appropriate to project 'green' to future cases as evidentially interrelated, as opposed to why we (should) accept this or that judgment in either set of terms.

Marc Lange's (2000a, 2000b, 2002, 2007, 2009) revisionist conception of natural laws can help here.[10] For Lange, the hypothesis that some statement is a law expresses what it would be for unexamined cases to behave in the same way as cases already considered. In taking a hypothesis to be a law, we commit ourselves to a set of inductive strategies and

9. The salience of a pattern encountered in a scientific phenomenon, then, should be sharply distinguished from the kind of formalism highlighted in Kant's (1987) account of judgments of the beautiful or the sublime (as opposed to the broader account of reflective judgment sketched in the first introduction to that work) or from any psychological account of how and why patterns attract our interest or appreciation in isolation.

10. Lange's account of laws and modalities is discussed more extensively in chapter 8.

thus to the inductive projectability of the concepts employed in the law. Since many inference rules are consistent with any given body of data, Lange asks which of these possible inference rules is *salient*. The salient inference rule would impose neither artificial limitations upon its scope nor unmotivated bends in its further extension. The salience of an inference rule, Lange argues, is not "something psychological, concerning the way our minds work. . . . [Rather] it possesses a certain kind of justificatory status: in the manner characteristic of observation reports, this status [determines] . . . what would count as an unexamined [case] being relevantly the same as the [cases] already examined" (2000a, 194).[11] Where Lange compares salient rules to observation reports, however, I compare them to the salient pattern of a phenomenon.[12] Its normative status as a salient pattern meaningfully articulates the world, helping render intelligible those aspects of the world that fall within its scope, albeit defeasibly so.

Such a role for meaningful patterns *in the world* does not steer us back onto the philosophical rocks of Scylla. The salient patterns displayed in natural or experimental phenomena are nothing Given but instead indicate the defeasibility of both the pattern itself and its scope and significance. One of Lange's examples illustrates this point especially clearly. Consider the pattern of correlated measurements of the pressure and volume of gases at constant temperature. Absent other considerations, their linear inverse correlation yields the familiar Boyle-Charles law. Yet couple this same phenomenon with a model—one that identifies volume with the free space between gas molecules rather than the size of their container and understands pressure as reduced by intermolecular forces that diminish rapidly with distance—and the salient pattern extension instead becomes the van der Waals law. What it would be for this pattern to continue "in the same way" at other volumes and pressures has shifted, such that the simple proportional extension of Boyle's Law now incorporates an "unmotivated bend." Moreover, when modeled and measured differently, all such general patterns dissipate in favor of ones specific to the chemistry of each gas.

Recognizing the inherent normativity of pattern recognition in experimental practice allows us to recover the requisite two dimensions of that normativity. I criticized Hacking and Cartwright for defining

11. I argue in chapter 8, however, that the role Lange here assigns to de facto agreement among competent observers is not appropriate in light of his and my larger purposes.

12. I nevertheless take partial issue with Lange's insistence that *agreement* among competent participants in a practice determines what would count as an unexamined case being relevantly the same as examined cases.

the scope and content of scientific concepts in terms of their successful application. Yet they were right to look to the back-and-forth between experimental phenomena and theoretical models as a locus for the articulation of conceptual content. Haugeland (1998, ch. 11) begins to point us in the right direction by distinguishing "two fundamentally different sorts of pattern recognition. On the one hand, there is recognizing an integral, present pattern from the outside—*outer recognition*. . . . On the other hand, there is recognizing a global pattern from the inside, by recognizing whether what is present, the current element, fits the pattern— . . . *inner recognition*. The first is telling whether something (a pattern) is *there*; the second is telling whether what's there *belongs* (to a pattern)" (1998, 285). A pattern is a candidate for outer recognition if what stands out as salient in context points beyond itself in an informative way. The apparent pattern is not just an isolated curiosity or spurious association. Consequently, there is something genuinely at stake in how we extend this pattern, such that it can be done correctly or incorrectly. For example, only if it *matters* to distinguish those motions that are caused by forces from those that would not be so caused is there anything in classical mechanics to be right or wrong about.[13] Whether a pattern actually does indicate anything beyond its own occurrence is defeasible, in which case it shows itself to be a coincidental or merely apparent pattern.

Inner recognition identifies an element in or continuation of a larger pattern. Inner recognition is thus only at issue if there is some larger pattern there with something at stake in getting it right. Inner recognition grasps how to go on in the right way consonant with what is thereby at

13. In the case of classical mechanics, we normally conclude that there are *no* motions within its domain that are not caused by forces (although of course quantum mechanics does permit such displacements, for example, in quantum tunneling, which are understood as outside the classical domain). That inclusiveness does not trivialize the concept in the way that Cartwright's restriction of scope to its approximately accurate models does, precisely because of the defeasible coincidence between inner and outer recognition (see below). There is (if classical mechanics is indeed a domain of genuine scientific understanding) a conceivable gap, what Haugeland (1998, ch. 13) calls an "excluded zone" between what we *could* recognize as a relevant cause of motion and what we *can* understand with the conceptual resources of classical mechanics. There are no situations that belong within the excluded zone because such occurrences are impossible. Yet such impossibilities must be conceivable and even recognizable. Moreover, if there were to be such impossibilities that could not be explained away, or isolated as a relevant domain limitation (as is done with quantum discontinuities), then what seemed like salient patterns in the various phenomena of classical mechanics would turn out to have been artifacts, curiosities, or other misunderstandings. On this conception, contra Cartwright, to say that a phenomenon belongs within the domain of the concept of force is not to say that we do or can understand how to model it in those terms; it does make the concept ultimately accountable to that phenomenon and empirically limited within its domain to the extent that it cannot be applied to that phenomenon to the degree of accuracy called for in context.

stake. So for classical mechanics, inner recognition is involved in identifying forces and calculating their contributions to an outcome. The existence of a pattern depends upon the possibility of recognizing how it applies. Haugeland thus concludes, rightly, that "what is crucial for [conceptual understanding][14] is that the two recognitive skills be distinct [even though mutually constitutive]. In particular, skillful practitioners must be able to find them in conflict—that is, simultaneously to outer-recognize some phenomenon as present (actual) and inner-recognize it as not allowed (impossible)" (1998, 286). Both dimensions of conceptual normativity, outer and inner, are needed to sustain the claim that the pattern apparently displayed in a phenomenon enhances the world's intelligibility. There must be something genuinely at stake in recognizing that pattern, and any issues that arise in tracking that pattern must be resolvable without betraying what was at stake.

III—Models and Conceptual Articulation

This two-dimensional account of the normativity of pattern recognition enables us to see Cartwright's and Hacking's discussions of the relation between laboratory phenomena and theoretical modeling in a new light. Cartwright can now be understood to challenge this two-dimensional approach to explicating the normativity of scientific understanding. We could think of her work from *How the Laws of Physics Lie* to *The Dappled World*, which emphasized trade-offs between explanatory power and empirical accuracy, as denying the compatibility of inner and outer recognition in physics. Expressed now in Haugeland's terms, her original (1983) argument was that the explanatory patterns expressed in the most fundamental laws and concepts of physics are mostly not candidates for inner recognition, since most events in the world cannot be accurately treated in those terms without ad hoc phenomenological emendation and ceteris paribus hedging. Such modifications of inner recognition belie the apparent clarity and systematicity of the explanatory pattern. Later (1999), she argued instead that the alleged universality of the

14. Haugeland actually talks about what is crucial for "objectivity" rather than for conceptual understanding. Yet objectivity matters to Haugeland only because it serves as the standard for understanding. In *How Scientific Practices Matter* (Rouse 2002, ch. 7–9), I argue that Haugeland's appeal to objectivity is misconstrued and that conceptual understanding should be accountable not to "objects" (even in the quite general and formal sense in which Haugeland uses that term) but to what is at issue and at stake in various practices and performances. See also the further discussion of this point in chapter 5.

fundamental laws is illusory. The scope of their concepts is restricted to those situations ("nomological machines") that actually generate more or less lawlike behavior and to the broader tendencies of their causal capacities. In the dappled world we live in, we need other, less precise concepts and laws to fill the gaps where the regularities expressed by supposedly fundamental laws dissipate. Scientific understanding is a patchwork rather than a conceptually unified field.

Cartwright is calling attention to two importantly connected features of scientific work. First, the concepts that express the patterns projected inductively from revealing experimental or natural phenomena often outrun the relatively limited domains in which scientists understand in detail how those concepts apply. Cartwright's examples typically involve mathematical theories in physics or economics where only a limited range of situations can be described and modeled accurately in terms of the theory, yet the point applies more generally. Classical genetics, for example, mapped phenotypic differences onto relative locations on chromosomes, but only a very few organisms were mapped sufficiently to allow genes correlated with various traits to be localized in this way. Moreover, for most organisms there were substantial practical barriers to establishing the standardized breeding stocks and a sufficiently wide range of recognized phenotypic mutations to allow for sufficiently dense and accurate mapping. Second, in the "gaps" where one set of theoretical concepts could not be applied in detail, other patterns could often be articulated as alternative ways to understand and predict behavior of interest. Cartwright (1999, ch. 1) uses the relation between classical mechanics and fluid dynamics to exemplify this apparent overlap. The motion of a paper banknote in a swirling wind does not allow a well-defined force function for the causal effects of the wind, but the situation may well be more tractable in the alternative terms, equations, and boundary conditions provided by fluid dynamics. This issue has been widely discussed in one direction in terms of reduction or supervenience relations between theoretical domains, but the conceptual relations go in both directions: the supposedly supervening conceptual domain might instead be said to "explicate" the concepts or events that cannot be accurately modeled at a more basic level of analysis.[15]

15. I use the term 'explicate' here for any set of domain relations in which the possibility of reduction or supervenience might be raised, even where we might rightly conclude that the explicating domain does not supervene on a base domain. Thus mental concepts explicate the domain of organismal behavior for some organisms with sufficiently flexible responsive repertoires, even if the mental concepts do not supervene upon physical states or nonmental biological functions of those organisms.

The issues and concerns Cartwright has identified are important, but her response to them remains not fully satisfactory. Her conclusion that "fundamental" concepts have a limited scope depends upon a familiar but untenable account of what it is to grasp a concept and apply it to worldly situations. On this view, grasping a concept is (implicitly) grasping what it means in every possible, relevant situation. Here, Cartwright actually agrees with her "fundamentalist" opponents that $F = ma$, the quantum mechanical formalism, and other theoretical principles provide schemata for applying their constituent concepts throughout their domains. She disagrees with them only, but quite dramatically so, concerning how far those domains extend. The fundamentalist takes the domain of these theoretical principles to be unrestricted, with only epistemic limits on our capacity to work out their application. Cartwright ascribes semantic and perhaps even metaphysical significance to those limitations, which she takes to display instead the inapplicability of those concepts outside a limited range.[16]

Mark Wilson's (2006) alternative approach to empirical and mathematical concepts helps show how to acknowledge and respond to Cartwright's concerns while also reconciling them with my concern about conceptual understanding. My concern has been to understand the conceptual significance of experimental phenomena and their relation to practices of theoretical modeling in ways that do not lose contact with what is at stake in the applicability of scientific concepts. Cartwright thinks the dappled, patchwork character of the world as we find it turns out to be unamenable to smooth, systematic inclusion within the supposedly regimented universality of fundamental physical concepts. Wilson instead rejects the underlying "classical picture of concepts" that Cartwright implicitly relies upon and treats empirical concepts as more complexly organized, for example, as akin to loosely unified patchworks of facades bound together into atlases or as overlapping patchworks pulled in different directions by competing "directivities."[17] A fully general concept

16. Cartwright does not assign this significance to de facto epistemic limitations that might merely reflect failures of imagination or effort. The concepts apply wherever more generally applicable models *could* be developed that would enable the situations in question to be described with sufficient accuracy in their terms, without ad hoc emendation. As she once succinctly put the relevant criterion of generality, "It is no theory that needs a new Hamiltonian for each new physical circumstance" (1983, 139).

17. Wilson identifies the classical picture of concepts with three assumptions expressible "within the homely vernacular of commonplace intellectual evaluation":

> "(i) we can determinately compare different agents with respect to the degree to which they share 'conceptual contents';

need not have any fully general way of applying it. As a telling example, he addresses "the popular categorization of classical physics as *billiard ball mechanics*. In point of fact, it is quite unlikely that any treatment of the *fully generic* billiard ball collision can be found anywhere in the physical literature. Instead, one is usually provided with accounts that work approximately well in a limited range of cases, coupled with a footnote of the 'for more details, see . . .' type. . . . [These] specialist texts do not simply 'add more details' to Newton, but commonly overturn the underpinnings of the older treatments altogether" (Wilson 2006, 180–81). In the case of billiard balls, a sequence of models treats them incompatibly first as point masses, then as rigid bodies, as almost-rigid bodies with corrections for energy loss, as elastic solids distorting on impact, as solids traversed by shock waves, as explosively colliding objects at high velocities, and so on. Some of these models also break down the response of the balls upon impact into stages, each modeled differently with gaps. Wilson concludes, "To the best I know, this lengthy chain of billiard ball declination never reaches bottom" (2006, 181).

Wilson provides extraordinarily rich case studies of disparate links among conceptual facades, patches, or platforms and the accompanying "property dragging" that sometimes shifts how the concepts apply in different settings. I only suggest one further distinction within that set of examples that might help indicate the extent to which empirical concepts need not be smoothly regimented or fully determinately graspable. Suppose we think of sequences of billiard ball collision models as exemplifying an *intensifying* articulation of concepts, with increasing precision and fine-grained detail, to the phenomena centrally at issue in their use. We then also need to acknowledge the *extensive* articulation required to adapt familiar concepts to unfamiliar circumstances. Wilson objects in the latter case to what he calls "tropospheric complacency" in our apparent grasp of familiar concepts: "We readily fancy that we already 'know what it is like' to be *red* or *solid* or *icy everywhere*, even in alien circumstances subject to violent gravitational tides or unimaginable temperatures, deep within the ground under extreme pressures,

(ii) that initially unclear 'concepts' can be successively refined by 'clear thinking' until their 'contents' emerge as impeccably clear and well defined;
(iii) that the truth-values of claims involving such clarified notions can be regarded as fixed irrespective of our limited abilities to check them" (2006, 4–5).

He also identifies that picture from a different direction under the rubric of "classical gluing," whereby predicate and property are reliably attached to one another directly or indirectly, for example, in the latter case, via a theoretical web and its attendant "hazy holism."

or at size scales much smaller or grander than our own, and so forth" (2006, 55). Thought experiments such as how to program a machine to find rubies on Pluto (Wilson 2006, 231–33) tellingly indicate the parochial character of our confidence that we already know how to apply familiar concepts outside their familiar settings or even that the correct application is determinate but unknown. The partial indeterminacy of further applications does not by itself impugn the scientific adequacy of the concepts or our grasp of them. Nor does it call for limiting the scope of the concept to their already-determinate or potentially determinate applications.

Joining Wilson in rejecting tropospheric and related forms of conceptual complacency lets me endorse Cartwright's denial that a general law–schema is sufficient to understand more complex or less accommodating settings while also rejecting her proposed limitations on the scope of the concepts employed in such schemata. We should instead recognize that concepts commit us to more than we know how to say or do. To adapt Cartwright's own terms, 'force' or 'gene' should be understood as dappled concepts rather than as more uniformly projectable concepts with limited scope in a dappled world. Brandom (1994, 583) suggests in this regard a telling analogy between conceptual understanding and grasping a stick. We may only firmly grasp a concept at one end of its domain, but we take hold of the entire concept from that end. We are also accountable for the sometimes unanticipated consequences of its use at the other end and in between. The same is true, however, for the pattern recognition displayed in experimental work, which I take to be integral to the articulation of conceptual understanding in the sciences. These patterns can be inductively salient far beyond what we know how to say or act upon, and it often takes extensive empirical work to figure out what our concepts say (not just whether they apply correctly) in more intensively articulated or further-extended contexts.

That is why I talk about inner and outer recognition in terms of what is at issue and at stake in concept use. 'Issues' and 'stakes' are fundamentally anaphoric concepts. They allow reference to the scope and significance of a pattern, a concept, or a practice (as what is at stake there) and what it would be for them to go on in the same way under other circumstances or more stringent demands (as what is at issue), even though those issues and stakes might be contested or unknown. As one illustration, recognizing the anaphoric character of conceptual normativity lets us see what is wrong with Lange's claim that inner recognition of conceptually significant patterns is shaped by disciplinary interests and concerns. He says that "a discipline's concerns affect what it takes for an inference rule to qualify as 'reliable' there. They limit the error that can be tolerated in a certain prediction . . . as well as deem certain facts to be entirely outside the field's range of interests. . . . With regard to a fact with which a discipline is *not* concerned, *any* inference rule is *trivially* accurate enough for that discipline's purposes" (Lange 2000a, 228). Lange makes an important point in this passage that is misleadingly expressed in terms of scientific disciplines and their concerns. First, what matters is not the de facto interests or concerns of a discipline but rather what is at issue and at stake in its practices and achievements. Members of a discipline can be wrong about what is at stake in their own work, and those stakes can shift over time as the discipline develops. Second, the relevant locus of the stakes in empirical science is not disciplines as social institutions but the domains of inquiry to which disciplined inquiry is accountable. The formation and maintenance of a scientific discipline is best understood as a commitment to the intelligibility and empirical accountability of a domain of inquiry with respect to what is at issue and at stake in that domain.[18]

These considerations about conceptual normativity also allow refinement of Hacking's notion of phenomena as salient, informative patterns. The concepts developed to express what is inductively salient in a phenomenon are always open to further intensive and extensive articulation. The same is true of the experimental phenomena themselves. The implicit suggestion that phenomena are stable patterns of salience thereby gives way to recognition of the interconnected dynamics of ongoing experimentation and model building.[19] Thus far, I have talked about experimental phenomena as if experimentation merely established or disclosed a significant pattern in the world, whose conceptual role would have to be further articulated by model building. That impression drastically oversimplifies the conceptual significance of experimentation. To begin with, we should think about systematically interconnected experimental capacities rather than distinct experimental phenomena. Salient patterns manifest in experimentation function

18. I discuss scientific domains of inquiry more extensively in chapters 8–10.

19. In my earlier work (Rouse 1996b) and elsewhere, I have argued for a shift of philosophical understanding from a static to a dynamic conception of epistemology. More recently (Rouse 2009), I suggest that the account of conceptual normativity in *How Scientific Practices Matter* (Rouse 2002) should be understood as a nonequilibrium dynamics of both conceptual and epistemic normativity. Brandom (2011) develops this analogy more extensively in his interpretation of Wilson (2006) as offering accounts of the statics, kinematics, and dynamics of concepts. Peschard (2010, 2011, 2012) emphasizes the interactive dynamics of theoretical modeling, experimental practice, and the identification and conceptualization of the target systems to be modeled theoretically and explored experimentally.

together to articulate whole domains of conceptual relationships rather than single concepts.[20] Moreover, what matters is not a static experimental setting but its ongoing differential reproduction, as new, potentially destabilizing elements are introduced into relatively well-understood systems. As Karen Barad noted, "[Scientific] apparatuses are constituted through particular practices that are perpetually open to rearrangements, rearticulations, and other reworkings. That is part of the creativity and difficulty of doing science: getting the instrumentation to work in a particular way for a particular purpose (which is always open to the possibility of being changed during the experiment as different insights are gained)" (2007, 170).

The shifting dynamics of conceptual articulation in the differential reproduction of experimental systems suggests the recognition that all scientific concepts are dappled—that is, always open to further intensive and extensive articulation in ways that might be only patchily linked together. That is not a deficiency. The supposed ideal of a completely articulated, accurate, and precise conceptual understanding is in fact far from ideal. Consider Lange's (2000a, 212–19) example of the conceptual relations among pressure, temperature, and volume of gases. Neither the Boyle-Charles law nor van der Waals's law yields a fully accurate, general characterization of these correlated macroproperties or the corresponding concepts. Yet each law brings a real pattern in the world to the fore, despite some noise that it cannot fully accommodate. We should not think of these laws as approximations to a more accurate but perhaps messy and complex relation among these macroproperties. Any treatment of pressure, temperature, and volume more precise than van der Waals's law requires attending to the chemical specificity of each gas, and since gases can be mixed in various proportions, there is no limit to the relevant variability. Insisting upon more precise specification of pressure-temperature-volume correlations thus requires abandoning any generally applicable conceptual relationship among these properties, whose more general relationships *only* show up ceteris paribus. Concepts that only apply with ineliminable imprecision or noise can be legitimate scientific concepts that articulate intelligible patterns in the world.

Hacking was nevertheless right to recognize the stabilization of some conceptual relationships in the sciences, even if such stability cannot be rightly regarded as self-vindicating. The patterns already disclosed and modeled in a scientific field are sometimes sufficiently articulated

20. See chapters 8 and 9 for further explication.

with respect to what is at stake in its inquiries. In such cases, the situations where inner recognition of those conceptual patterns might falter if pushed far enough do not matter to scientific understanding, and those divergences can be rightly set aside as noise. That is why Lange (2000a, ch. 8; 2007) indexed natural laws and their component concepts to scientific disciplines, or as I prefer, to their domain-constitutive stakes. The scientific irrelevance of some gaps or breakdowns in theoretical understanding can hold even when more refined experimental systems or theoretical models are needed in engineering or other practical contexts.[21] At other times, however, seemingly marginal phenomena, such as the fine-grained edges of shadows, the indistinguishable precipitation patterns of normal and cancerous cells in the ultracentrifuge, the discrete wavelengths of photoelectric emission, or subtle shifts in the kernel patterning of maize visible only to an extraordinarily skilled and prepared eye turn out to matter in ways that conceptually reorganize a whole region of inquiry.[22] That is why, contra Cartwright, the scope of scientific concepts extends further and deeper than their application can be accurately modeled, even when the current articulation of those models seems sufficient to their scientific stakes.

IV—Experimentation and the Scientific Image

In the previous chapter, I proposed a reconception of "the scientific image" as naming what a scientific understanding of the world amounts to as a whole. I argued that attention to scientific practice encourages us to think of scientific understanding as yielding not a single "image" of

21. It is thus not surprising that the vast majority of Wilson's (2006) examples are drawn from materials science, engineering, and other forms of applied physics that have had to pay close attention to the behavior of actual materials. Brandom (2011) suggested that this domain exemplifies the pressure that can be put on concepts when they are routinely examined and developed by professionals through multiple iterations of a feedback cycle of extensions to new cases. He notes that jurisprudence may provide a similar case in which concepts like contract or property that have perfectly acceptable uses in political thought are similarly put under pressure by their application in case law. In both cases, we have domains whose concepts are generally in good order for what is at stake in some contexts but open to indefinitely extendable intensive and extensive articulation for other purposes.

22. These examples are chosen as prominent illustrations of initially obscure or marginal aspects of a scientific domain that turned out to point toward major conceptual reorganizations of those domains: shadow-edges for geometrical optics; the similar precipitation patterns of normal and cancerous cells in the ultracentrifuge pointing toward conceptualization of the relation between cellular structure and function in modern cell biology; the discrete wavelengths of photoelectric emissions displaying the quantum behavior of light and electrons; and the kernel patterns in maize displaying the repositioning of supposedly stable genes on chromosomes.

the world within the space of reasons but an ongoing reconfiguration of the space of reasons itself. Unificationist and disunificationist accounts of scientific knowledge disagree about how the different aspects of a single scientific image relate to one another, but they share a concern with how the components of an overall conception of the world fit together as a single "image." I have been arguing that the sciences neither produce nor aim for such an image, even as a disunified patchwork. The sciences instead refine and articulate the conceptual space within which we understand, reason about, and act with respect to many aspects of the world, including allowing aspects of the world to become newly manifest within the space of reasons and to show up intelligibly in new ways. The point was not to deny that scientific fields sometimes arrive at widespread consensus about some aspects of their domain. We should instead recognize that even when consensus is attained, it presupposes a transformation of a materially situated scientific practice within which reasoned consensus can emerge. Scientific practice is also always directed ahead toward further research possibilities that can transform the space within which current conceptions are understood and assessed.

My argument in this chapter makes clear that this "space of reasons" should not be considered an ethereal domain of disembodied propositions and inferences. Scientific practice articulates the world conceptually not merely by developing new theories and models but also by building and developing experimental systems. Conceptual understanding is partially embodied in words and mathematical expressions but also in the instruments, skills, practices, and material phenomena that are integral to any understanding of what theories and models say and what they are about and thereby accountable to. Theoretical language is doubly mediated by models and experimental practices in ways that contribute to the requisite dual normativity of conceptual understanding: both how we understand what happens in a scientific domain and the correctness or incorrectness of that understanding are mutually interactive in their empirical accountability. Most philosophical reflections upon scientific understanding take the material mediation of concepts for granted and presume a determinate interpretation of theoretical concepts to which we are not philosophically entitled. These philosophical accounts exceed their entitlement both by leaving out the material mediation of conceptual content and by overlooking the ways in which scientific understanding is always conceptually open to further intensive and extensive articulation and refinement. What is thereby reconfigured is not merely talk and reasoning about the world but the world itself in ways that allow for its intelligibility and consequent openness to reasoned understanding.

The remaining three chapters of this part of the book develop further this account of how scientific understanding is embedded in mutually interactive practices of verbal or mathematical modeling and broadly experimental articulation of the world. In chapter 8, I build upon Marc Lange's and John Haugeland's reconception of scientific laws and nomological necessity to show how the normativity of conceptual understanding in the sciences has an ineliminable alethic-modal dimension. Lange and Haugeland do not begin with a philosophical conception of what scientific laws or laws of nature are in order then to ask which sciences discover such laws and what roles they play there. They first consider how alethic and normative modalities are integral to scientific understanding in practice, in complementary ways, and identify laws and their necessity with what actually plays these roles in scientific practice. The resulting account of laws as "constitutive standards" governing the intelligibility of scientific domains displays a mutual interdependence between the holistic counterfactual stability of sets of laws and the skills and commitments through which scientific practice allows laws and their constituent concepts to bear upon and be accountable to relevant aspects of the world. These modal aspects of scientific understanding thereby turn out to play indispensable roles in allowing for the dual normativity of conceptual understanding, through which scientific understanding is both contentful and empirically accountable.

Recognition of the holistic interrelations among domain-constitutive scientific laws, and their mutually constitutive involvement with experimental practice, may nevertheless seem to pose difficulties in accounting for how scientific understanding ever gets off the ground or how new aspects of the world or new scientific domains ever come into the space of reasons. Chapter 9 takes up this issue and emphasizes the role of new experimental *systems* in opening domains of systematically interrelated concepts with intelligible empirical content. I argue that these systems should be understood as fictional constructions, where the relevant sense of "fiction" does not involve false claims or nonreferring terms but instead refers to relatively self-contained "worlds," whose limited and controlled patterns of interaction enable the development of conceptual relations (laws) as constitutive standards for scientific domains. These standards can then be extended outside of the "fictional" domain with the requisite dual-normativity characteristic of all conceptual understanding. These considerations reinforce the central lesson of the current chapter: that experimental practice is integral to the opening and further articulation of scientific understanding.

The final chapter in this part of the book shows how the normative authority of the conceptual understanding worked out in scientific practice arises from its belonging to larger patterns of human life, through which there is something at issue and at stake in scientific practice. Scientific practices are never self-contained but are always situated within what thereby becomes a broadly scientific culture. Moreover, these more extensive interactions are not external impositions upon the supposed conceptual and empirical autonomy of scientific practice but are instead integral to the conceptual normativity of scientific understanding itself. Only because the sciences matter to who we are and how we live is there a contentful configuration of scientific understanding. That conceptual space is in turn a condition for the possibility of scientific judgments being correct or incorrect, according to standards that govern their conceptual coherence and empirical accountability. We should not mistake this insistence upon the wider embedding of scientific understanding as a social or cultural explanation of scientific concepts and their normative authority, however. Such efforts at social scientific explanations of scientific understanding are triply mistaken, as we shall see. First, they take scientific practices as a relatively self-contained explanandum, such that the forms of human life within which they are embedded or entangled are understood as "external" to science. Scientific performances, skills, concepts, and domain articulations are instead integral to broader patterns of human life. Second, social or cultural "explanations" of scientific understanding mistakenly identify "society" or "culture" as something separable from or opposed to the material, "natural" world to which human life integrally belongs. Third, they fail to recognize that the accountability of scientific practice within broader patterns of human life and understanding is mutual, such that scientific work often reconfigures the very issues and stakes to which it is held accountable. In this respect, as we shall see, the future-directed open-endedness characteristic of scientific research practice turns out to be exemplary of the larger patterns of normative accountability that constitute human life as a space of reasons.

This conclusion to my discussion of scientific practice and its transformation of the scientific image brings back the considerations from the first part of the book, which addressed the problem of how to understand conceptual normativity naturalistically. As maintained throughout the book, the most basic criterion for the coherence of any putatively naturalistic philosophical understanding is its ability to show that and how the authority and force of scientific understanding are situated within nature as scientifically understood. This second part of the book shows how to begin to accommodate this account of scientific practice and its normativity as intelligible within the conceptual space opened and articulated by evolutionary biology and related scientific domains.

EIGHT

Laws and Modalities in Scientific Practice

Laws of nature have had a complex and contested role within philosophical reflection on the sciences. Laws were prominent in canonical presentations of scientific work, from Newton's and Boyle's Laws, through the classical nineteenth-century laws of electromagnetism and thermodynamics, to efforts to extend the notion beyond physics with the periodic law and law of definite proportions in chemistry, or Mendel's laws, the Hardy-Weinberg equilibrium, and the central dogma of early molecular biology in the life sciences. Hume's challenge to attributions of causal connection or natural necessity also gave laws a central role in empiricist conceptions of science, albeit under a more constrained interpretation of lawfulness as empirically discernible regularity.

Emphasis upon laws has shaped recent philosophical conceptions of science in at least three ways. First, Hempel's (1965) and Goodman's (1954) treatments of the role of counterfactuals in explanation and inductive reasoning were among the earliest significant challenges to strict empiricist conceptions of science and to less robust conceptions of laws as contingent regularities. Explanation, empirical confirmation, and the meaning of theoretical concepts were the central topics in mid-twentieth-century empiricist philosophy of science, and arguably each of them turned on counterfactual reasoning that involves laws or natural necessity. Second, recognition of these roles for modal concepts within the sciences contributed to the revival of philosophically respectable metaphysics, often taken as allied to a scientific or naturalistic standpoint. Sellars and other advocates of scientific realism emphasized that scientific understanding outruns what is empirically discernible in order to uncover deeper or more fundamental structures that explain what can be observed and account for observed data's departures from simple or strict regularities. Outside of philosophy of science, naturalistic conceptions of intentionality have also increasingly relied upon laws or nomological necessity to underwrite their attributions of intentional content. Third, the prominence of laws in the physical sciences and their apparent universality and strictness seemed to many philosophers to vindicate a distinctive role for physics among the sciences, relegating other disciplines to the secondary status of "special sciences" whose law attributions were looser, less robust, or dependent upon their derivation from physical laws.

The more recent resurgence of philosophical interest in scientific practice has in turn seemed to challenge both the centrality of laws to the sciences themselves and many of the philosophical projects advanced by attributing prominence to laws of nature. These challenges have come from three distinct but mutually reinforcing directions. First, close attention to the so-called special sciences, especially the life sciences whose achievements seemed exemplary in the late twentieth century, seems to disconnect scientific understanding from any demand to discern underlying laws. Pervasive evolutionary contingency; the complexity of life processes that consequently display few simple, invariant regularities; and the context-sensitive functionality of biological mechanisms seem to block any central role for laws in biology. Physical or chemical laws may still contribute to scientific understanding in the life or earth sciences, but their significance is mediated by contingent histories and functional complexities, which resist reduction to physical or chemical processes without loss of their biological significance or conceptual coherence (Beatty 1995; Brandon 1997; Mitchell 2003, 2009). Molecular genetics, once the centerpiece of reductionist projects in biology, has instead seen the proliferation of molecular complexity and dependence upon cellular and larger functional contexts. Genetics has been increasingly overtaken by genomics, proteomics, epigenetic regulation of gene expression, and even the resurgence of developmental biology and comparative morphology as background for understanding genes as resources for cells and organisms rather than as their lawful determinants.

A different challenge to the philosophical significance of laws of nature emerged from their heartland in the physical sciences. As philosophers have shifted attention from the structure of scientific theories to their uses, laws of nature have sometimes been viewed as a vestige of the sciences' origin in a theologically situated natural philosophy (Giere 1999, ch. 5). A now-widespread philosophical strategy de-emphasizes laws in favor of the relative autonomy of diverse models, which highlight more localized patterns of similarity or approximation to idealized or simplified systems (Morgan and Morrison 1999). Theoretical understanding is located in the details of the models and their analogical extension rather than in general principles taken as laws. Model-based views have challenged the centrality of laws in diverse ways: as not describing actual physical systems (Cartwright 1983, 1999; Giere 1988, 1999; Teller 2001), as principles that merely guide and loosely unify model building (Giere 1988), as conceptually gerrymandered claims of empirically limited scope (Cartwright 1999), or as loosely bound atlases of discontinuous "theory facades" (Wilson 2006).[1] Whereas focus on laws was often associated with a unifying "scientific image," the mediating-model approach to theoretical understanding suggests a pervasive disunity to scientific understanding at all levels. Pluralistic modeling challenges any "Perfect Model Model" of scientific understanding (Teller 2001) and highlights scientific recognition of "the richness and variety of the concrete and particular [such that] things are made to look the same only when we fail to examine them too closely" (Cartwright 1983, 19).

A third challenge to nomocentric conceptions of scientific understanding has emerged in renewed emphasis upon causality, causal modeling, and causal explanation (Salmon 1999; Cartwright 1999, 2003; Pearl 2000; Woodward 2003; Hitchcock 2003). The Humean tradition treated laws as surrogates for the causal connections that strict empiricism cannot countenance, but many of the prominent counterexamples to deductive-nomological accounts of explanation suggested that causal relevance and causal asymmetry cannot readily be captured by

1. Thinking about the role of models in mediating theoretical understanding emerged alongside and in close conversation with a different conception of "models," deriving from model-theoretic work in logic and mathematics, as part of a "semantic conception" of theories (e.g., Suppes 1967; Suppe 1977; van Fraassen 1980, 1989; Lloyd 1988). The semantic conception also de-emphasizes the role of natural laws (see especially van Fraassen 1989) but retains emphasis upon theories as structured products of scientific research rather than upon the process of theorizing. My discussion primarily concerns models-as-mediators rather than a model-theoretic conception of theoretical structure.

laws.[2] The challenge of differentiating causal connections from acausal correlations in multifactorial contexts such as the relations between smoking and lung cancer or contraceptive pills and thrombosis highlighted an indispensable role for causal understanding. Efforts to cash out causal relations in terms of changes in probabilities instead seemed to make causal terms irreducible due to the need to conditionalize probabilities on other possible causal factors. Once causal relations are accepted at face value, the modal character of laws may seem increasingly superfluous. That dismissal was reinforced by the suggestion that alternative accounts of causality might indicate a plurality of distinct causal relationships rather than a failure to understand "the" relation between causes and their effects (Cartwright 2003).

These three challenges to nomocentrism can be mutually reinforcing or overlapping, exemplified by the widespread use of mechanistic models in biology or chemistry (Bechtel 2006). They also play variations on common themes of disunity, concreteness, and the nonlinguistic character of scientific understanding, since explanatory models often take the diverse forms of diagrams, tables, simulations, three-dimensional graphic representations, or physical models, as well as algebraic equations. These challenges also suggest that a philosophical focus on laws might be an undesirable vestige of outmoded views. Natural laws still carry some theologically inspired overtones of a divine legislator, but they also suggest a philosophical orientation toward scientific knowledge as a product, extracted from its "natural habitat" in scientific research. My reconception of the scientific image is indebted to the philosophical considerations motivating these challenges to nomocentrism—theoretical modeling, causal analysis, the autonomy of the special sciences, and the apparent disunity of scientific understanding in practice. The import of these considerations for understanding laws and nomological necessity has nevertheless been widely misunderstood. Attention to scientific practice and the diversity of the sciences challenges familiar *conceptions* of laws and necessity but does not diminish their importance for scientific understanding when lawfulness is more adequately conceived.

2. Among the now standard causal-oriented counterexamples to nomological conceptions of scientific explanation were (1) the failures of asymmetry in purported explanations of the height of a flagpole by the length of its shadow, the expanding universe by the red-shift of stellar spectra, or measles by Koplik's spots; (2) the gerrymandered subdivision of laws exemplified by the dissolution of salt in holy water as uncomfortably parallel to more standard partitions such as Kepler's laws, Galileo's law of free fall, or sexual selection and predation as subclasses of evolutionary changes in gene frequencies; (3) the conjunction problem for explaining laws (e.g., the conjunction of the Boyle-Charles law and Mendel's laws explaining Mendel's laws); and (4) the probabilistic explanations of improbable events such as paresis or slow radioactive decay.

The remainder of this chapter works out the beginnings of a more adequate conception of laws and their necessity.

I—Lange on Laws and Natural Necessity in Scientific Practice

Philosophical disagreement about whether natural laws are important for science is matched or exceeded by disagreement about what laws are. Humeans (e.g., Hempel 1965; Lewis 1973) identify laws with occurrent regularities or regularities that extend counterfactually or subjunctively. Necessitarian conceptions (e.g., Armstrong 1983; Dretske 1977) identify laws with relations among universals or properties. Causal conceptions regard laws as generalizations of causal patterns, taking their modal or counterfactual import from their causal instances. Each of these familiar conceptions nevertheless agrees in taking laws as components of scientific knowledge or structural features of the world to be understood. What laws are is thereby taken as independent of and philosophically prior to whatever role they have within scientific practice.

Recent work by Marc Lange (2000a, 2000b, 2002, 2007, 2009) and John Haugeland (1998, 2000, 2007, 2013) proceeds differently. Both Lange and Haugeland take the role of laws within scientific research as the basis for clarifying and assessing conceptions of what laws are.[3] The primary contributions they emphasize are familiar. Laws play central roles in inductive confirmation and measurement, in scientific reasoning (which often proceeds subjunctively or counterfactually), and in the integrally conjoined roles of conceptual articulation and explanation. Lange and Haugeland also ascribe a less familiar role for laws in accounting for a familiar feature of scientific work. Scientific research is organized into disciplines that apply their own methods and standards to distinct domains of inquiry. Lange and Haugeland appeal to laws to explicate the boundaries of scientific disciplines, their partial autonomy, and how a discipline's

3. Mitchell (2009) also argues that philosophers should develop accounts of what laws are that are guided by the functions of laws in scientific practice. She nevertheless remains committed to a conception of laws as regularities and emphasizes variations in the degree of contingency that attaches to laws in different scientific domains. Apart from thereby drawing too sharp a distinction between logical and nomological necessity, and overlooking the role of modalities in scientific reasoning, her focus on laws as contingent regularities overlooks the holistic, domain-constitutive role of laws and their role in guiding research practice through their combination of systematicity and open-endedness. The importance of these aspects of Lange's and Haugeland's accounts of laws emerges in the remainder of the chapter.

claims are or are not accountable to other disciplines.[4] Although these issues are mostly familiar, Lange's and Haugeland's emphasis on the laws' roles in scientific practice yields a novel and powerful conception of what laws are and how they modalize scientific understanding. Their conceptions were developed independently but are complementary and mutually reinforcing. In this section, I explicate the core elements of Lange's view, which provides a more detailed and sophisticated conception of what laws are and how they are "necessary." The next section shows how Haugeland's account both contributes to and constructively revises Lange's approach. The concluding sections of the chapter consider how further refinement of their conjoined accounts contributes to reconceiving "the scientific image" and conceptual understanding more generally.

Lange begins with a constitutive difference between laws and accidents: their range of invariance under counterfactual suppositions. The laws would still have held even under different circumstances, whereas accidents are circumstantially contingent.[5] This difference is subtle, however. Many accidents remain invariant under wide ranges of counterfactual conditions: had I worn sandals this morning, or taken a different route to work, most contingencies would be unaffected. On other counterfactual suppositions, some accidents might have held even though some laws do not. It is presumably a law of chemistry or materials science that copper conducts electricity, but if copper had been an insulator, or had no free electrons in its outer shell, copper would not have been electrically conductive. On that counterfactual supposition, however, I might still have worn sneakers this morning. Lange responds to the overlapping range of invariance between laws and accidents by arguing that laws have *maximal* counterfactual invariance, whereas accidents hold within a more limited counterfactual range.

Maximal invariance is not displayed by laws individually, however, but only as a holistic feature of sets of laws. Lange introduces this holistic conception of laws via the notion of "subnomic" truths.[6] A statement

4. Strictly speaking, as discussed below, what the laws explicate is the autonomy of domains of scientific inquiry rather than scientific disciplines and subdisciplines. Scientific disciplinary boundaries often track what turn out to be autonomous domains of inquiry, but institutional, economic, and other social-practical issues also bear on the organization of scientific disciplines. Lenoir (1997, ch. 2) provides a thoughtful historical reflection on the multifarious aspects of discipline formation, although he did not have available the conception of domains of inquiry discernible in Lange and Haugeland.

5. This initial discussion of laws and accidents deliberately preserves an ambiguity between whether these are different kinds of truths, different kinds of events or patterns in the world, or both. This ambiguity is resolved in subsequent discussion.

6. For ease of exposition, I will follow Lange's exposition in distinguishing nomic from subnomic *truths*. I nevertheless still leave open the possibility that the appropriate distinction is

is subnomic if its truth or falsity does not depend upon what the laws are. The subnomic truths include subnomic correlates to the laws themselves, understood as truths regardless of their range of counterfactual invariance (thus, the subnomic correlate to the second law of thermodynamics correctly describes the de facto, probabilistic increase in entropy in any closed system without asserting or denying any counterfactual import to this regularity). Laws are distinguishable from accidents by their membership in a *set* of statements that is subnomically stable: all members of the set remain true under any subnomic counterfactual supposition that is consistent with the set, in which case each member of the set is a law.[7] Moreover, this criterion requires no prior, independent determination of what the laws are: any nonmaximal set of subnomic statements (or patterns) is a set of laws if the set is subnomically stable.

The intuitive import of subnomic stability can be made clear quickly. Add even one accident to a set of subnomic truths that also express laws, and this enlarged set would not have been counterfactually stable. Take the set composed of the subnomic regularities expressed by the laws of physics, whatever they are, and the accidental truth that I arrived at my office this morning. Nothing in this set is *logically* inconsistent with the possibility that I had a fatal bicycle accident on route to work, yet in that eventuality, one member of the enlarged set would not have held. For any accidental truth, there will be some nomologically possible conditions under which it would not have held (adding additional accidents to the set would not help, since additional accidents provide additional opportunities for counterfactual instability). The one exception would be the set of all subnomic truths, which is trivially stable, since *no* counterfactual possibility is consistent with that set. That is why we need consider only nonmaximal sets of subnomic truths.

An important aspect of the laws' role in scientific practice emerges directly from the constitutive holism of the laws. Philosophers often discuss laws from an implied position outside the world, from which one could specify what the laws are independent of what is actually the

between nomic and subnomic events or patterns of events. In that case, the subnomic domain would be those occurrences or patterns whose identity does not depend upon which ones are lawful. In that case, just as there would be truths that are subnomic correlates to the laws, we would talk about events or patterns without consideration of whether and how that pattern or kind of event extends counterfactually.

7. If one were to construe laws and accidents as events or patterns in the world rather than statements, then we would talk about "compatibility" with the set of laws rather than "logical consistency." Statements whose truth values belong to a counterfactually stable set might, in that case, turn out to be a special case of laws.

case.[8] Lange's conception (along with Haugeland's) eschews any such standpoint. Whether a pattern is lawful depends upon whether it belongs to a stable set of laws, and arguably no such sets have been extensionally determined. Lange (2000a, 2000b) instead focuses initially upon the prospective *commitment* expressed in *taking* a hypothesis as a law. Laws are not just important achievements in retrospect but also integral to further exploration in research. Indeed, the difference between laws and subnomic truths primarily concerns their prospective rather than retrospective import. A long empiricist tradition asserts that a retrospective assessment of empirical data could justify no claim stronger than the subnomic counterpart to a law; empiricist skepticism reduces laws to regularities (i.e., to subnomic truths) as all that can be retrospectively justified empirically.[9] Lange notes instead that "a basic presupposition of scientific research is that we do not need to examine everything in order to know everything. Rather, a few observations, restricted in space, time, and other respects, sometimes suffice to render salient a hypothesis that is accurate to all unexamined cases in a remarkably wide range of cases" (2000b, 240–41). Laws thereby express inductive-inferential norms of reasoning within scientific practice. In taking a hypothesis to be a law, scientists implicitly claim that the best inductive *strategies* to pursue in this context will vindicate the reliability of the inference rule

8. In counterfactual contexts, however, the issue of "standpoint" might seem to go the other direction. Many philosophers presume that what the laws are in a possible world supervenes on the subnomic truths in that world. Lange (2000a) uses the example of the nearest possible world that consists only of a lone proton to argue the contrary. What the laws are in the nearest lone-proton world depends upon what the laws are in our world and not upon the subnomic truths in that possible world. Later, I argue that such priority for the actual world as context for conceptual norms and alethic modalities is a requisite commitment for a naturalistic understanding of laws.

9. The crucial difference between laws and their subnomic counterparts thus turns on their relation to counterfactual conditionals. The subnomic counterpart to a law is a regularity (a regularity in what happens but without the modal qualifier "actually"). Taken as a law, however, the regularity extends to cover the conditions expressed by counterfactual antecedents as well. When the distinction between laws and subnomic facts is understood in this way, the empiricist tradition concerning laws that stems from Hume must deny that there are any laws (instead acknowledging only subnomic regularities). Goodman's (1954) classic *Fact, Fiction and Forecast* powerfully argued that empirical confirmation cannot be understood unless it extends counterfactually (e.g., canonically, such that the empirical evidence for emeralds being green also must confirm counterfactual claims such as "had this emerald first been discovered much later, it would still have been green"). Lange (2009) then argues that subjunctive facts (the facts expressed by counterfactual or subjunctive conditionals) should be understood to be the "lawmakers," in parallel to Plato's famous question about piety and the gods in the *Euthyphro*: it is the truth of the subjunctive facts that determines which other truths are laws rather than the laws that determine which subjunctive conditionals are true. In appropriating Lange's conception of laws, I will not utilize his later conception of subjunctive facts as lawmakers. This issue implicitly arises below in my reassessment of whether laws are best understood as a distinctive kind of truth or as a counterfactually reliable pattern in the world.

corresponding to that hypothesis.[10] A law expresses what it would be for unexamined cases to behave "in the same way" as those already considered. The familiar difficulty, of course, is that many inference rules are consistent with any given body of data. Lange therefore asks which inference rule is *salient* in this context. The contextually salient inference rule would impose neither artificial limitations in its scope nor unmotivated bends in its subsequent extension.[11] For example, inferring from nearby electrical experiments that "all copper objects *in Connecticut* are electrically conductive" would be an inappropriately narrow scope limitation. Absent further considerations, geographic location is not salient for those experiments. In the other direction, "grue" (Goodman 1954) and "quus" (Kripke 1982) are infamous examples of unmotivated bends.

As I noted in the previous chapter, Lange rightly insists that inductive salience is a normative matter rather than a psychological or sociological propensity. Salience indicates what extension of a set of data one *ought* to take as indicating that it continued "in the same way" as the original data. Data alone are not sufficient to determine what is inductively salient, however. Different extensions of the same data might be salient in different contexts or from different "outlooks."[12] Lange first uses the Boyle-Charles and van der Waals laws to indicate how different outlooks might yield different patterns of inductive salience. Different outlooks suggest different inductive strategies to pursue from initial data about the covariation of the pressure and volume of gases at a constant

10. The point also applies in reverse: in pursuing an inductive strategy, scientists thereby implicitly commit themselves to taking a hypothesis that expresses that strategy to be a law. Inductive strategies can vary in both scope and content. Consider the inductive strategies Mendel might have pursued from his data about inheritance patterns in peas. One strategy (not a wise one!) would have been to limit his inductive inferences geographically: his experiments confirm that Mendelian inheritance patterns hold in Brno or in the Austro-Hungarian empire. Another might be to limit them taxonomically: the experiments justify inferences to Mendelian ratios in peas or in plants. Clearly these would have been inappropriate scope restrictions to impose at the outset in the absence of independent reason to impose those limits.

11. When Lange speaks of "unmotivated" bends, "motivation" is a normative matter of reasons or justification rather than a de facto psychological inclination or social expectation.

12. Lange (2000b) takes the term 'outlook' from McDowell's (1984) discussion of moral realism to highlight a common strategy underlying Sellars's arguments against phenomenalism, McDowell's arguments against naturalistic reduction of moral categories, and Fodor's and others' arguments against the reduction of folk-psychological categories to neuroscientific kinds. The core argument is that even if the "higher-level" categories were to supervene on the "lower-level" kinds, the lower-level categories corresponding to those higher-level classifications would be gerrymandered in ways that would block their inductive salience. The point is not just that we couldn't discover these categories without an outlook that makes them salient: we couldn't even confirm their reliability inductively. I return to these issues below in discussing the implications of this view for conceptions of the unity or disunity of the sciences. Alongside McDowell's notion of what is salient from an outlook, one could also cite Dennett's (1987) conception of a "stance"; I shortly make further use of Haugeland's (1998, ch. 11) extension of Dennett's stances to similar ends.

temperature. Depending upon which strategy one adopts, the initial data drawn from gases at relatively low pressure would have different implications for what to expect at very high pressures and low volumes. In this case, the difference arises from different *conceptual* outlooks: with only minimal assumptions about what gases are, or how they "occupy" a volume and exert pressure on a container, the Boyle-Charles law is salient. A model that instead identifies gases as composed of energetic particles of low volume and relatively weak mutual attractive forces points toward the van der Waals law; from the latter outlook, the straight-line extension projected by the Boyle-Charles law would be an "unmotivated bend."

Conceptual or theoretical commitments are not the only possibly relevant differences in outlook that would yield different statements or patterns as saliently stable. One relevant consideration is the requisite degree of accuracy; different inductive strategies may then be called for, yielding different laws. Lange's primary examples of variations in inductive salience that result from different outlooks come from different scientific disciplines rather than from intradisciplinary conceptual or methodological orientations: "A discipline's concerns affect what it takes for an inference rule to qualify as 'reliable' there. They limit the error that can be tolerated in a certain prediction . . . as well as deem certain facts to be entirely outside the field's range of interests. . . . With regard to a fact with which a discipline is *not* concerned, *any* inference rule is *trivially* accurate enough for that discipline's purposes" (2000a, 228). Evolution introduces historical contingency into the life sciences, for example, since most of the patterns they discern are not physically necessary and many tolerate exceptions: reverse transcriptase, meiotic drive, syncytial development, or transposons are prominent examples of exceptions to important biological generalizations, and endosymbiosis or Hox gene duplication exemplify initially idiosyncratic cases that evolved (literally) into pervasive and reliable biological patterns. Such exceptions do not refute biological generalizations. In some contexts, they amount to tolerable noise that, in other contexts, instead becomes the relevant signal. Such noise-disrupted patterns are "real patterns if anything is" (Dennett 1991, 31). Some patterns that contribute to biological functioning have a counterfactual stability that constitutes a relevant form of biological invariance. Moreover, discerning such patterns is crucial to scientific practice in biology. Biologists regularly reason from evidence about a very small number of organisms to ascribe characteristic traits to species and larger taxonomic units, despite the recognition of variation within populations. How else could one sensibly speak of sequencing *the* genome of *C. elegans, D. melanogaster*, or *H. sapiens* or even establish a stable background against which the emergence of evolutionary novelties could be discerned? Lange (2000a, ch. 8; 2007) argues that functional biological research requires commitment to patterns of relevant counterfactual stability concerning biological species or other taxa, even though those patterns are noise-laden with intrapopulational variation, and the relevant domains are both evolutionarily contingent and historically circumscribed.

Evolutionary biology requires different patterns of reasoning, however. In evolutionary contexts, variation within populations is signal rather than noise, and thus on Lange's view, evolutionary biology appeals to a different set of laws. Medical reasoning, psychological reasoning, geological reasoning, or reasoning in other contexts of inquiry might in turn display their own characteristic patterns of subnomic stability. Lange highlights several suggestive examples. Internal medicine or cardiology requires reasoning about reliable patterns of responsiveness of heart function to the injection of epinephrine or ingestion of nitroglycerin (with some known exceptions). In medical contexts, Lange argues, evolutionary counterfactuals about how the human heart might have evolved differently exemplify cases in which it doesn't matter medically how such cardiological responses to pharmaceutical intervention would have changed under those circumstances. The relevant medical laws still hold regardless of what is to be said in such cases (Lange 2000a, 231–32). The inductive strategies undertaken in clinical trials carry their own distinctive nomological commitments:

> The 1987 edition of the American Psychiatric Association's *Diagnostic and Statistical Manual of Mental Disorders* . . . co-classifies 5,860 different combinations of symptoms as "autism." In adopting this classification, psychiatrists have agreed to pursue certain inductive strategies—to regard one case of autism as bearing confirmationwise on each other "suitable" kind of unexamined case of autism, no matter which one of the 5,860 combinations of autistic symptoms it displays. The significance accorded to a researcher's observations of certain autistic patients, the sorts of epidemiological studies that researchers decide to pursue, and so on, depend on the psychiatric community's commitment to these strategies. (Lange 2000a, 209–10)

Absent a defeasible commitment to the collective counterfactual stability of these different combinations of psychiatric symptoms, such a clinical trial would make no sense, and its results would not be evidence for any intelligible generalizations.

The appeal to disciplines or disciplinary interests can only be heuristic, however. What matters is not what the members of a discipline happen to be interested in or concerned about but what is at stake in their inquiries. Here is an instructive example. Circa 1950, counterfactual hypotheses about how the history of the earth might have been different would have plausibly been regarded as outside the range of interests of geology, while evidence about "martiological" or "venusiological" phenomena would have had minimal evidential bearing on geological hypotheses. Those evidential relations dramatically changed once plate tectonics and evidence about the earth's mantle and core yielded a deeper understanding of its physical dynamics; extraterrestrial probes to Mars, Venus, and the moon provided relevant new data; and astronomical discovery and analysis of other solar systems generated theories about the processes that form rocky inner and gaseous outer planets. Geology as a discipline has, to say the very least, evinced considerable interest in these connections, and comparative planetary science is now a lively area of inquiry. Yet if geologists had resisted such efforts, they would have been mistaken: in the context of this new theoretical and evidential background, counterfactual histories of the earth and the comparative dynamics of planetary formation do importantly bear on geological understanding, whether or not geologists had recognized and responded to them. Perhaps an indifferent response would not have been mistaken in all aspects of the field, however: the inferential connections among stratigraphic sites that contribute to geological history might remain largely impervious to comparative planetary understanding. What matters in either case is not the psychology of scientists or the sociology of disciplines but the intelligibility of a domain of inquiry and the practices that disclose it. Scientists could be wrong about what is at stake in their own work, how those stakes govern their practice, or indeed about whether there is an intelligible domain of inquiry there at all (i.e., one with its own counterfactually stable set of laws) from an outlook or scientific practice that could make such laws salient.

With this revision in place, Lange's account of the disciplinary autonomy of sets of laws as relevantly counterfactually stable patterns provides a richer perspective on familiar issues concerning the unity or disunity of the sciences. Lange primarily emphasizes nomological disunity. Although the existence and boundaries of counterfactually stable sets that constitute laws and domains of inquiry are an empirical matter, there does seem to be a prima facie plurality of sets of laws, roughly aligned with the boundaries of some scientific disciplines. These nomological domains are literally autonomous at three distinct levels. First, their very character as laws is constituted by their mutual inferential interrelations, without reference to other domains. Second, the standards that determine whether the laws in a domain do indeed hold under all relevant counterfactual considerations are themselves domain-relative: different domains require different degrees of accuracy or precision, across different ranges of relevant counterfactual suppositions, with varying tolerance for exceptions and ceteris paribus qualifications. Those normative differences might remain autonomous even if the events in one domain of inquiry were also subsumed within the domain of another. For purposes of illustration, assume that all genes that code for proteins are physical entities, and indeed, are composed of segments of DNA molecules that incorporate discrete sequences of complementary base pairs.[13] Even though all genes are then physical, chemical, and molecular-biological entities, the constitutive differences among sequences that compose genes and those that are not genes (not to mention pseudogenes) would not be physically, chemically, or molecular-biologically salient, and hence the relevant patterns could not be empirically confirmed by their instances when characterized at those "lower" levels. Third, and most striking, the laws in one domain may remain relevantly stable under counterfactual suppositions that violate the laws of other domains, even of domains upon which they seem to supervene. Under some circumstances that violate laws of physics, for example, many laws in the chemical, life, or psychological sciences might have remained invariant. In Lange's example, the laws of island biogeography (if there be such) might still have held under some suppositions that contravene the laws of physics, such as some birds having evolved "modest anti-gravity organs, assisting in takeoffs," or "had material bodies consisted of some continuous rigid substance rather than corpuscles" (Lange 2007, 499). In that case, the laws of physics would not even be a subset of the set of laws that stably define the counterfactual invariance of island biogeographic phenomena.

Physics is not utterly irrelevant to island biogeography or other supposedly special sciences, but often only the gross features of physical laws matter in other domains and not the more richly or finely

13. On this simplifying supposition, we ignore RNA viral genes and also rule out as constituents of "genes" the genetic-regulatory and epigenetic and functional determinants of differential gene expression. "Genes" are then the DNA segments actually expressed by causal correspondences in organisms between DNA base pairs and amino acid sequences of synthesized proteins. Crucially, we are *not* ruling out the *effects* of differential transcription and translation that identify and extract exons or methylation and other epigenetic patterns that block or promote gene expression.

articulated patterns of counterfactual stability that constitute *physical* invariance. Put another way, different scientific domains are mutually accountable in their characterizations of what actually happens—that is, their subnomic truths: chemistry, biology, or psychology cannot license claims about the actual course of events that entail any claims that contradict what physics says about those events, and vice versa. Their domains are nevertheless modally autonomous. They may differ in the circumstances under which one *ought* to say that their claims are true or false (differing in their requisite degree of accuracy or precision, for example) and in their ranges of *counterfactual invariance*. As Lange (2007, 499) concludes, they neither inherit their counterfactual stability as a domain from the stability of other domains nor possess the same ranges of stability. These modal dimensions of scientific understanding nevertheless constitute the conceptual articulation and intelligibility of what happens within those scientific domains.

These modal variations among scientific domains point toward the final theme I take directly from Lange's account of laws. So far, I have carefully avoided the familiar language of necessity and possibility and spoken only of laws as having holistic invariance or stability under relevant counterfactual perturbation. That modal reticence was only strategic, however. Just as Aristotle noted that "there are many senses in which a thing may be said to 'be', but all that 'is' is related to one central point" (Aristotle *Metaphysics* 1941, bk. IV, ch. 2, 1002), so Lange's account reconstitutes the different alethic modalities to show how and why there is an underlying unity to the different senses of necessity rather than a debilitating ambiguity.[14] Relations among logical necessity, metaphysical necessity, natural necessity, and whatever other kinds of necessity there are (conceptual necessity? moral necessity? practical necessity? institutional or bureaucratic necessity? and so forth) have often seemed perplexing. There are perfectly straightforward senses in which natural laws are not logically, conceptually, or metaphysically necessary, and hence might have been different. Not only can we intelligibly ask what the world would have been like had some of the natural laws been different (e.g., if the gravitational force had diminished by the inverse cube of the distance between two masses rather than by its inverse square), but we can often provide determinate, justifiable answers to such questions. In borderline cases, the grounds for determining whether there

14. Haugeland (2013) directly connects the plurality of domains of scientific laws with Aristotle's and especially Heidegger's ([1927] 1962) insistence upon the manifold sense of being. See section II of this chapter.

"are" such kinds of necessity have also been unclear. We might characterize other senses of the term as metaphorical, but that just restates that we do not yet know what the relevant form of "necessity" is said to consist in.

Lange's account of subnomic stability attractively resolves these perplexities. He proves that if two distinct sets of statements within a single domain are subnomically stable, one must be a subset of the other.[15] Instead of asking in each case what we (should) mean by logical, metaphysical, natural, or other varieties of necessity, we can ask instead which sets of statements or patterns remain stable under relevant counterfactual suppositions. We could then assign familiar terms for different varieties of necessity to the most plausibly appropriate stable set or introduce new terms that better indicate their relevant mode of counterfactual invariance. Logical necessity would correspond to the minimal stable set (with consequently the widest range of invariance). Logical laws always hold, except under circumstances that would violate another logical law, in which case anything follows inferentially. That minimal form of invariance then defines the stability relation for other sets: the *minimal* consistency relation ("logical" consistency) must hold between the members of another set and the counterfactual hypotheses relevant to its stability. What kinds of necessity do hold would not be determined by fiat but by which statements or patterns are indeed counterfactually stable. This hierarchical array of levels of inferential stability allows for a straightforward sense in which "lower" levels of necessity are contingent at higher levels. We can understand how the natural laws might have been different, for example, because maintaining the subnomic stability of higher grades of necessity allows determination of what would then have happened under "counternomic" suppositions. As Lange notes, a single domain may incorporate different levels of natural necessity. In physics, for example, the fundamental dynamical law and the conservation laws seem "necessary" in a stronger sense than do the laws governing specific forces: we can determine how the world would have been different had different force laws obtained but only by utilizing the conservation and dynamical laws in the calculation. Along with the hierarchically arrayed necessities of different levels of lawfulness, Lange's conception

15. Although Lange's proof turns on the concept of truth, I invite continuing agnosticism about whether the relevant subsumption relation is primarily a semantic relation among sets of subnomic truths or a different kind of inclusion among patterns in the world (of which the semantic relation may or may not be a special case).

straightforwardly interprets a different kind of modal variation. What the laws are is determined by which sets of statements or patterns are subnomically stable; we can similarly understand metalaws, such as symmetry principles that "govern" the laws, as patterns that remain *nomically* stable under relevant counternomic suppositions, in parallel to the *subnomic* counterfactual stability of the first-order laws (Lange 2009, sec. 3.4–3.5).

Clarifying and ordering different levels and kinds of necessity in a coherent and informative way is an important achievement of Lange's account. He provides a common core meaning to the varieties of necessity, a principled, defeasible basis for understanding which varieties obtain and why and an informative understanding of their interrelations. No other account of multiple modalities provides a comparably nonarbitrary basis for modal concepts. In this context, however, the relations among different domains of natural necessity stand out as somewhat anomalous. Physical necessity, chemical necessity, functional-biological necessity, evolutionary-biological necessity, and whatever other forms of natural necessity turn out to display the requisite counterfactual stability are not related to one another with the same clarity. Lange is clear about their mutual independence but has little to say about whether and how they are mutually accountable, especially since what are salient patterns from one nomologically constitutive outlook are usually conceptually gerrymandered and thus indiscernible from another. Each form of necessity incorporates the same "higher" levels of necessity as stable subsets, and perhaps might be incorporated as subsets of "lower" levels, but these interrelations have no clear bearing upon the distinct forms of natural necessity.

II—Haugeland on the Normativity of Law-Governed Domains

Lange and Haugeland approach the role of laws in scientific practice from opposite directions. Lange began with the laws' constitutive forms of counterfactual or subjunctive invariance, in contrast to those of mere accidents. This holistic invariance is the key to understanding their contribution to inductive strategies, counterfactual reasoning, explanation, and a disciplinary division of labor. Haugeland begins instead from the laws' role in constituting and sustaining scientific inquiry. The laws' distinctive forms of alethic-modal invariance then emerge as indispensable to their normative role for scientific understanding. Despite differences in emphasis, these two projects are complementary approaches to understanding the modal character of conceptual understanding in scientific practice. That a similar conception of laws and modalities emerges from either direction is not coincidental; it indicates, I argue, the coconstitutive roles of conceptual normativity and alethic modalities.

Since Lange developed many of the basic elements of this conception of laws and their necessity, I will only briefly characterize Haugeland's alternative approach. I focus instead on Haugeland's distinctive contributions to an amalgamated view. Haugeland introduces scientific laws as a special case of domain-constitutive rules. Constitutive rules were originally understood through reflection upon games, although games were stand-ins for social practices more generally, including linguistic practices (Rawls 1957; Searle 1969, ch. 2). The rules for how rooks move in chess, for example, are not merely regulations that constrain already-extant entities ("rooks"). There are no rooks apart from the institution and maintenance of the authority and force of these rules, and for rooks, the rules are not optional: not to move in the appropriate ways within the right kind of setting (a chess game in this case) is not to be a rook. Haugeland's point in this approach was not to treat science or nature as a game; the differences matter along with the similarities. The artificial setting of socially instituted practices nevertheless usefully shows how object-constitutive "rules" or norms are more complex than is usually recognized.

Constitutive rules were traditionally conceived as regulations governing the players of games: to say that rooks move only along ranks and files was shorthand for saying that players must not move rooks in any other way. Haugeland argued instead that regulations are a special case of more general standards governing all the entities in a domain. We readily overlook the standards governing rooks themselves because their compliance is normally "built in," but a rook that was immovably massive, autonomously self-moving, indistinguishable from bishops, or randomly changing color would violate the standards of chess as much as any player could.[16] He then argues that intelligible domains of entities also presuppose two additional forms of constitutive normativity: *skills* of discernment and performance that are integral to the domain and

16. Haugeland does insist upon a distinct role for the "players" in such a domain. Rooks may self-effacingly violate their constitutive standards (in which case they are no longer rooks), but they cannot be held accountable for upholding the integrity of the standards. That difference indicates the importance of what Haugeland calls "constitutive commitment" to domain-constitutive standards for sustaining the intelligibility of the domain.

commitment to uphold the standards throughout the domain (Haugeland 1998, ch. 13). Scientific practices articulate and enforce norms that govern the intelligibility of entities within their domains, and laws acquire their characteristic necessity as integral to the empirical accountability of scientifically disclosed domains.

What does Haugeland's account distinctively contribute to Lange's conception of laws? This section highlights three important ways in which Haugeland further develops our understanding of laws in scientific practice. The following sections add two further considerations that raise more fundamental questions about what laws are and how they are authoritative in scientific practice.

Haugeland's first contribution is to relocate laws. Where Lange talks about the laws of various *disciplines* such as physics, evolutionary biology, island biogeography, or cardiology, Haugeland connects laws with the intelligibility of *domains* of entities. We saw that Lange had relatively little to say about the relations among various disciplinary sets of laws compared to his precise account of different levels of law and their necessity. He only recognizes their source in "disciplinary concerns," while acknowledging that whether any laws satisfy those concerns is an empirical matter. For Haugeland, by contrast, a set of laws "constitutes" a domain of entities by holding them to defeasible standards.[17] Domains of entities are normatively constituted by the laws as constitutive standards rather than by disciplinary interests. Laws nevertheless cannot serve as constitutive standards *for* entities unless the projected laws are also accountable to the very entities whose intelligibility they make possible. Haugeland's central concern is to understand the conditions for that mutual accountability. Conjoining Lange's work with Haugeland's clarifies these conditions in turn by showing how a *set* of putative laws might be collectively accountable to the entities they govern, via the requirement that the set remain stable under relevant counterfactual perturbation. Haugeland's account of scientific laws presumes that various laws function as standards for a domain of entities, but he has no obvious way to say why, for example, the force laws and their symmetries, the conservation laws, the fundamental dynamical law, and so forth belong together as laws of physics or how they are related in hierarchical

17. "Constitute" is a technical term for Haugeland. It marks out a relation weaker than "creating" or "instituting" but stronger than merely letting an already extant entity "count as" something else. By establishing, recognizing, and critically reflecting upon constitutive standards, our activities "let" the relevant entities be, in an enabling sense of that term. For more extensive discussion of constitution as letting be, see Haugeland (1998, ch. 10–13; 2007; 2013).

"grades" or levels of necessity.[18] Lange's nomological holism shows *how* laws collectively constitute intelligible domains of entities.

Haugeland's second contribution is to show how laws play their characteristic roles in scientific practice as empirically defeasible "constitutive standards." They do so as components of a larger nexus of constitutive skills and commitments. The important issue in the background is how to understand the laws' two-dimensional normative role in measurement and inductive reasoning. Philosophers of science recognized long ago that scientific practices of observation and measurement are "theory-laden," such that theoretical considerations can serve as standards for the assessment and revision of empirical data (Hanson 1958; Feyerabend 1962; Kuhn 1970; Boyd 1973). A "crisis of rationality" in the philosophy of science (Hacking 1983, ch. 1; Zammito 2004) initially arose from the recognition that the coherence and empirical adequacy of a whole network of concepts might be at issue together with the skills, methods, and norms that supposedly enable their empirical defeasibility. The sense of crisis abated as philosophers proposed ways to accommodate the interdependence of standards and skills, whether by insisting that the acceptance of a "linguistic framework" or a whole theory is entirely pragmatic (Carnap 1950; Quine 1953), allowing room for reasonable judgment rather than procedural rationality in choosing between "paradigms" (Kuhn 1970), acknowledging the sciences' allegedly ineliminable recalcitrance to methodological constraint (Feyerabend 1975), identifying the emergence of a referentially successful theory as "the beginnings of successful methodology within a scientific field" rather than its consequence (Boyd 1990, 366), or finding yet other ways to set the issue aside.[19] Instead of setting the issues aside, Haugeland shows how to *understand* the mutual accountability of data and methods or skills to

18. Strictly speaking, symmetry principles are not laws but metalaws. Where laws indicate the stability of sets of subnomic truths under relevant counterfactual perturbation, metalaws indicate the stability of sets of nomic truths (laws) under relevant counternomic perturbation.

19. There is more than one way of "setting an issue aside," and these need not be problematic in themselves. In this case, one way of doing so is to focus on issues concerning confirmation, explanation, the structure of theories, and the like within accepted theoretical contexts. Another way to do so is to engage philosophically with particular scientific areas whose conceptual structure is at issue (evolutionary biology, interpretation of the quantum mechanical formalism, or the role of folk psychological concepts in cognitive science are some prominent cases in which philosophical work has been conducted within a space of conceptual controversy in the sciences). Such philosophical work takes seriously the scientific background to such controversies, while also implicitly suggesting that the scientific background alone is not sufficient to settle them, by seeing a usefully contributory role for philosophy in those contexts. These are important and valuable projects, yet they do not directly address the general issue of how to understand the multidimensional normativity of scientific research.

theoretical "frameworks" and of theoretical understanding to empirical findings. Such accountability requires both something akin to Lange's account of the laws' holistic counterfactual stability and some further articulation of the normativity of scientific practice. Lange himself had already claimed, following Goodman (1954), that the inductive reasoning through which data bears confirmationwise on other actual or possible cases implicitly invokes the lawfulness of the concepts inductively projected. John Roberts (2008) further emphasizes the internal relation between lawful invariance in Lange's sense and norms of measurement: "If you want to engage in empirical science at all, you must be committed to acknowledging the counterfactual reliability of everything you acknowledge to be a legitimate measurement procedure" (2008, 288). Haugeland then analyzes more deeply this bearing of laws on scientific practice.

Haugeland argued that four mutually responsive aspects of conceptual understanding in scientific practice are needed to understand its multidimensional normativity.[20] Functioning together, these four kinds of "norms" allow specific performances or judgments in the course of scientific work to be normatively accountable, even while the norms or standards to which they are held to account are empirically defeasible in turn. Along with the laws governing a scientific domain as "constitutive standards," he argues for recognition of two different levels of scientific skill ("mundane" and "constitutive" skills) and a constitutive commitment to the authority of the laws. I will introduce each of these elements in turn before considering how and why they must function together in scientific practice and why they require the laws' distinctive forms of invariance or "necessity."

Haugeland's central concern is how to understand the empirical accessibility and accountability of scientific phenomena through observation and measurement, broadly construed: "Observation and measurement only make sense if there is, in principle, some way to distinguish between *correct* and *incorrect* results" (2007, 98). This truism is more challenging, however, once one rejects the Myth of the Given and recognizes that both the results themselves and the standards that govern them are empirically

20. As noted above, Haugeland takes these considerations to apply to any constituted region of entities and not just scientific domains. He insists that scientific domains have a more wide-ranging empirical accountability, however (nothing in science is entirely at scientists' discretion, whereas he thinks that games and other social practices are to some extent accountable only to their players/practitioners collectively). I will set this issue aside here and consider only the intelligibility of scientific domains. In my previous book (Rouse 2002, 244–46), however, I argue that even games and other social practices are accountable to stakes that are up to not just their participants.

defeasible. Empirical data can be criticized on theoretical or methodological grounds, but the theoretical and methodological norms themselves also need to be under empirical control. That mutual accountability is the context for Haugeland's claim that "the only fundamental way to establish that something must be wrong is to show that some plurality of results is not mutually compatible. And that, finally, presupposes antecedent constraints on what combinations would and would not be possible—which is to say, *laws*" (2007, 100–101). Understanding this role for laws as "constitutive standards" for a scientific domain requires the other three elements of Haugeland's account, which together will show in turn why only a domain-constitutive set of laws could serve as the relevant constitutive standards.

The first step toward this more complex analysis is to recognize that laws would exercise no authority over scientific practice and hence would play no normative role there, unless scientists could tell whether some phenomenon actually accords with or violates the laws. Those tellings (as "constitutive skills") are themselves normative; they can be exercised correctly or incorrectly and well or poorly in various other respects.[21] Yet their normativity takes a different form than that of the laws themselves since they involve capacities or skills for discernment. As Haugeland comments about the constitutive skills of chess, his initial case for expository simplicity: "The rules that are being followed in exercising [constitutive skills] are not at all the same as the rules—namely the constitutive standards—[whose] compliance is being monitored. . . . Accordingly, the normative authority of these rules, and the way it is brought to bear, must also be distinct from the standards. . . . A player need have no further knowledge of the constitutive standards beyond this ability to tell whether they are being followed in practice" (1998, 323). A second step recognizes that the constitutive skills for discerning whether or not some event does or would violate the laws in a given scientific domain presuppose a wide range of other "mundane" scientific skills. The mundane skills in a science are the familiar abilities acquired through training and experience: applying concepts and drawing inferences from them, using equipment, knowing when it is working properly, maintaining and manipulating experimental systems, observing or discerning relevant events, theoretical calculation, designing and implementing experiments, and much more. Distinguishing

21. Lange (2000b, 227) also calls attention to the role of scientific skills in constituting the justificatory salience of laws, but he does not develop this theme in a way comparable to Haugeland's account of the relations among mundane and constitutive scientific skills.

the two kinds of skill is important, however. Haugeland points out that the constitutive skills "cannot be just further mundane skills. They must be 'meta' or 'monitoring' skills vis-à-vis the results of mundane performances; for their essential exercise is to watch out for incompatibilities among those results—an exercise of *vigilance*" (1998, 335). What one is vigilant *for* is not a conflict between constitutive and mundane skills but an apparent conflict among the mundanely skillful performances. Moreover, skills of both kinds must be *reliable* and *resilient*. Reliability is familiar: unless scientists could *consistently* perform experimental and theoretical activities in a research field and adjudicate between proper and improper performance, with outcomes mostly in accord its governing standards, scientific inquiry could not proceed effectively. Resilience is the ability to respond appropriately to apparent breakdowns or failures of reliability or compatibility of the results:

> *Resilience* as here intended, is . . . a kind of perseverance born simultaneously of adaptability and self-assurance. . . . A paradigm of resilience [is] an expert who "knows full well" that he or she can do something—and *so* is not turned aside or discouraged at the first, or even second, sign of recalcitrance. . . . If performances or skills are "revised" or "repaired" casually, at the first sign of trouble, then nothing is seriously excluded, and all "testing" is a farce. That is *why* the skills must be *resilient*: they must be able to stand up to one another, and hold their ground, lest any contentions among them be hollow and inconsequential. (Haugeland 1998, 322, 334)

A reliable and resilient skillfulness, not only in undertaking basic preparations, procedures, and assessments in science but also in determining whether their outcomes are consistent or inconsistent with the possibilities marked out by the laws, is indispensable for allowing a set of laws to *govern* the phenomena within a scientific domain.

The final element in Haugeland's account of constitutive normativity is what he calls a "constitutive commitment" on the part of at least some scientists within a field to hold its entire practice accountable to the laws as constitutive standards by not tolerating violations of them. This commitment has disjunctive import. The first disjunct is straightforward: it is simply the commitment to exercise the relevant mundane and constitutive skills conscientiously, holding them accountable to the relevant norms (if there are mistakes in performance, or if equipment or other materials do not meet the requisite standards, those mistakes should be corrected and the original results discounted). The second disjunct concerns what to do when the relevant mundane skills all seem to have been properly performed yet yield a result in apparent conflict with the domain's constitutive standards (the laws). The demand for *resilient* skill comes into play here. Laws would not allow for the intelligibility of a scientific domain if scientists were to give up on them too early. So once further testing and rechecking confirms that the various mundane skills and performances were properly done, scientists make adjustments in their skills and standards to bring them all back into line: "In contrast to rectifying particular performances, repairing and improving the skills themselves is a matter of changing how they are performed in general, altering the relevant abilities and dispositions. As resilient, objective skills must be resistant to repair, just as they are to revision, and for the same reason. Repairs, unlike revisions, are prompted not by isolated discrepancies, surrounded by results that agree, but rather by persistent and recalcitrant patterns of discrepancy. If things keep going wrong, maybe the problem is not individual errors in performance, but deficiencies in the skills themselves" (Haugeland 1998, 335). The third disjunct, which comes into play only in the face of inability to repair a persistent incompatibility among the performances, standards, and skills is to give up the entire domain of research. In section IV, I give a different account from Haugeland of just what that would mean, but additional considerations must be in place before we can grasp the significance of the difference.

Haugeland's third major contribution is to show more clearly why the constitutive standards governing scientific domains have to be laws. To understand his point, we need to go back to basics and think about the typical aims of scientific research. As Lange had pointed out, a limited range of observations must suffice to make salient a hypothesis that accurately extends more widely to unexamined cases. The accuracy of a scientific hypothesis in unexamined cases, however, subjunctively presupposes the proper performance of any assessment of that accuracy. The mundane skills that together constitute scientific conceptualization and measurement must thus be under normative control. Not just any norm will do. As Erwin Schrödinger once noted in reflecting upon the perplexities introduced by the then-new quantum mechanics, "There must still be some criterion as to whether a measurement is true or false, a method is good or bad, accurate, or inaccurate—whether it deserves the name of measurement process at all. Any old playing around with an indicating instrument in the vicinity of another body, whereby at any old time one then takes a reading, can hardly be called a measurement of this body" (1983, 158). What makes a performance a measurement is its dual normativity: the *proper* (i.e., skillful) performance of measurements normally produces the *correct* outcomes.

The crucial point leading Haugeland to the need for *laws* as constitutive standards is that only holistic sets of laws would let the two levels of normative constraint be mutually constitutive without thereby becoming conceptually empty. The constitutive standards mark out a significant and revealing pattern in the world (a "salient hypothesis that is accurate to all unexamined cases in a remarkably wide range of cases," as Lange put it in a passage I quoted previously). But such a pattern only shows up if scientists' mundane skills enable the discernment of the elements that collectively compose that pattern. The mundane skills are in turn *skills* (rather than just a strange de facto regularity in what some people, "scientists," do) because their correct exercise displays a significant and revealing pattern:

> If the independent identifiability of the elements of an orderly-arrangement pattern is problematic, and if, at the same time, the identity of a recognition pattern can be context dependent, then the one hand may wash the other. . . . [The] "elements" of an orderly arrangement need no longer be thought of as *simple* ("elementary"), like bits or pixels, or even as independently identifiable. On the contrary, they might be quite elaborate, elusive, and/or subtle—so long as some relevant creatures are (or can learn to be) able to recognize them. This recognizability, in turn, can perfectly well depend, in part, on their participation in the arrangements (= the context) of which they are elements. (1998, 275)

This mutual relation has to occur in the right way, however, on pain of triviality and consequent conceptual emptiness (in which case, the correct exercise of the skills would *not* yield a significant and revealing pattern). It is not sufficient, for example, that proper performance of the mundane skills in a scientific domain always *does* yield correct outcomes. That became clear in the previous chapter in my criticism of Hacking on self-vindication and Cartwright on the scope of scientific concepts. In their accounts, the two levels of normative accountability (proper performance of the mundane skills and correct outcomes of proper performance) collapsed into one another, such that "correct" outcomes were simply the ones that proper performance reliably produces and any phenomena outside that range of mutual adjustment were thereby excluded from the domain of the concepts applied by the skills. The collapse of the two levels into one does present a locally stable pocket of order but one that consequently did not *say* anything beyond itself. In that eventuality, we *would* have to examine everything (i.e., produce local pockets of order that included each thing) in order to know everything. To avoid this collapse of scientific concepts into emptiness, the proper exercise of the mundane skills must be able to extend beyond what the constitutive standards would permit. Properly exercised skills *would* discern an incorrect result if one *were* to occur. That they do not *and would not* occur (*if* that is so) is what makes the pattern expressed by the laws significant and revealing (i.e., having conceptual content *about* some domain).

No single dimension of measurement could mark out such a domain, for it could provide no basis for this dual normativity. One could set up standards for proper performance of the relevant skills of measurement and take their results as defining the application of a concept, but the result is then trivial. Hasok Chang (2004) nicely illustrates the relevant problem in the case of measuring temperature. Consider the "two-point method" exemplified by calibrating mercury thermometers in equal units of length between two fixed points (such as the freezing and boiling points of water):

> The procedure operates on the assumption that the fluid expands uniformly (or linearly) with temperature, so that equal increments of temperature results in equal increments of volume. To test this assumption, we need to make an experimental plot of volume vs. temperature. But there is a problem here, because we cannot have the temperature readings until we have a reliable thermometer, which is the very thing we are trying to create. If we used the mercury thermometer here, we might trivially get the result that the expansion of mercury is uniform. And if we wanted to use another kind of thermometer for the test, how would we go about establishing the accuracy of the thermometer? (Chang 2004, 59)

What Chang calls the "problem of nomic measurement" here is a special case of the more general issue of understanding the two-dimensional normativity of any empirical concept: the same data set by itself cannot specify standards for both proper performance of the relevant measurements and their correct outcome.

Haugeland's point is that one needs an interlocking set of results whose defeasible collective invariance across its domain (their "comparability" in the terms Chang takes from the French physicist Regnault) then provides a standard for determining that any results not fitting the pattern *must* be incorrect. The force of this "must" is double-edged. On the one hand, any result that does not fit into the larger pattern should be regarded as incorrect, and the procedures and performances that produced it must be corrected to arrive at the correct result—that is, one fitting the larger pattern that constitutes the lawfulness of that domain. On the other hand, an inability to revise or repair the relevant skills or standards to sustain the counterfactual stability of the pattern would in turn put the intelligibility of the entire domain into doubt. Whereas a single dimension of measurement could only trivially define a concept even when applied and measured carefully, it is a significant empirical achievement that multiple results can be successfully aligned throughout an open-ended domain. Only the reliable mutual adjustment of a plethora of skills and outcomes such that they *might* have diverged incompatibly, but *did not* do so, could allow for the nonarbitrary determination of conceptual norms whose content and application are both empirically accountable. Hence the empirically defeasible counterfactual stability of a set of laws, which collectively rules out otherwise possible events or states, is essential to any empirically answerable conceptual understanding.

III—Law-Patterns and Pattern Recognition

So far, Haugeland develops or expands upon Lange's conception of scientific laws in three ways: relocating the normativity of scientific practice from discipline-constitutive concerns to the intelligibility of domains of entities; recognizing the integral role of scientific skills and commitments in establishing and sustaining the domain-constitutive role of laws; and showing why the two-dimensional normativity of laws as empirically defeasible constitutive standards for domains of inquiry requires their counterfactual and thus alethic-modal significance. A fourth contribution requires a more striking alteration in familiar conceptions of laws.

The first step toward this reconception is the recognition that laws must be scientifically accessible. Scientific inquiry requires laws whose collective stability in some empirical domain *governs* scientific (and other) comportment toward that domain by recognizably and authoritatively revealing what can and cannot occur. Lange referred to this aspect of lawfulness as the salience of the laws from some outlook, while Haugeland characterized it as the essential intertwining of lawful patterns and scientific pattern recognition.[22] For Lange, the salience of laws from some theoretical or practical outlook plays a "justificatory" role: any body of data instantiates many patterns, most of which could not be justifiably projected onto subsequent, unexamined cases.

22. In this respect, both Haugeland and Lange are committed to thinking of laws as *scientific* laws rather than laws of *nature*. Laws of nature would not authoritatively bind scientific practice unless they were appropriately salient (recognizable) from the relevant scientific outlook.

Such projection is justified only if it is intelligible and salient what it would be for evidence from new cases to go on "in the same way" as the cases already examined. In the absence of any "justifiably" projectable concepts, however, there is also no possibility of scientific inquiry. Haugeland therefore insists that scientific pattern-recognition skills play a *constitutive* role for the intelligibility of scientific domains. The mutual intertwining of mundane and constitutive scientific skills secures the defeasible claim that there is indeed an intelligible scientific domain here at all. Mundane skills are resilient capacities for "inner" recognition of how to extend the putatively lawful pattern to previously unexamined cases, thereby enabling the appropriate application of concepts and the recognition of phenomena in their terms. Constitutive skills allow "outer" recognition that there is in fact a significant pattern discernible in some domain of events and processes, marked by the actual and counterfactual compatibility of the deliverances of what only thereby become mundane *skills*. Lange and Haugeland make what is effectively the same point. When Lange talks about the "justificatory" role of inductive salience, he is not talking about what justifies a scientific claim but instead about what makes it even a *candidate* for inductive justification. The justification would then come from the evidence actually discerned by further investigation as confirming or disconfirming the pattern inductively projected from the original data. Unless there is a projectable (conceptual) pattern displayed in some domain, however, there is nothing for evidence to bear on one way or the other.

Lange nevertheless joins much of the tradition in thinking of laws as a special kind of truth claim ("necessary" truths). Haugeland's expansion of the notion of a domain-constitutive rule shows why that is too narrow a conception. Law-statements and models articulate and express a pattern in the world, for which the skillful recognition of its counterfactual stability is an *understanding* of what that pattern encompasses. Yet such expression is only an element in the larger configuration of pattern-cum-pattern-recognition; to identify the laws with their expression in words or models is to mistake part for whole. Haugeland's reasoning for this claim is that two distinct notions of "pattern" are at work in our conception of laws, and both are needed together to account for what laws contribute to scientific reasoning and understanding. We most commonly think of laws as expressing "some sort of orderly or nonrandom arrangement—the opposite of chaos." If aspects of the world behave lawfully, there is an orderly pattern to them, regardless of whether anyone or anything notices, responds to, or understands that order. In a different sense, "patterns are 'by definition' candidates for discernment or recognition" (Haugeland 1998, 273). In this second sense, intelligibility is always intelligibility to someone or from some outlook. Haugeland's point is that these two senses of order or pattern need to function together in scientific laws and other domain-constitutive norms.

Both senses of order are needed because each by itself displays a characteristic insufficiency that the other then resolves. The notion of order or pattern as nonrandom arrangement presupposes a prior specification of the elements that are arranged in an orderly way. Yet the articulation of those basic elements is itself a form of order or pattern; hence, there is a sense in which the notion of pattern as orderly or nonrandom arrangement cannot be the most basic notion. Davidson (1984, ch. 13) cogently expressed this point long ago in his criticism of the scheme/content distinction as the "third dogma of empiricism": "We cannot attach a clear meaning to the notion of organizing a single object (the world, nature, etc.) unless that object is understood to contain or consist in other objects. Someone who sets out to organize a closet arranges the things in it. If you are told not to organize the shoes and shirts, but the closet itself, you would be bewildered" (1984, 192). The same point is also expressed in Sellars's (1997) rejection of the Myth of the Given. While his primary target was epistemological foundationalism, for which the "Given" elements of intelligibility would be experiential, the point is more general. The mere existence or presence of an entity, whether it be a sensation, a particle, a bit or pixel, or a quantum of energy, carries no normative authority. Only through its contribution to a larger, defeasible pattern do the pattern elements mean or justify anything, including any intelligible order among them. Karen Barad (2007) has recently argued in some detail that a distinctive version of this theme is integral to quantum mechanical understanding: there *are* no inherent boundaries to objects except as integral components of larger phenomena. Moreover, "phenomena" in her sense embody conceptual norms as communicable standards for the reproduction of the same phenomenon in other circumstances. In this respect, Barad's understanding of phenomena as embodying concepts is the flip side of Lange's and Haugeland's conceptions of laws as constitutive standards and/or patterns salient from an "outlook."[23]

23. For Barad, the role of a larger "outlook" or way of life in constituting the communicability and reproducibility of phenomena is expressed by noting that phenomena, as embodying a "constructive cut" between objects-in-phenomena and the "agencies of observation" to which they are manifest, have no back boundary. I discuss Barad's conception more extensively, including its implications for philosophical naturalism, in my other works (Rouse 2002, ch. 8–9; 2004; 2014a).

A conception of patterns as recognizabilia is also not sufficient by itself, for a different reason. By itself, our differential responsiveness to some features or aspects of our surroundings, even if learnable by and communicable among us, is only informative about us. It picks out/constitutes some features of our surroundings as relevant to our ongoing way of life, perhaps even to the point of being transformative of that way of life in some cases. That responsiveness then has no further significance beyond its articulation or development of our ongoing way of life as an organismic lineage (except to the extent that it, and the further capacities it enables, indirectly becomes part of the taxon-relative environment of some other organisms). Put another way, our response pattern by itself involves no *recognition* of anything but only a normal-functional responsiveness to what organisms of our kind normally do respond to (whether as an "innate" or learned pattern of response). There is and can be only one-dimensionality to such differential normativity: the normal-functionality of organism-cum-selective-environment and what deviates from it, whether those deviations are adaptive, maladaptive, or selectively neutral. It matters that these latter differences belong to a single dimension of biological normativity, since what is initially adaptive may be maladaptive in the long run; in either case, it merely changes the goal-directed pattern of the taxon to the extent that it is reproduced in subsequent generations.

Matters change fundamentally when the two senses of "pattern" function together in a more complex kind of pattern—it is this kind of worldly, two-dimensional pattern that I identify as a *conceptual* articulation of the world. We have already seen Haugeland's insistence upon this mutual reinforcement of "inner" and "outer" recognition, but we can now better understand the point. On the one hand, informative patterns in the world (i.e., conceptually articulative patterns) need not be composed of basic elements whose role as contributors to intelligibility are just "Given," brute facts. The elements of those patterns (particles, genes, chemical elements, traits, fields, strata, tectonic plates, molecules, and so much more) are what they are, as "elements" in a lawful pattern, only through their contribution to a larger arrangement and often only recognizable according to communicable standards that express their contribution to that arrangement. The arrangement itself, however, is not just a contingent fact about our discriminations and responses to our surroundings, because it is held accountable to and revisable for the sake of some further difference it makes to the larger pattern of which it is a component. As I put it in chapters 5 and 7, there is something at issue and at stake in how those distinctions and discriminations are made and revised. At the ground level, such further accountability differentiates the flexibly goal-directed capacities of other organisms from those capacities that are rightly regarded as conceptually responsive skills.

Andrea Woody's (2004a, 2004b, 2014; Woody and Glymour 2000) work on explanation in chemical practice illuminates clearly why the "laws" that do this explanatory work—which I would redescribe as conceptual-articulative work—encompass the practical skills for the discernment of the lawful pattern and its "elements." Woody is primarily concerned with the laws' explanatory function, and its importance in scientific practice, whereas I am interested in what performs that function.[24] I deliberately choose several of Woody's examples to make a further point. Woody highlights their collective challenge to familiar attempts at a general account of explanation:

> The diversity of our examples makes it implausible that we might conceive of explanatory power as a straightforward property of a theory or model. . . . In each of our examples the explanatory structures do not supply the most theoretically principled account of the phenomena available. Conceptions are idealized, quantitative relations are approximated or even transformed into qualitative relations, and information is represented in ways that exploit our visual and spatial capacities. . . . "Explanatory power" appears to be shorthand terminology that may indicate any among a vast array of distinct properties that facilitate our ability to reason with and by our theories. (2004b, 34–35)

Such a pragmatic conception of scientific understanding diverges from familiar conceptions in which the sciences aspire to transcend our all-too-human limitations in order to uncover laws *"of nature"* whose lawfulness or necessity is independent of anyone's capacities to recognize them. The conjoining of counterfactually stable patterns with the scientific skills and commitments that constitute relevant capacities for pattern recognition offers a different model for how this divergence from transcendent conceptions of natural laws does not thereby make these capacities arbitrary or anthropocentric.

Woody's first example, the ideal gas law expressed mathematically as $PV = nRT$, seems to exemplify a traditional conception of laws as general relations among a small set of variables, albeit one that holds only

24. Woody herself only describes two of the three cases as laws (the ideal gas law and the periodic law), but part of what I am doing is expanding the conception of what laws are to encompass whatever performs the explanatory role that she focuses upon, and I take her to agree that all three cases have the kind of counterfactual stability that is central to Lange's and Haugeland's accounts of laws.

approximately and ceteris paribus due to its constitutive idealizations. We already saw that this pattern does not stand alone: the van der Waals equation and various templates for more chemically specific equations of state also indicate intelligible patterns in the familiar phenomenological properties of gases. Woody's point is that the conceptual role of the ideal gas law in chemistry cannot be understood solely due to its approximate accuracy, even when conjoined with its mathematical simplicity. The idealization involved in the gas law picks out this approximation from others as conceptually significant for chemical practice and understanding: "The law [expresses] a conception of the core theoretical properties of gases. The relevance of an ideal gas model for understanding actual gas behavior would not depend directly, then, on the empirical accuracy of approximations included in the ideal gas description. Not merely an inaccurate description, the law provides selective attention to certain gas properties and their relations by ignoring other aspects of actual gas phenomena. It instructs chemists in how to think about gases *as gases*" (Woody 2004b, 21). That role alone does not account for its prominence in chemistry, however, for the conceptual role of gas behavior extends throughout a broader domain of inquiry. Gas behavior highlights the conception of chemical substances as composed of molecules whose intermolecular forces provide the basis for understanding their chemical properties. Thus Woody concludes, "Discussion [of the ideal gas law] functions not only to investigate the behavior of gases but to orient the entire field of chemistry, both conceptually and methodologically. The ideal gas law serves as a bridge between the realm of bulk substances, the traditional subject matter of chemistry, and the realm of atoms and molecules, the discipline's endorsed theoretical framework [and] provides a concrete example of how these two realms should be joined" (Woody 2004b, 24). The ideal gas law thus exemplifies what may on the surface seem to be a general descriptive equation yet whose conceptual role far outruns its descriptive application and even its approximate accuracy. Understanding the conceptual role of the gas law includes recognition of the ways in which the behavior of gases and of their constituent molecules in other phases can depart from the "ideal" expressed by the equation: "By trying to account for the failure of gases to obey Boyle's law exactly, we can learn about the size of molecules and the forces that they exert on one another" (Mahan 1975, 33, cited in Woody 2004b, 24). It also recognizes that chemists routinely take measurements of molecular properties in the gas phase as characterizing "the molecule itself," even for molecules that more commonly appear as condensed matter (e.g., Gu, Trindle, and Knee 2012).

The periodic law departs much more substantially from a conception of laws as modalized truth claims. There are some verbal expressions of the periodic law (e.g., "the properties of the elements are periodic functions of their atomic numbers" Mahan 1975, 569), yet the vagueness of that general claim contrasts strikingly with the depth and significance of the periodic relations among the elements that articulate chemical understanding. The periodic law articulates a pattern in part by suppressing confounding detail, and it does so in a different way than the gas law. The relations among the properties of elements found "nearby" on the periodic table reflect a qualitative ordering rather than an idealizing approximation of the quantitative relations among the various properties ordered. The periodic law implies no claim about any finer-grained patterns among these properties or any more-detailed parallels among the multiple properties ordered by the periodic scheme.

One might object that the periodic table nevertheless roughly or vaguely approximates a more theoretically precise scheme expressed by a quantum theoretical understanding of the electronic structure of atoms. Woody explains clearly why the periodic ordering schema remains indispensable: quantum theory treats electrons as mutually indistinguishable, whereas the periodic law articulates an ordered relationship among electron shells that are successively filled as one "moves" along the spatial rows of the table. Similarly, the vertical columns that mark the most distinctive chemical classifications expressed as a periodicity are not quantum-theoretically reducible. She comments that "the extra information in a full-blown quantum mechanical treatment, in addition to introducing electron indistinguishability, would obscure many of the patterns clearly revealed by the periodic table. The periodic law gives us 'halide' while the Schrödinger equation does not. At best we may read 'halide' into the formalism *post hoc*—something akin to seeing objects in the patterns of clouds" (Woody 2004b, 28). An understanding of atomic structure further legitimates the significance of the periodicities displayed on the table, without supplanting them or removing their distinctively qualitative and relational character.

The two-dimensional spatial display that quickly became the iconic expression of the periodic law is also indispensable to its articulative role: the ability to track the digitized spatial patterns that display these periodic relations is integral to the pattern it articulates. The periodicity expressed by the law is embedded in an understanding of how columns and rows signify conceptually significant relations. These relations in turn draw upon extensive capacities to track the quite diverse properties implicitly ordered by this spatial array, as well as to isolate the elements themselves to explore *their* constituent properties in the first place. The isolation, classification, and denumeration of chemical elements as the conceptually significant contributors to chemical materials and interactions is, after all, elegantly embedded in the tabular display in the form of conventional symbols for the elements. The periodic law is important precisely because it encapsulates in readily surveyable and comprehensible form so much about the material targets, practices, skills, and understanding that demarcate the domain of chemistry. The structure and content of this pattern as it contributes to the counterfactual stability of the laws of chemistry outruns any verbal specification of the periodic law, but it also encompasses far more than just the tabular representation of its most familiar expression.

We are accustomed to thinking of laws as *representations* of patterns that hold counterfactually. For many of the most basic and far-reaching scientific laws, however, we cannot understand their conceptual content, or what it is for that content to "hold" counterfactually, except as encompassing a larger pattern of material transformation, skillful discernment, and inferentially extended use. Their content is not self-contained as a representation whose accuracy is then assessed by other means: the counterfactually stable pattern that makes up a holistic, domain-constitutive set of laws incorporates the skills of experimental practice and the conceptual-articulative determinations of when and how to apply their characteristic expressions as laws. The reliability and resilience of scientists' mundane and constitutive skills are *part* of what it is for a set of laws to remain stable under relevant counterfactual perturbation. Absent such reliability and resilience, laws would typically display misleading instability (as putative counterexamples to the laws were too readily taken at face value) or superficial stability (as telling counterexamples to what the laws exclude are obscured or overlooked).

That one could instead formulate further truth claims about the reliability and resilience of these scientific skills might seem to be an attractive alternative that would retain a more traditional conception of laws as truth claims or their truth conditions. Incorporating those verbal formulations among the laws of a scientific domain would nevertheless mischaracterize the distinctive contribution of scientists' constitutive and mundane skills to the counterfactual stability of the overall pattern that constitutes and articulates a scientific domain conceptually. The resilience of scientific practice is not a ceteris paribus truth but a flexible skill. Scientists' skillful ability to respond appropriately to apparent violations of the laws is what sustains the counterfactual stability of the pattern to which they contribute, not the truth of claims about those skills.

This difference shows up especially clearly once we consider the modal character of the laws: resilience *is* the relevant mode of counterfactual stability for scientific skills. Under relevant counterfactual perturbations of scientific practice, scientists may well be capable of responding resiliently to apparent violations of the law. Yet the *claim* that scientists' skills are resilient is not counterfactually stable under all circumstances consistent with the laws. The resilience of scientific skills is an integral component of chemical necessity, but claims about that resilience are not chemically necessary truths. The laws of chemistry would still hold, for example, on the counterfactual (one hopes!) supposition of a cataclysm obliterating all human life, and the supposition of such a cataclysm would not relevantly change the resilience of chemical skill, but on that supposition, the truth claim about that resilience would no longer hold in the absence of the skills. The modality of the contribution of scientific skills to the counterfactual stability of the laws is affected by semantic ascent in ways that would undercut that contribution.

IV—Constitutive Commitments and Conceptual Normativity

A potentially more troubling objection arises if we juxtapose the resulting conception of laws as encompassing scientists' constitutive and mundane skills with Hacking's account of self-vindication criticized in the preceding chapter. Hacking also proposed a constitutive interconnection ("self-vindication") between experimental practice and the content and empirical accountability of scientific domains. Hacking argued that the development of theories or models in the experimental sciences typically involves a process of mutual adjustment between theoretical modeling and the instrumentally mediated data domains to which they are empirically accountable. I then argued that Hacking's proposal would untenably collapse the two-dimensional normativity that lets scientific understanding be a conceptually articulated engagement with the world. The worry I now consider is that Haugeland's implicit extension of lawful patterns to encompass their correlated skillful capacities for pattern recognition might also eviscerate the conceptual contentfulness of the laws.

This issue is in fact central to Haugeland's project. Skillful scientific responsiveness to the lawful patterns that constitute intelligible domains of entities on his account exemplifies the difference between genuinely conceptual capacities and the merely "normal" discriminative capacities of organisms, or people's capacities to participate in merely socially instituted practices. As chapter 2 showed, the "problem" with biological responsiveness or social institution is that what organisms or social instituted practices do cannot mean anything other than what such organisms normally respond to or what a particular community actually accepts as proper performance.[25] Haugeland takes scientific inquiry to differ from such cases through what I have been calling its two-dimensional normativity. For Haugeland, the difference is that conceptual capacities enable the recognition of objective features of the world rather than just a one-dimensional, biological-functional, or social-institutional responsiveness. What then accounts for and explicates this difference are the constitutive commitments noted above as the final element in Haugeland's expansion of the constitutive normativity of scientific laws. Sciences are skillful discoveries of things as they are (rather than merely what we make of them) to the extent that their practitioners can undertake the kind of dual-edged constitutive commitment described above. What is needed is a resilient effort to develop and extend the mundane skills that articulate an intelligible domain of inquiry, even in the face of apparent anomalies or other internal incompatibilities, coupled with a resolute willingness to give up the entire enterprise—including the whole domain of entities, skills, and laws that it articulates—if those efforts cannot appropriately reconcile the incompatibilities.

Haugeland was right to raise this issue about the normative authority and force of scientific skills and the lawful patterns they articulate. I take this issue to be his fifth major contribution to a reconception of scientific laws. If sciences were just activities that some historically situated cultures or social groups within them happened to engage in, or preferred to other ways of life, those practices or preferences would primarily characterize *them*, as a particular kind of organism or as a social group within that kind. The features or patterns to which scientific practices are responsive would then just be part of the way of life of those organisms or those groups in their surroundings. Haugeland argues that sciences and other characteristically human practices have a different normative character because those entire ways of life are held accountable to *objects* themselves and not merely to our conception of or behavior toward those objects. "Objects" are constituted by those practices as loci of potential incompatibility, and it is the resolute insistence not to

25. "Problem" is in scare quotes because there is nothing problematic about these features of either organismic behavior or social institutions for organisms or social groups. They are only problematic if they are mistaken for instances or models of conceptually articulated normativity.

tolerate such incompatibilities that supposedly constitutes law-governed scientific domains. We supposedly transcend the local, all-too-human-animal particularity of our species precisely through resolute and resilient determination to hold ourselves and our performances accountable to their constituted objects as objectively governing *standards*.

Here the modal character of scientific understanding intersects the more general account of conceptual normativity developed in the first part of the book. In chapter 5 (and also in Rouse 2002, ch. 7–9), I challenged Haugeland's conception of the normative authority and force of conceptually articulated domains as dependent upon a first-person, "subjective" commitment to hold one's performances accountable to objects as constituted by those very performances. His account of existential commitment as a self-binding taking-over of responsibility for the intelligibility of an entire conceptual domain failed to account for the normative force of scientific intelligibility. As Kierkegaard (1954, 203) tellingly put it, self-binding commitments have no more authority than would a monarch in a country where revolution is legitimate. Nor did this failure merely display inadequacies in Haugeland's arguments. The central theme of my previous book (Rouse 2002) was that philosophical debates about naturalism have been framed by underlying conceptions of nature and of ourselves as knowers and concept users that make those debates unresolvable. Conceptions of nature as anormative, and of the normative authority and force of concepts and knowledge claims as somehow introduced or imposed by us (e.g., by our existential commitments on Haugeland's version), make intractable how normative authority could intelligibly bind us as natural beings.

This book's account of conceptual normativity as a form of biological niche construction resolves those problems by showing how conceptual normativity is authoritative, as integral to our inherited developmental and selective environment, in a way that is temporally open-ended. Through mutual dependence upon one another and our surrounding environment, we are accountable to what is at issue and at stake in the practices that continually reconstitute our biological niche as a conceptually articulated space of reasons. The two-dimensional normativity of our practices and our biological way of life allows a biological goal-directedness toward both *whether* and *how* our way of life will maintain itself. The "objects" to which we are thereby accountable, anaphorically characterized as what is at issue and at stake in our environmentally situated way of life, are the configuration of that way of life as directed toward intelligible possibilities. Their normative force arises in the same ways in which other organisms are dependent upon and responsive to their environments but with this uncharacteristic two-dimensional articulation as temporally extended possibilities.

Such temporal extension shows up in the recognition of scientific understanding as situated in ongoing research. Research traditions are configured conceptually by their own historically shaped direction toward further issues and possibilities. The remaining chapters of the book further consider how the temporality of conceptual normativity is evident within scientific practice. Chapter 9 addresses the initial opening of scientific domains as intelligible fields of research possibilities. The "fictional" constitution of a law-governed conceptual domain is the scientifically specific form of what Kukla and Sellars conceived as the "mythical" constitution of normative authority, which was discussed in chapter 5. Haugeland's account of existential commitment has no obvious way to account for how law-governed domains are initially constituted, except by appealing to a willful commitment.

Haugeland's account is even more clearly incompatible with the futural dimension of temporally extended conceptual normativity and hence with important features of scientific practice. What Haugeland cannot readily account for in this respect are the selective priorities of scientific research and the consequent complexity of its conceptual norms. Chapter 10 takes up this issue, but for current purposes, the point can be seen at two levels. On the one hand, not all aspects of the world are scientifically significant. Haugeland's account of conceptual normativity seems bereft of resources for understanding why some object domains serve as standards appropriately governing a domain of research practices and others do not.[26] On the other hand, an appeal to a constitutive, "existential" commitment to objective authority cannot account for the detailed articulation of standards for scientific understanding. Chapter 7 showed that scientific concepts are always open to further intensive and extensive articulation. Sciences vary in the extent to which their concepts require more fine-grained intensive or extensive articulation, such that the conceptual adequacy of the field is at issue

26. Haugeland could argue that the present configuration of scientific fields to some extent reflects actual successes in demarcating domains of law-governed invariance; or he might instead argue (extending his partial appeals to Popperian falsification as a predecessor) that there are no philosophical constraints upon which scientific domains to pursue, only a demand for resilient, resolute accountability to objects within whatever domains scientists choose to investigate. Neither approach, I would argue, comes close to capturing adequately the ways in which scientific inquiries have been prioritized, pursued, or focused within specific disciplinary domains, although I will not try making that historically detailed argument here.

in its further development. Haugeland's insistence that one must either revise or repair discrepancies within a conceptual field, or else give up the entire domain, does not readily allow for such variance.

A failure to capture the selective and differential significance of scientific work also "reflexively" shows up in Haugeland's own appeal to objectivity to express what is at stake in sustaining a scientific practice.[27] Haugeland's specific account of the normative authority of scientific understanding, as *objective* accountability to the world, is a synecdoche: it conflates one particular answer to the question of how and why scientific understanding matters with the configuration of the whole conceptual field.[28] In appealing to the objectivity of conceptual norms, he draws upon an influential and constructively important aspect of our conceptual heritage, to which I also appealed in a more limited way in chapter 5. The adequacy of that answer remains seriously contested in ways that leave the question of how scientific understanding matters still partially up for grabs. To see that the issue remains unsettled here, we only need consider some vigorous defenses of alternative conceptions: Nancy Cartwright (1999, 2007) or Richard Rorty (1991, 1998, 2007) appeal to the same practices and history to argue for abandoning any quest for objectivity and transcendence in favor of the amelioration of human life; Jürgen Habermas (1971a, 1971b, 2003), Philip Kitcher (2001), and Sheila Jasanoff (2005) argue that scientific inquiry and its constitutive standards should be accountable within a broader democratic politics;

27. Optical metaphors of reflection have played a familiar role in philosophical accounts for how the conceptual "illumination" of some domain of objects also bears on its own self-understanding. Barad (2007, especially ch. 2) has argued, rightly, that diffraction provides a better metaphoric resource for this purpose. A geometrical optics of reflection does not take into account the *material* intra-action of light and matter, and the reflective metaphor thereby mistakenly encourages overlooking the materiality of our conceptual-articulative practices, a central theme of my account in this book. I continue to use the word 'reflexive' rather than speaking of the diffractive insights gained by thinking of Haugeland's account in his own terms for two related reasons. First, I have nothing substantial to add to Barad's insightful discussion of why diffraction provides a better conceptual resource for this borrowing from physics. I could not use the term 'diffraction' here without having to explain it, and a detailed explanation would simply recapitulate Barad's point while also digressing from mine. Second, retaining the term is appropriate once one has grasped Barad's point, since there is no relevant *contrast* between reflection and diffraction. The reflection of light by matter *is* a diffraction intra-action, and as I briefly noted in chapter 7, the geometrical optics of reflection must therefore be recognized as a misconception of the entire optical domain, despite its residual utility for some purposes. Hence I take Barad to have shown what a more adequate grasp of a "reflexive" understanding would involve while also invoking my own further elaboration in chapters 7 and 9 of why conceptual articulation cannot be merely intralinguistic.

28. This conflation reenacts a familiar pattern. Sellars's and other identifications of a scientific understanding of the world with a particular position in the space of reasons, rather than with the conceptual configuration of the entire space, is also a synecdoche. See chapter 6 for further discussion.

and in some cultural and religious contexts, others offer very different arguments for the subordination or rejection of scientific practices and norms in deference to faith in a divine order. My point is not to defend any of these views, nor to advocate a relativistic indifference among them, but only to recognize that the standards appropriate for assessing scientific significance belong to a contested conceptual field with live issues and stakes. A philosophical account of conceptual understanding in science must recognize that scientific practices *and our philosophical conceptions of them* are situated within those ongoing debates and cannot be an external authority or trump card that might settle them preemptively. Scientific research and its constitutive standards and norms are not indifferent to how those debates develop and set the context for the conduct of further scientific work.

NINE

Laboratory Fictions and the Opening of Scientific Domains

I—Introduction

This part of the book develops an alternative conception of what Wilfrid Sellars called "the scientific image" as an idealized composite of what the sciences achieve. On this alternative conception, sciences belong to the discursive practices through which the world is conceptually articulated. Discursive practices sustain what Sellars called the "space of reasons," within which we can talk about, act upon, recognize, and reason about aspects of the world in normatively accountable ways. Sellars took the scientific image to be a position within that space, a composite, more or less unified representation of the world whose empirical success and explanatory power confer a comprehensive epistemic authority: "In the dimension of describing and explaining the world, science is the measure of all things, of what is that it is, and of what is not that it is not" (1997, 83). I argue instead that scientific practices continually reconfigure the space of reasons itself from "within," without needing to converge on a determinate "image" within it. The sciences open, sustain, and often expand the range of intelligible possibilities for describing, explaining, and acting within the world, including possibilities for intelligible disagreement and practical conflict. They also refine that space by correcting or replacing patterns of discursive interaction that do not stand up to ongoing critical reflection.

The Sellarsian space of reasons is often misconceived as an ethereal region of claims and counterclaims, a linguistic or thinly social space of reasoning distinct from the causal events in the world that we often reason about. Such misconceptions are more difficult to sustain after recognizing, as chapters 7 and 8 showed, that the material practices of experimentation and other forms of empirical research are integral to the space of reasons. Scientists must act upon and within the world to let it be intelligible conceptually, and instruments, experimental systems, and skills help sustain the intelligibility and the content of scientific claims and reasoning. It would still be a misconception, however, to think that experimental practices add worldly substantiality to an otherwise ideal or ethereal space of disembodied propositions or to a domain of representations "in the head" or "in language" that is even notionally distinct from the world "outside" language. As chapters 3 and 4 showed, language itself is a worldly phenomenon that only exists through its continual reproduction as materially part of the biological environment in which human beings develop and reproduce. Scientific practices, and the more extensive field of discursive practices to which they belong, *articulate* the world conceptually rather than representing it. A more subtle misconception would then contrast articulation "from within" the world and representation "from without." Spatiality is itself at issue in the contrast between articulation and representation; we do better to think of the world as also spatially articulated in part through scientific and other conceptual practices.

This chapter explores further how the sciences articulate the world conceptually by considering how scientific research opens and sustains conceptual domains. Chapter 2 showed that conceptual articulation allows for intentionally directed comportments, which are normatively accountable in two distinct but interconnected dimensions. Conceptual understanding involves an active capacity to track and adjudicate performances within a social practice, both for their appropriateness and significance within the practice and for their broader practical and perceptual significance. For declarative linguistic performances, these two dimensions correspond to their meaningfulness and their justification or truth;[1] for equipment, two-dimensionality connects the appropriateness

1. Whether justification or truth is the primary measure of success for declarative utterances along that dimension depends upon one's other philosophical commitments. Kukla and Lance (2009) remind us, however, that declarative utterances are far from exhaustive of discursive practice,

and purposiveness of its use and the resulting success or failure with respect to what is at issue and at stake in its use.[2] The two-dimensionality of such comportments enables a possible divergence between how one takes some aspect of the world to be and how it shows itself in response to one's engagement with it.

Philosophical interest in the sciences often compresses these two dimensions. One-dimensional assessments of the justification or truth of scientific claims subordinate the articulation of conceptual domains, which establishes and refines intelligible possibilities for making and justifying claims.[3] A telling example of such compression is that philosophical conceptions of science have sometimes been troubled by the recognition that scientific theorizing often departs from veridical representation of the world. Scientific conceptualizations are replete with idealizations, approximations, ceteris paribus clauses, metaphors, and other figurative expressions and even characterizations of entities well understood not to exist.[4] Many philosophers of science have been especially troubled by fictional scientific representations or the possible assimilation of idealizations and approximations to fictions (Suárez 2009a). Advocates of instrumentalism, fictionalism, pragmatism, social constructivism, and other sophisticated antirealist conceptions of the aims of scientific theorizing instead embrace such "trouble." These antirealists nevertheless also suppress the two-dimensionality of scientific conceptual articulation. They substitute a more proximate goal, such as empirical adequacy or prediction and control, for a realist insistence upon scientific aspirations to truth as correspondence. Both realists and antirealists, however, focus on the fulfillment of their proposed goals

even in the sciences and philosophy, which are often presumed to aim solely at producing declarative knowledge claims. Scientific papers often invite consideration of alternative hypotheses, remind readers of relevant background considerations, question familiar assumptions, and so forth. For reasons of brevity I note only that these alternative pragmatic performances also have their own two-dimensionality, which concerns the appropriateness, meaningfulness, or relevance of the performances themselves and their success or failure in fulfilling their characteristic pragmatic aims.

2. We normally speak of the purposes or goals of equipment use rather than what is at issue and at stake there since the proximate goals of equipment use are often taken for granted at the time. The question of just what one is doing or trying to accomplish in using equipment can nevertheless arise, even when one least expected it.

3. Logical empiricist philosophers of science often did address scientific conceptual articulation but did so in problematic ways that could not readily reconcile the roles they ascribed to linguistic frameworks and to supposedly "Given" empirical content. Subsequent work in the philosophy of science has too often presumed that conceptual development is intralinguistic.

4. A classic example of scientific models that knowingly refer to and describe nonexistent kinds of entities (rather than just idealized properties or behavior of extant kinds) are the "silogen" atoms that average over properties of hydrogen and silicon atoms in some materials science models of semiconductor materials. For discussion, see Winsberg (2010).

rather than the expansion and reconfiguration of the possibilities with respect to which those goals can be specified and pursued.

This compression of the two-dimensional normativity of scientific understanding, with consequent inattention to the sciences' contribution to conceptual articulation, are thus manifest both in philosophical anxieties about "fictional" contributions to scientific understanding and in antirealists' countervailing embrace of those anxieties. Typically, conceptions of "fictions" and fictional contributions to scientific understanding also betray an inappropriately narrow conception of scientific accomplishments. Reconsidering what we should mean by "fictions" and how they contribute to scientific understanding not only broadens our grasp of scientific achievements and what different aspects of scientific work contribute to them but also allows us to gain a better grip on what it means to articulate the world conceptually and how such articulation transforms the world.

II—Fictions and Scientific Understanding

Philosophers who consider fictions or fictional representations in science typically address some canonical cases: idealized models, simulations, thought experiments, or counterfactual reasoning. The philosophical issues raised by these cases may also seem straightforward. Sciences aim to discover actual structures and behaviors in the world and to represent and understand them accurately. Scientific fictions provoke the question of how fictional representations or, more provocatively, misrepresentations could contribute to understanding how the world actually is.

I take up the role of fictions in science to address a different issue, concerning conceptual meaning and significance rather than either truth and falsity or justification. These issues are closely connected, but it is a mistake to conflate them. Nancy Cartwright collected her important early essays under the provocative title *How the Laws of Physics Lie* (1983). Laws lie, she argued, because they do not accurately describe real situations in the world. Descriptions of actual behavior in real situations would require supplementing the laws with more concrete models, ad hoc approximations, and ceteris paribus provisos. Cartwright suggested that the fictional character of physical laws is analogous to literary and, specifically, theatrical fiction; like film or theatrical productions, the genre of physical law demands its own fictive staging. I argued at the time (Rouse 1987, ch. 5) that Cartwright had mischaracterized the import of her concerns. Her arguments would challenge the *truth* of law statements only if their *meaning* were fixed in ways at odds with how scientists actually use such expressions in practice. The need for models, provisos, and ad hoc approximations to describe the actual behavior of physical systems in theoretical terms is no surprise to physicists. The models were integral to their education in physics, and the open-ended provisos and approximations needed to apply them were implicit in their practical grasp of the models. The "literal" interpretation of the laws that Cartwright once took to be false thus does not accurately express what the laws mean in scientific practice.[5]

The previous chapter already developed this point in more detail. The necessity expressed by scientific laws is entangled with the normativity of scientific practices. What an expression such as $F = ma$ or one of Maxwell's equations says is a normative matter, expressing a connection between how and when it is appropriately employed in scientific practice and the appropriate consequences of its employment.[6] To that extent, understanding laws and other verbal or mathematical expressions cannot be easily disentangled from understanding the circumstances to which they apply. As Donald Davidson noted, we thereby "erase the boundary between knowing a language and knowing our way around in the world generally" (2005b, 107).

Cartwright (1999) now also recognizes trade-offs between truth and meaning. What concerns her more recently is not the laws' truth but their scope: which events they are informative about and accountable to. Moreover, she equates the scope of the laws with the scope of their concepts. The laws of classical mechanics, for example, apply wherever the causal capacities that affect motions are appropriately characterized as "forces." My concern here is not Cartwright's answers to the questions of which circumstances fall within the domains of scientific concepts. I only insist that questions of meaning and of truth must remain

5. Cartwright (1999) now recognizes that what was at issue in her concerns about the accuracy of the laws is not their truth but their meaning (and more specifically, the scope of their application). Her revised view still differs from mine in at least two important respects, however. First, her account relies upon only one of two aspects of conceptual content (Dummett 1975; Brandom 1994, 117–18): she determines the scope of laws from the evidential circumstances of their application, without regard to the connection between application and inferential consequences. Second, she takes the empirical adequacy of the models as the only relevant criterion for assessing the applicability of laws, whereas I think empirical adequacy is one among multiple relevant considerations. See Rouse (2002, 319–34).

6. I adapt this two-dimensional account of conceptual meaning from Brandom 1994, who in turn adapted it from Dummett 1975.

distinct even though interconnected. We cannot ask whether a theory, a law, or any other hypothesis is true without some understanding of what it says and to which circumstances it should apply.

This chapter takes up conceptualization and meaning by exploring a different kind of phenomenon than the canonical "scientific fictions." Its focus is the development and exploration of laboratory "micro-worlds" (Rouse 1987) or "experimental systems" (Rheinberger 1997). "Micro-worlds" are "systems of objects constructed under known circumstances and isolated from other influences so that they can be manipulated and kept track of, . . . [allowing scientists to] circumvent the complexity [with which the world more typically confronts us] by constructing artificially simplified 'worlds'" (Rouse 1987, 101). Illustrative experimental microworlds include the Morgan group's system for mapping genetic mutations in *Drosophila melanogaster*, the many setups in particle physics that direct some form of radiation toward a shielded target and detector, or the work with alcohols and their derivatives that marked the beginnings of experimental organic chemistry (Klein 2003). These are not verbal, mathematical, or pictorial representations of some actual or possible situation in the world. They are not even physical models, like the machine-shop assemblies Watson and Crick manipulated to represent three-dimensional structures for DNA. They are instead novel, reproducible arrangements within the world.

Associating experimental systems with the canonical scientific fictions may seem strange. Philosophical discussions of scientific fictions normally take experimentation for granted as well understood. The question can then be raised whether thought experiments or computer simulations relevantly resemble experimental manipulations as "data-gathering" practices.[7] I proceed in the opposite direction; I ask whether and how the development of experimental systems resembles the formulation and use of the canonical "scientific fictions." The issue is not whether simulations or thought experiments contribute data but whether and how laboratory work joins thought experiments in articulating and consolidating conceptual understanding. Chapter 7 challenged philosophers' tendency to exclude experimentation from processes of conceptual development. To caricature a complex tradition, the logical empiricists confined experimentation to the context of justification rather than discovery; scientific realists and other metamethodological postempiricists emphasized that experimentation *presupposes* prior theoretical articulation of concepts;

7. See Humphreys (2004), Hughes (1999), Norton and Suppe (2001), and Winsberg (2003, 2010).

and reaction to the excesses of both traditions insisted that experimentation has a life of its own apart from developing or testing concepts and theories. None of these traditions said enough about experimental contributions to conceptual articulation.

As background to examining experimental systems, consider briefly Kuhn's classic account of the function of thought experiments. Thought experiments become important when scientists "have acquired a variety of experience which could not be assimilated by their traditional mode of dealing with the world" (Kuhn 1977, 264). By extending scientific concepts beyond their familiar uses, he argued, thought experiments *bring about* a conceptual conflict rooted in those traditional uses rather than *finding* one already implicit in them. Kuhn insisted upon that distinction because he took the meaning of concepts to be open textured rather than fully determinate. By working out how to apply these concepts in new, unforeseen circumstances, thought experiments retrospectively transformed their use in more familiar contexts, rendering them problematic in illuminating ways.

Thought experiments could only play this role, however, if their extension to the newly imagined setting genuinely extended the original, familiar concepts. Kuhn consequently identified two constraints upon the imaginative extension of scientific concepts, "if it is to disclose a misfit between traditional conceptual apparatus and nature." First, "the imagined situation must allow the scientist to employ his usual concepts in the way he has employed them before, [not] straining normal usage" (1977, 264–65). Second, "though the imagined situation need not be even potentially realizable in nature, the conflict deduced from it must be one that nature itself could present; indeed, . . . it must be one that, however unclearly seen, has confronted him before" (1977, 265). Thought experiments, that is, are jointly parasitic upon the prior employment of concepts and the world's already-disclosed possibilities; like Davidson, Kuhn found it hard to disentangle our grasp of concepts from "knowing our way around in the world" more generally. Against that background, thought experiments articulate concepts by presenting concrete situations that display differences that are intelligibly connected to prior understanding. In Kuhn's primary example, the difference between instantaneous and average velocity only becomes conceptually salient in circumstances where comparisons of velocities in those terms diverge. My question, however, is how scientific concepts come to have "normal usage" in the first place, not merely describing what actually happens but providing a grasp of the possible situations "that nature itself *could* present" (my emphasis).

The novel circumstances of experimental systems or thought experiments are important because they make salient conceptually significant differences that do not show themselves clearly in more "ordinary" circumstances. Experimental systems can also play a pivotal role in making possible the conceptual articulation of a domain of phenomena in the first place. A postempiricist commonplace rejects "Whig" histories of science that narrate relatively seamless transitions from error to truth. In many scientific domains, however, earlier generations of scientists could not have erred because the relevant errors were not yet even conceivable. In the most striking cases, scientists' predecessors either had no basis whatsoever for making claims within a domain or could only make vague, unarticulated claims. In Hacking's (2002, ch. 11) apt distinction, they lacked not truths but possibilities for truth or error: they had no way to reason about such claims and thus could not articulate claims that were "true-or-false." Allowing new aspects of the world to show up as conceptually articulable is thus a distinctive feat of laboratory and other experimental sciences.

III—Phenomena and Conceptual Articulation

Chapter 7 called attention to Hacking's suggested parallel between creating experimental phenomena and discovering scientific laws as integral to how scientists come to "know their way around in the world generally": "In nature there is just complexity, which we are remarkably able to analyze. We do so by distinguishing, in the mind, numerous different laws. We also do so by presenting, in the laboratory, pure, isolated phenomena" (1983, 226). Phenomena in Hacking's sense are events in the world rather than appearances to the mind, and for the most part, scientifically informative phenomena are now created in the laboratory rather than found in the world. Experimental work does not strip away confounding complexities to reveal underlying nomic simplicity; it creates new complex arrangements as indispensable background to any foregrounded simplicity. Most philosophical readers have not taken Hacking's parallel between phenomena and laws as modes of analysis sufficiently seriously. We tend to think only laws, models, or theories analyze and enable understanding of nature's complex occurrences. Creating phenomena may be an indispensable means to discerning relevant laws or constructing illuminating theories but could only indicate possible directions for analysis, which must be then worked out theoretically. Limiting laboratory phenomena in this way as mere means to the verbal or mathematical articulation of theory is nevertheless a mistake, even if one acknowledges that experimentation also has ends of its own. Experimental practice is integral to conceptual understanding, and not merely instrumental.

As created artifacts, laboratory phenomena and experimental systems have distinctive purposes. Most artifacts, including the apparatus within experimental systems, are used to accomplish some end. The goal of an experimental system itself, however, is not what it does but what it shows. Experimental systems are novel rearrangements that allow some aspects of the world that are not ordinarily manifest and intelligible to *show* themselves clearly and evidently. Such arrangements may isolate and shield relevant interactions or features from confounding influences. They also introduce signs or markers into the experimental field, such as radioactive isotopes, genes for antibiotic resistance in the presence of antibiotics, or correlated detectors for signals whose conjunction indicates events that neither signifies alone. Creating experimental phenomena reverses the emphasis from traditional empiricism: what matters is not what the experimenter observes but what the phenomenon shows.[8]

We have already encountered Catherine Elgin's (1991, 1996, 2009) distinction between the features or properties an experiment *exemplifies* and those that it merely *instantiates*.[9] In her example, rotating a flashlight ninety degrees instantiates the constant velocity of light in different inertial reference frames, whereas the Michelson-Morley experiment exemplifies it.[10] Elgin thereby emphasized the symbolic function of experimental performances and the parallels between their cognitive significance and that of paintings, novels, and other artworks. A fictional character such as Nora in *A Doll's House* strikingly exemplifies a debilitating situation that the lives of many women in traditional marriages have merely instantiated. A well-formed fly leg where an antenna would normally grow similarly exemplifies the modularity of development. Elgin nevertheless distinguished scientific experimentation from both literary and scientific fictions. An experiment actually instantiates the features

8. The point of the contrast is not to excise the experimenter or her perceptual capacities but to emphasize that experiments arrange meaningful, discernible patterns in the world. To the extent that the experimenter is involved, it is as agent and skillful participant rather than external, passive observer.

9. See chapter 7.

10. Although I will not belabor the point here, it is relevant to my subsequent treatment of experimental systems as "laboratory fictions" that, strictly speaking, the Michelson-Morley experiment does not instantiate the constant velocity of light in different inertial frames because the experiment is conducted in a gravitationally accelerated rather than an inertial setting.

it exemplifies, whereas thought experiments and computer simulations join many artworks in exemplifying features that they only instantiate metaphorically.

Elgin's distinction between actual experiments and fictional constructions prioritizes instantiation over exemplification. Nora's life is fictional and thus only metaphorically constrained, whereas light within the Michelson interferometer really does travel at constant velocities in orthogonal directions, and homeotic mutants really do grow appendages in the "wrong" place. Thought experiments, computer simulations, and novels are derivative, fictional, or metaphorical exemplifications because exemplifying a conceptually articulated feature requires instantiating that feature. The feature is already 'there,' awaiting only the articulation of concepts that allow it to be recognizable. Unexemplified features of the world that thereby remain unconceptualized would be like the statue of Hermes that Aristotle said exists potentially within a block of wood, awaiting only the sculptor's or scientist's trimming away of extraneous surroundings (Aristotle, *Metaphysics* IX, ch. 6, 1048a).[11]

In retrospect, with a concept clearly in our grasp (or better, with ourselves already gripped by what is at issue in its application), the presumption that the concept applies to already-extant features of the world is unassailable. There were mitochondria, spiral galaxies, polypeptide chains, and tectonic plates before anyone discerned them or even conceived their possibility. This retrospective standpoint, where the concepts have already been developed and the only question is where they apply, nevertheless crucially mislocates important aspects of scientific research. In Kantian terms, researchers initially seek reflective rather than determinative judgments. Scientific research must articulate concepts with which the world can be perspicuously described and understood rather than simply apply those already available. To be sure, conceptual articulation does not begin de novo but extends a prior understanding that indispensably guides inquiry. Yet in the sciences, such prior articulation is tentative and open textured, at least in those respects that the researcher aims to explore.

11. Hacking's initial discussion of the creation of phenomena criticized this very conception of phenomena as implicit or potential components of more complex circumstances: "We tend to feel [that] the phenomena revealed in the laboratory are part of God's handiwork, waiting to be discovered. Such an attitude is natural from a theory-dominated philosophy. . . . Since our theories aim at what has always been true of the universe—God wrote the laws in His Book, before the beginning—it follows that the phenomena have always been there, waiting to be discovered. I suggest, in contrast, that the Hall effect does not exist outside of certain kinds of apparatus. . . . The effect, at least in a pure state, can only be embodied by such devices" (Hacking 1983, 225–26).

The dissociation of experimental work from conceptual articulation reflects a tendency to think of conceptual development as primarily verbal, a matter of gaining inferential control over the relations among our words. We have already seen that tendency encapsulated in Quine's (1953, 42) influential images of scientific "conceptual schemes" as self-enclosed fabrics or fields that accommodate the impact of unconceptualized stimuli at their boundaries solely by internal adjustments in the theory. Both Donald Davidson (1984) and John McDowell (1994) have criticized the Quinean image, arguing that the conceptual domain is unbounded by anything "extraconceptual." I agree. Reflection on the history of scientific experimentation nevertheless strongly suggests the inadequacy of Davidson's and McDowell's own distinctive ways of securing the unboundedness of the conceptual.[12] Against Davidson, that history reminds us that conceptual articulation is not merely intralinguistic, as we saw in chapter 7.[13] Against McDowell, the history of experimentation reminds us that conceptual articulation incorporates causal interaction with the world and not just perceptual receptivity.

Both points are highlighted by a series of examples in which experimentation opened whole new domains of events to conceptual understanding where previously there was, in Hacking's apt phrase, "just complexity." Wilhelm Johannsen introduced the word 'gene' in 1905 for a hypothetical entity. There was not yet a recognized region of distinctively genetic phenomena, however, before the Morgan group's correlations of crossover frequencies with variations in chromosomal cytology (Kohler 1994). "Genes" were postulated elements in theories of heredity rather than a distinctive domain of phenomena open to experimental exploration and understanding.[14] From a later biological perspective, what we now distinguish as genetics and development were thoroughly entangled in early twentieth-century conceptions of heredity.

In a similar way, the intercalibrated practices of thermometry enabled the differentiation of temperature from quantity of heat by establishing

12. I have developed these criticisms of Davidson and McDowell more extensively in my previous book (Rouse 2002), as well as in chapters 5 and 7.

13. Davidson (2001) would also argue that the "triangulation" involved in the interpretation and articulation of concepts is not merely intralinguistic. He nevertheless sharply distinguishes the merely causal prompting of a belief from its rational, discursive interpretation and justification. For discussion of why his view commits him despite himself to understanding conceptual articulation as intralinguistic, see chapters 2 and 6 of Rouse (2002).

14. Mendel's celebrated experiments with pea plants only displayed discrete, phenotypic differences that were heritable in integer ratios across generations. Genes were posited to explain such heritable patterns, but classic Mendelian experiments did not make genes themselves manifest except indirectly through associated patterns in phenotypes.

norms for the quantitative identification of temperature differences on a common scale (Chang 2004). Henrietta Leavitt's and Harlow Shapley's tracking of period-luminosity relations in Cepheid variable stars likewise brought interstellar relations within the scope of quantitatively differentiated distances. The deployment of the ultracentrifuge and the electron microscope not only gave further articulation to the structure of cells but more important connected the spatial differentiation of cells' physical components with their functional significance for cellular life (Bechtel 1993; Rheinberger 1995). The use of radioactive decay products to probe seemingly indivisible and unarticulated atoms, most strikingly in Rutherford's targeting gold leaf with beams of alpha particles, disclosed the subatomic domain as accessible and comprehensible. These aspects of the world had previously been less ineffable than the "absolute, unthinkable, and undecipherable nothingness" that Hacking (2002, ch. 6) memorably ascribed to anachronistic human kinds. They nevertheless lacked the articulable differences that sustain conceptual understanding and enable its further development. What changed the situation was not just new kinds of data, or newly imagined ways of thinking, but new interactions that articulated the world differently. For example, almost anyone in biology prior to 1930 would have acknowledged that cellular function must involve a fairly complex internal organization of cells for them to perform their many roles in the life of an organism. Nevertheless, such acknowledgment was inevitably vague and detached from any determinate understanding, and from any intelligible program of further research, apart from visual identification of some static structures such as nuclei, cell walls, mitochondria, and in vitro exploration of a few biochemical pathways. Without further *material* articulation and rearrangements of cellular components, there was little one could say or do about the integration of cellular structure and function. Such material articulation was integral to the introduction, stabilization, and inferential deployment of descriptive terminology for functionally significant structures and structurally situated functional processes.

The construction of experimental microworlds thus plays a distinctive and integral conceptual role in the sciences. Heidegger, who was among the first to ascribe philosophical priority to scientific research over the retrospective assessment of scientific knowledge, forcefully characterized the role I am attributing to some experimental systems: "The essence of research consists in the fact that knowing establishes itself as a 'forging-ahead' *(Vorgehen)* within some realm of entities in nature or history. . . . Forging-ahead, here does not just mean procedure, how things are done. Every forging-ahead already requires a circumscribed domain in which it moves. And it is precisely the opening up of such a domain that is the fundamental process in research" (Heidegger 1950, 71; 2002, 59, translation modified). The creation of laboratory microworlds is often indispensable to opening domains in which scientific research can proceed toward articulated comprehension of circumscribed aspects of the world.

IV—Laboratory Fictions

What does it mean to open up a scientific domain, and how do experimental systems accomplish it? Experimental systems always have a broader intentional significance. Biologists speak of the key components of their experimental systems as model organisms, and scientists more generally speak of experimental models. The crossbreeding of mutant strains of *Drosophila* with stock breeding populations, for example, was not of interest for its own sake, but it was also not merely a peculiarity of this species of fruit fly. The *Drosophila* system was instead taken to show something of fundamental importance about *genetics* more generally and helped constitute genetics as a research field. Elgin already called attention to the symbolic role of experiments in exemplifying rather than just instantiating some of their features. Domain-opening experimental systems "exemplify" in a stronger sense because they help open a conceptual space within which their own features can show up intelligibly.

We can extend Elgin's parallel between experimentation and aesthetic understanding by considering a sense in which domain-constitutive experimental systems are "laboratory fictions." It is no objection to this claim that laboratory systems are existing phenomena in their own right. The symbolic role of experiments as exemplifications already shows that laboratory phenomena are intentionally directed beyond themselves. Yet talking about actual laboratory settings as "fictions" may still seem strange, even when their creation and refinement are unprecedented. By definition an experimental setup is an actual situation, in contrast to thought experiments or those theoretical models that represent idealized or even impossible situations. Ideal gases, two-body universes, "silogen atoms" (Winsberg 2010), or observers traveling alongside electromagnetic waves nicely exemplify the familiar contrast between fictional representations and experimental facts. Even among this class of theoretical representations, Winsberg argued persuasively that some idealized constructions, such as ideal gases or frictionless planes, serve as reliable guides to an actual domain and hence should not be regarded as fictional representations, in contrast to fictional posits such as silogen atoms, whose functional role in a larger representational system requires giving up their more localized representational reliability.

Winsberg's distinction between fictional and nonfictional elements of theoretical models turns on how these theoretical constructions fulfill a representational function. Yet representation is not the sole or even the primary function of scientific work. Scientific work also has a discursive function that articulates conceptual patterns. I have argued elsewhere (Rouse 2002, ch. 5–9) that scientific models and theories only acquire a representational function through their articulative and inferential role in discursive practices. My (1987) criticism of Cartwright's early interpretation of fundamental laws as fictions also situated the representational role of laws within the context of their discursive use within scientific practice. When we address this issue from the other direction, moreover, we can see that serious scholarly discussion of literature nowadays likewise gives priority to the discursive role of literature, including literary fictions.[15] We understand literary fiction too narrowly in thinking of fictions as imaginary constructions that represent nonexistent situations. Such representationalist conceptions do not take account of how fictional work draws upon, plays with, and is ultimately accountable to the larger material-discursive setting within which it works. Fictional writing explores the world by working out discursive variants that disclose and foreground some aspects of it. Such writing is not self-contained; its effects draw upon and are understood against the background of actual settings and practices, including other textual constructions. Moreover, its significance arises from what it can disclose beyond its own discursive construction. Fictional writing and experimental science draw upon different resources, and work in different ways, disclosing different aspects of the world within which they are situated. For some scientific and philosophical purposes, it is appropriate to take for granted the discursive context that lets a representational relation between a theoretical model or concept and some situations in the world seem transparent. This chapter instead highlights aspects of scientific practice that condition such representational treatments of scientific work.

15. For a useful discussion of how and why philosophical conceptions of literary language need to take into account the conceptions of discourse that now guide contemporary literary theory, see Bono (1990).

Two interconnected features of experimental systems that bring an entire field of phenomena "into the open" or "into the space of reasons" guide consideration of them as laboratory fictions in instructive parallel to the literary concept. First is the systematic character of experimental operations. Fictions in the relevant sense are not just any imaginative construction but have sufficient self-enclosure and internal complexity to constitute a situation whose relevant features arise from their mutual interrelations. Second, fictions thereby constitute their own "world." The point is not that fictional constructions do not resemble or otherwise correspond to anything outside the situations they construct; many fictional constructions explicitly invoke "real" settings. Even then, however, they constitute a world internally rather than by their external references. In this sense, we can speak of Dickens's London in much the same way that we speak of Tolkien's Middle Earth.

My first point, that scientific experimentation typically requires a more extensive experimental system and not just individual experiments, is now widely recognized in philosophy and science studies, albeit in ways that limit its import.[16] Ludwik Fleck (1979) was among the first to highlight the priority of experimental systems to experiments, and most later discussions follow his primary concern with the *justificatory* importance of experimental systematicity: "To establish proof, an entire system of experiments and controls is needed, set up according to an assumption or style and performed by an expert" (Fleck 1979, 96). Hacking's (1992) discussion of the "self-vindication" of the laboratory sciences also concerned epistemic justification. The self-vindicating stability of the laboratory sciences, he argued, is achieved in part by the mutual adjustment of theories and the experimental systems that generate the relevant data.

I endorse the underlying recognition that experiments cannot be adequately understood in isolation but make a different point than Fleck or Hacking. The systematic interconnectedness of experimental systems does not just serve a justificatory role. The establishment of systematically intraconnected "microworlds" plays a primary role in disclosing and constituting aspects of the world as a coherent domain open to contentful conceptual articulation at all.[17] I noted above that "genes,"

16. Notable defenses of the systematic character of experimental systems and traditions include Rheinberger (1997), Galison (1987, 1997), Klein (2003), Kohler (1994), and Chang (2004).

17. Fleck, at least, was not unaware of the role of experimental systems in conceptual articulation, although he did not quite put it in those terms. One theme of his study of the Wassermann reaction was its connection to earlier vague conceptions of "syphilitic blood," both in guiding the subsequent development of the reaction and also thereby in articulating more precisely the

for example, were transformed from merely hypothetical entities to the locus of a field of inquiry ("genetics") via the development of pure-bred experimental lineages, hybridizations of those "pure" lines, and the subsequent correlation of the crossover frequencies of mutant traits with visible transformations in chromosomal cytology in fruit flies crossbred to a standardized breeding population. Carbon chemistry, understood as distinct from earlier phenomenological descriptions of organically derived materials due to its compositional orientation, likewise became a domain of inquiry through the systematic, conceptually articulated tracking of ethers and other derivatives of alcohol (Klein 2003).[18] Leyden jars and voltaic cells played similar roles for electricity. What is needed to open a novel research domain is typically the ability to create and display an intraconnected field of reliable differential effects: not merely creating individual phenomena but situating them within a systematic but open-ended experimental practice.

My shift of focus from justification to conceptual articulation and domain-constitution differs from Hacking's account of experimental phenomena creation in a second way. Hacking (1992) later sought to understand the eventual stabilization of fields of laboratory work as the result of an eventual mutually self-vindicating adjustment of theories, apparatus, and skills. I instead call attention to the beginnings of this process of developing an experimental field, the opening of new domains for conceptual articulation rather than their eventual practical and conceptual stabilization. My argument then shows Hacking's account of self-vindication in a new light. We need to consider not just the supposedly self-vindicating justification of experimental work but its scientific significance more generally. Taken by themselves, mutually adjusted experimental phenomena and theoretical models may seem self-referential and self-vindicating. Their place within an ongoing scientific enterprise

conceptual relations between syphilis and blood. He did not explicitly connect the systematicity of experimental practice with its conceptual-articulative role, however. Hacking was likewise also often concerned with conceptual articulation (especially in some of Hacking's collected papers [1999, 2002]), but this concern was noticeably less evident in his discussions of the laboratory sciences (e.g., Hacking, 1983, ch. 12, 16; 1992).

18. Klein (2003) also emphasizes the development of a systematic symbolic nomenclature for chemical components of organic material, which she tellingly describes as "paper tools," but this role for explicitly theoretical articulation of a conceptual domain does not undercut my point. Experimental construction of systematically related phenomena need not function in isolation from the verbal and other symbolic innovations often needed to track their components, features, and interrelations. My point is instead that systematically interconnected experimental systems function together with the articulation of a theoretical vocabulary to constitute a law-governed scientific domain. Without the experimental syntheses and subsequent chemical analyses of various ethers and alcohols, the nomenclature could not have been developed or interconnected.

nevertheless depends upon their being informative beyond the realm of their self-vindication. Hacking's emphasis upon self-vindicating stabilization thus subtly downplays the intentional and conceptual character of experimental systems.[19]

To understand the intentional directedness of experimental systems, I turn to the second feature of "laboratory fictions," their constitutive character.[20] The constitution of a scientific domain accounts for the conceptual character of the distinctions that articulate the associated field of scientific work. Consider further what it means to say that the *Drosophila* system developed in Morgan's group was "about genetics." We need to be careful here, for we cannot presume the identity and integrity of genetics as a domain. The word 'gene' predates Morgan's work, and the notion of particulate, germ-line "units" of heredity emerged even earlier in work by Mendel, Darwin, Weismann, Bateson, and others. The conception of genes as the principal objects of study within the domain of genetics marks something distinctive, however. Prior conceptions of heredity did not and could not distinguish genes from the larger processes of organismic development within which they functioned. What the *Drosophila* system initially displayed, then, was an open-ended field of distinctively genetic phenomena, for which the differential development of organisms was part of the experimental apparatus. Against that background, genes could be jointly indicated by their relative chromosomal locations and characteristic patterns of meiotic crossover, as well as their correlated phenotypic outcomes. This example also highlights the significance of shifting attention from the representational to the discursive functions of experimental systems. Morgan's *Drosophila* system does not *denote* the domain of genetics any more than Austen's *Pride and Prejudice* denotes the traits of character that it articulates within the fictional context of 'three or four families in a country village' (Elgin 2009). Austen rightly titled her novel *Pride and Prejudice* rather than

19. In chapter 7, I criticized Hacking's notion of experimental self-vindication on different grounds. I extended McDowell's (1994) criticism of Davidson to show how Hacking's account of self-vindication could similarly serve its manifest purpose of undercutting skeptical doubts about the epistemic justification of experimental science only by losing its grip on the semantic contentfulness of the concepts deployed in those self-vindicating domains. The two criticisms are complementary: in chapter 7, I argued that Hacking's account of the systematicity of experimental systems and their self-vindicating mutual adjustment with theories detached experimental practice from its conceptual-articulative role. Here I am exploring how a different model of experimental systematicity might account more successfully for that role.

20. John Haugeland (1998, ch. 13) suggests the locution "letting be" to explicate what we both mean by the term 'constitution.' Constitution of a domain of phenomena ("letting it be") must be distinguished from the alternative extremes of creation (e.g., by simply *stipulating* success conditions) and merely taking antecedently intelligible phenomena to "count as" something else.

Elizabeth Bennet and Mr. Darcy; its primary function as a fictional construction is not denotative, and to read it as primarily about these characters would be to miss the point. One would similarly miss the point if one regarded the Morgan group's research as a study of artificially standardized fruit flies. The discursive role of both "fictional" constructions is articulative rather than denotative; they reveal an interconnected domain of entities, properties, and relations rather than being about any particular entities or properties within that domain. They do so in part by providing standards for the appropriate application and use of its key concepts within that confined and controlled experimental setting.

The *Drosophila* system allowed a much more extensive inferential articulation of the gene concept. Concepts are marked out by their possible utilization in contentful judgments, which acquire their content inferentially as well as referentially. For example, a central achievement of *Drosophila* genetics was the identification of phenotypic traits, or more precisely trait *differences*, with chromosomally located "genes." Such judgments cannot simply be correlations between an attributed trait difference and what happens at a chromosomal location because of their inferential interconnectedness. Consider the judgment in classical *Drosophila* genetics that the Sepia gene is *not* on chromosome 4.[21] This judgment does not simply withhold assent to a specific claim; it more specifically indicates either that the Sepia gene has some other chromosomal locus or that no single locus can be assigned to the distinctive trait differences of Sepia mutants—that is, such judgments indicate a more or less definite space of alternatives through their place in a practical-inferential domain. Yet part of the content of the "simpler" claim that Sepia is on chromosome 3 is the consequence that it is not on chromosome 4. Thus any single judgment in this domain presupposes the intelligibility of an entire conceptual space of interconnected traits, loci, and genes, including the boundaries that mark out what is or is not a relevant constituent of that space.[22]

21. The importance of negative descriptions for displaying conceptual holism was clearly indicated by Hanna and Harrison (2004, ch. 10).

22. Classical genetic loci within any single chromosome are especially cogent illustrations of my larger line of argument, because prior to the achievement of DNA sequencing, any given location was only identifiable by its relations to other loci on the same chromosome. The location of a gene was relative to a field of other genetic loci, which in turn are only given as relative locations. DNA sequencing did not directly allow for a more specific, and specifically individuated, referential specification of classical gene talk, however. The identification of coding regions of DNA not only required a different experimental system but also complicated the identification of trait-differential "genes" with coding regions: only in conjunction with a complex regulatory and

Inferential relations play a central role in the opening of conceptually articulated domains, but inferential relations are not entirely linguistic.[23] Conceptual distinctions can function implicitly in practice without being expressed in words at all. Scientific work is normally sufficiently reflective and communicative that important conceptual distinctions are eventually marked linguistically. A central point of this chapter, however, is to argue that the inferential articulation of scientific concepts is part of the systematic development of an interconnected domain of phenomena that can display the appropriate conceptual differences. The experimental practices that open such a domain make it possible to form judgments about entities and features within that domain, but the practices themselves already articulate implicitly "judgeable contents" prior to the explicit articulation of judgments. Verbal articulation in the sciences would be idle and empty without extensive, systematic, skillful involvement with experimental practices in settings that implicitly regulate the appropriate invocation of verbally explicit conceptual relations.

These points about the relation between scientific domains and entities disclosed within those domains can also be usefully illustrated by comparison to Hanna and Harrison's (2004) discussion of the conceptual space of naming, which I discussed briefly in chapter 3. Proper names are often taken as the prototypical case of linguistic expressions whose semantic significance is directly conferred by their relationship to entities bearing those names. Hanna and Harrison's counterclaim is that proper names themselves do not *directly* refer to persons, ships, cities, and the like but only do so through the mediation of a larger framework of practices that they identify collectively as the "name-tracking

---

transcriptional-translational apparatus within a cell and its genome could a specific DNA sequence be a gene "for" an element within an intelligible field of trait differences.

23. This emphasis upon inferential articulation as the definitive feature of conceptualization is strongly influenced by Brandom (1994, 2000, 2002), albeit with some important critical adjustments (see Rouse 2002, ch. 5–7). My semantic inferentialism also has important affinities to Suárez (2009b) but with one very important difference. Suárez emphasizes the inferential *uses* of scientific representations, including fictional representations. I emphasize instead that representational relations are themselves inferentially constituted, and I go even beyond Brandom in emphasizing how causal interaction with the world in experimentation is caught up within the discursive, inferential articulation of conceptual understanding. Brandom thinks of inferential relations as systematic intralinguistic relations, which then interface with perception and action as "language entrances" and "language exits," whose conceptual role is indirect. I am arguing that experimental practice is more "intimately" entangled with the verbal aspects of conceptual articulation. For an extended account of the difference between localized interaction at an interface, and a more intimate, "broad-band" entanglement, see Haugeland (1998, ch. 9).

network":[24] "Such [name-tracking] practices are mutually referring in ways that turn them into a network through which the bearer of a given name may be tracked down by any of dozens of routes" (Hanna and Harrison 2004, 108).

We can use names to refer to persons, cities, or ships but only via the maintenance and use of such an interconnected field of naming and name-tracking practices.[25] Such practices constitute people and some objects as trackable through the *accountable* use of names for their reidentification as the same individual person or thing.

Experimental systems play a role within scientific domains comparable to the role of the name-tracking network in constituting people and things as nameable individuals. They mediate the accountability of verbally articulated concepts to the world, such that the use of those concepts is more than just a "frictionless spinning in a void" (McDowell 1994). Chapter 7 presented this role for experimental systems by extending Morgan and Morrison's (1999) account of theoretical models as partially autonomous *mediators* between theories and the world, recognizing that scientific understanding is often *doubly* mediated. Experimental systems and theoretical models function together to "mediate" between theoretically interrelated concepts and the circumstances to which they ultimately apply. It is often only through initial application of these models within the microworld of one or more experimental systems that they come to have an intelligible application anywhere else. Moreover, in many cases, the experimental model comes first; it introduces relatively well-behaved circumstances that can be tractably modeled in other ways

24. Hanna and Harrison (2004) present the role of the "name-tracking network" in a way that I do not fully endorse. They argue that naming practices constitute a distinct domain of "nomothetic" objects ("name-bearerships"). They then contrast nomothetic objects, such as chess pieces and name-bearerships, to natural objects. Coming to recognize the "nomothetic" character of scientific domains as amounting to the conceptual articulation of an intelligible objective domain undercuts any distinction between their "nomothetic objects" and objects more generally. The practical systematicity of the name-tracking network as a constitutive condition for persons and objects to be nameable as distinct individuals does not depend upon their conception of nomothetic objects, however.

25. "Reference" in this sense is a relation that also presupposes an understanding of the possibility of judgment, typically expressed in sentences. The sentence is the minimal unit of linguistic expression, with reference to objects playing a "subsentential" role. The point I take from Hanna and Harrison is that *proper* names require dual contextualization: within a discursive field of possible judgments that constitute the possibility of reference as an aspect of predicative judgment and also a social-material practice of naming and name-tracking that allows for reference to reidentifiable individuals. This emphasis upon sentences as minimal units of linguistic expression should not be misunderstood as another example of what Kukla and Lance (2009) call the declarative fallacy. On the one hand, not all sentences or judgments are declaratives. On the other hand, there are important pragmatic and semantic differences between sentences of only one word ("Rabbit!" or "Go!") and words as iterable elements usable in various sentences.

(e.g., by a *Drosophila* chromosome map).[26] These mediating systems constitute their larger domain as answerable to conceptual norms and thus as conceptually articulated.

We need to ask what "well-behaved circumstances" means here in order to understand this claim and its significance. Cartwright (1999, 49–59) introduced similar issues by talking about mediating models in physics or economics as "blueprints for nomological machines." Nomological machines are arrangements and shielding of various components, such that those components' capacities reliably interact to produce regular behavior. I am expanding Cartwright's conception to include not just regular behavior but conceptually articulable behavior more generally. I nevertheless worry about the metaphors of blueprints and machines. The machine metaphor suggests an already determinate purposiveness, something the machine is a machine *for*. With purposes already specified, the normative language permeating Cartwright's discussion of nomological machines then becomes straightforward. She speaks of *successful* operation, of running *properly*, and of arrangements that are fixed or stable *enough*. Yet where do the purposes and norms come from, and how do they acquire authority and force within that context? Those questions point toward the most basic reason to think about experimental systems as laboratory fictions mediating between theoretical models and worldly circumstances. Experimental systems exemplify what it would be for circumstances to be "well behaved" or for nomological machines (or experiments with them) to run "properly" or "successfully." Scientific concepts, then, both express and are accountable to patterns of intelligible interrelations with respect to which notions of proper behavior or successful functioning could be understood.

Cartwright and Ronald Giere (1988) have each tried to regulate the normativity of theoretical models and the concepts they employ, appealing respectively to their empirical adequacy or their systematic resemblance to real systems. In considering the appropriate domain for the applicability of the concept of 'force,' for example, Cartwright claims, "When we have a good-fitting molecular model for the wind, and we

26. Marcel Weber (2007) argues, for example, that the classical gene concept as it was materially articulated in *Drosophila* chromosome maps played an important role in developing the molecular gene concept that is often taken to have supplanted it. The extensively interconnected chromosome maps available for *Drosophila* were a resource allowing the identification of the relevant regions of the chromosome from which the coding and regulatory sequences could be isolated. As Weber succinctly concludes, "The classical *gene* may have ceased to exist, but the classical gene *concept* and the associated experimental techniques and operational criteria proved instrumental for the identification of molecular genes" (2007, 38).

have in our theory . . . systematic rules that assign force functions to the models, and the force functions assigned predict exactly the right motions, then we will have good scientific reason to maintain that the wind operates via a force" (1999, 28). Giere instead argued that theoretical models like those for a damped harmonic oscillator directly characterize abstract entities of which the models are strictly true, and their relation to real situations is one of relevant similarity: "The notion of *similarity* between models and real systems . . . immediately reveals—what talk about approximate truth conceals—that approximation has at least *two* dimensions: approximation in *respects*, and approximation in *degrees*" (1988, 106). To answer my concerns about how one could *initiate* a conceptually articulated field of modeling and experimental practice, however, both empirical adequacy and systematic similarity come too late. What is at issue is the ability to identify relevant *respects* of possible resemblance and what differences in degree are degrees *of*.[27]

Elementary mechanics serves as the proximate example for both Cartwright's and Giere's discussions, and the prior articulation of its domain is easily taken for granted in that context. The domain of mechanics seems well marked out, because the relevant experimental or observation-tracking systems were long ago established and stabilized in mutual adjustment with idealized models. The phenomena that initially allowed the grasp of mechanics as an intelligible domain—pendula, springs, levers, free-falling objects, and planetary trajectories, along with their conceptual characterization—are now familiar and well regimented

27. McManus (2012) argues that we should not think of measurement practices as discovering a "con-formity" between categories of thought and features of the world:

> By introducing the "technology" of tanks with uniform cross-sections and equally-spaced gradations, and the practice of using a particular container so as always to add what we come to call a "standard unit" volume of water, we are now treating water in a way which allows us to apply arithmetical rules to it meaningfully. . . . What we discover when we develop the technology of measuring cups, and so on, is not the way in which liquids have hitherto hidden their "arithmeticality." Instead, we have reconceptualized liquids so as to foreground something about them that we can describe in arithmetical ways. . . . [Similarly,] to learn how to measure heights by learning to lay the measuring rod straight along the body is not so much a matter of learning *how* to measure heights as how to measure *heights*. It is not that one learns something *about* heights; rather one learns *of* heights. (McManus 2012, 142–44)

McManus thus argues that one can thereby have standards for the correctness of measurements of volume or height but cannot take the measures of volume or height to be correct or incorrect in the same way. McManus's point is important, but he oversimplifies it by considering only single measures of single features. It matters, for example, that different ways of measuring distances (with rulers, surveying instruments, clocks and sextants, period-luminosity relations in Cepheid variable stars, or various measures of small distances such as wavelengths of light or molecular bond lengths) function together coherently and that measures of one concept be compatible with measures of others that can be inferentially related to it. These considerations show from a different direction why conceptual normativity and alethic modality are connected.

in their canonical cases. Cartwright's and Giere's presumption of that familiarity also overlooks the extensive practical and theoretical work required to conceptualize that domain more intensively. I previously called attention to Mark Wilson's (2006) exploration of the conjoined practical and theoretical work needed to bring the motions of nonrigid materials into the more narrowly articulated domain of mechanics alongside other extensions of the canonical phenomena and models that initially opened the domain.[28] Wilson's examples from applied mechanics and materials science show how much intertwined experimental and mathematical work was needed to extend the domain of mechanics beyond its initial idealizations of rigid-body interactions. Further in the background, Andrew Warwick (2003) identified a differently taken-for-granted conceptual system in the historical introduction and regimentation of paper-and-pencil calculation, through which the familiar mediating models of mechanics and classical electrodynamics were mathematically developed and extended. Both books show in exquisite detail the complexity and the extent of practical work needed to let classical trajectories be intelligible as a law-governed conceptual domain.

Acknowledgment of the extensive work required to articulate classical mechanics intelligibly brings this seemingly straightforward space of mathematical reasoning into contact with other cases when scientists began to formulate and explore new domains of phenomena. To return to the case of classical genetics, for example, Mendelian ratios of inheritance obviously predated Morgan's work, but spatialized "linkages" between heritable traits were novel. The discovery that the white-eyed mutation was a "sex-linked" trait provided an anchoring point within the emerging field of mutations, much as the stabilization of freezing and boiling points of water helped anchor the conceptual field of temperature differences. Hasok Chang (2004, ch. 1) has shown in the latter case that these initially familiar phenomena of phase transitions in water could not be taken for granted scientifically; in order to anchor a conceptual space of temperature differences, the phenomena of boiling and freezing required canonical specification. Such specification required

28. For the reasons indicated in my critical discussion of Hacking on geometrical optics in chapter 7, the domain of mechanics already incorporated the motions and interactions of nonrigid substances, but it did so in a mostly vague, unarticulated way. The further mathematical and experimental work indicated by Wilson and Warwick (2003) established more specifically articulated norms for the correct or incorrect application of the central concepts of mechanics within less "well-behaved" regions of its domain, thereby constituting a field of judgments that are "true-or-false," in ways that could be reasoned about and engaged experimentally and observationally.

practical mastery of techniques and circumstances as much or more than explicit definition.[29] Indeed, the point is that practical articulation and verbal articulation of the phenomena proceed together. That process also includes the development and refinement of the instruments that allow such phenomena to become manifest, such as thermometers for temperature differences or breeding stocks for trait linkages.

Experimental systems function in this domain-opening, concept-defining way, without having to be typical or representative of the features they exemplify. Consider again *Drosophila melanogaster* as an experimental organism. As the preeminent model system for classical genetics, *Drosophila* was highly *a*typical; as a human commensal, it is more cosmopolitan and genetically less diversified than most plausible alternative model organisms. More important, however, for *D. melanogaster* to function as a model system, its atypical features had to be artificially enhanced, removing much of its residual "natural" genetic diversity from experimental breeding stocks (Kohler 1994, ch. 1, 3, 8). *Drosophila* is even more anomalous in its later incarnation as a model system for evolutionary developmental biology. *Drosophila* is now the textbook model for the development, developmental genetics, and evolution of animal body plans (Carroll, Grenier, and Weatherbee 2001, esp. ch. 2–4), and yet the long syncytial stage of *Drosophila* development is extraordinary even among arthropods. These are model systems *not* in the sense of being "representative" of a domain of phenomena but in their constitutive role as concept articulative and domain opening.

One might nevertheless object that conceiving these possibly anomalous yet domain-constituting experimental systems in parallel with literary fictions renders the constitution of these domains merely stipulative and thus not answerable to empirical findings. Set aside any concerns about the objection's implicit presumption that literary fictions are "stipulatively" constructed with no normative accountability beyond the author's imagination, although I take that presumption to be clearly false.[30] We can still ask directly whether this role for experimental systems in constituting normative standards for the application of con-

29. Chang (2004) argues that in the case of state changes in water, ironically, ordinary "impurities" such as dust or dissolved air and surface irregularities in its containers helped *maintain* the constancy of boiling or freezing points; removing the impurities and cleaning the contact surfaces allowed water to be supercooled or superheated. My point still holds, however, that canonical circumstances needed to be defined in order to specify the relevant concept, in this case temperature (in part by spelling out its inferential and practical relations to concepts of boiling and freezing).

30. Anyone who doubts that fictional constructions can fail in their own terms should read as many bad novels as are required to overcome their doubts.

cepts would remove their accountability to assessment. Chang's (2004) study of the practices of thermometry shows one important reason such norms are not and could not be merely stipulative. There are many ways to produce regular and reliable correlates to increases in heat. Much work went into developing mercury, alcohol, or air thermometers, along with their analogues in circumstances too hot or cold for these canonical thermometers to register. Yet it was not sufficient merely to establish a reliable, reproducible system for identifying degrees of heat by correlating them with the thermal expansion or contraction of some canonical substance. The substantial variance in measurement among alternative possible standard-constituting systems pointed toward a conceptualization of temperature independent of any particular measure, however systematic and reproducible. Such a conceptualization, if it could be coherently worked out, would introduce order into these variations by establishing a standard against which the correctness of the measuring systems could themselves be assessed. That the development of a standard is itself normatively accountable is clear from the possibility of failure. Perhaps there would turn out to be *no* coherent, systematic way to correlate the thermal expansion of different substances on a single scale that would consequently spell out the concept of temperature.[31]

I previously cited Chang's (2004, 59–60) identification of the deeper issue here as "the problem of nomic measurement": identifying some concept X (e.g., temperature) by some other phenomenon Y (e.g., thermal expansion of a canonical substance) presupposes what is supposed to be discovered empirically—namely, the form of the functional relation between X and Y. The more basic underlying issue is not the identification of the *correct* functional relation, however, but the projection of a concept to be right or wrong about in the first place. The issue therefore does not apply only to quantitative measurement. It affects nonquantitative concepts like the relative location of genes on chromosomes or the identification of functionally significant components of

31. The systematic empirical and inferential accountability of temperature extends well beyond the experimental domain of thermometry in ways that highlight the complex, systematic empirical interdependence of scientific concepts. In chapter 8, I called attention to Andrea Woody's (2004b) claim that the ideal gas law is not just an approximately accurate empirical description of macroscopic properties of gases within a limited range but a conceptual indicator of the domain of chemistry as composed of interactions of energetic molecules via their intermolecular forces. An important part of that conceptually articulative role is that the model of an ideal gas, and its applicability to real gases via recognized ceteris paribus provisos, also further develops the concept of temperature and its systematic contribution to understanding the domain of chemical bonds and reactions, with additional inferential and practical-experimental relations of mutual accountability with thermometry.

cells as much as it applies to measurable quantities like temperature or electrical resistance.[32]

The normativity of experimental domain constitution is more dramatically displayed, however, when domain-constituting systems are abandoned or transformed by a constitutive failure or by a reconceptualization of the domain. Consider the abandonment in the 1950s of the *Paramecium* system as a model organism for microbial genetics.[33] *Paramecium* genetics was dealt a double blow. Its apparently distinctive advantages for the study of cytoplasmic inheritance, which was its primary raison d'être, became moot with the dissolution of supposed differences between nuclear and cytoplasmic inheritance due to the discovery of DNA in cellular organelles (including the *Kappa* particle in *Paramecium*). More important from my perspective, however, is the biochemical reconceptualization of genes through the study of auxotrophic mutants in organisms that could grow on a variable nutrient medium. Despite extensive effort, *Paramecium* would not grow on a biochemically characterizable medium and hence could not clearly display the auxotrophic mutations that allowed direct correlation of gene products with biochemical pathways. In Elgin's terms, cytogenetic patterns in *Paramecium* could no longer exemplify the (newly) distinctive manifestations of genes but only instantiate them, due to the intersection of its physiological idiosyncrasies with the further development of the gene concept.

A different kind of failure occurs when the "atypical" features of an experimental system become barriers to further conceptual development. For example, the very standardization of genetic background that made the *D. melanogaster* system the exemplary embodiment of chromosomal genetics also blocked any display of population-genetic variations and their significance for evolutionary genetics. Theodosius Dobzhansky had to adapt the techniques he learned from classical *Drosophila* genetics to the much more genetically varied *D. pseudoobscura* in order to exhibit the genetic diversity of natural populations (Kohler 1994, ch. 8).

32. One might argue that chromosomal locations in classical genetics were quantitative properties also, either because they were assessed statistically by correlations in genetic crossing over or because spatial location is itself quantitatively articulable. Yet crossover correlations were only measures of relative location, and because locations were only identifiable through internal relations on a specific chromosome map, I would argue that these were not yet quantitative concepts either; only the evidence for their application was quantitative.

33. For detailed discussion, see Nanney (1983). Sapp (1987) sets this episode in the larger context of debates over cytoplasmic inheritance.

More recently, Jessica Bolker (1995) has suggested that the very features that seem to recommend the standard model organisms in developmental biology as tractable model systems for the laboratory may have systematically misconfigured the domain conceptually. Laboratory work encourages use of organisms with rapid development and short generations; these features in turn correlate with embryonic prepatterning and developmental canalization, both of which insulate development from environmental fluctuation. The choice of these experimental systems thereby *materially* conceives development as a relatively self-contained process for which ecological interaction is "external" to embryonic development.[34] Any reconceptualization of development as ecologically mediated would likely require exemplification in different experimental practices employing different organisms, much as *Paramecium* was bypassed in microbial genetics.

V—Conclusion

I have been examining experimental systems as materialized fictional "worlds" that are integral to scientific conceptualization. The experimental practices that establish and work with such systems not only exemplify conceptualizable features of the world but help constitute the fields of possible judgment and conceptual normativity that allow those features to show themselves intelligibly. The sense in which these systems are "fictional" is threefold, and these three aspects of their "fictionality" function together in shaping the content and authority of the conceptual field they help articulate. First, exemplary experimental systems are simplified and rearranged as "well-behaved" circumstances that allow some features of the world to be more readily manifest. The world as we find it is often unruly and unarticulated. By arranging and maintaining more clearly articulated and manifest differentiations, we establish conditions for conceptual understanding that can then be applied to more complicated or opaque circumstances.[35] Second, these artificially regulated circumstances and the differentiations they display

34. For more extended discussion of the reconception of biological development as ecologically mediated, see Sultan (2007, forthcoming).

35. It is no part of my claim to suggest that experimental practices stand alone in this role. Mark Wilson (2006), for example, shows in exquisite detail how mathematics often plays an important role in further articulating concepts beyond the range of their initially specified applications, often in patchwork ways.

are sufficiently interconnected to allow for systematic interrelations among those differences. The resulting interrelations, if they can be coherently sustained and applied in other contexts or to new issues, demarcate norms for conceptual intelligibility rather than merely isolated, contingent correlations. What might otherwise be merely localized empirical curiosities instead become scientifically significant because they allow those situations to manifest an interlocking field of conceptual relationships. By clarifying relevant conceptual norms, experimental systems join theoretical models in mediating jointly between scientific theory and the world it thereby makes comprehensible. Third, and crucially, the establishment and institution of these systems, and the conceptual relations they exemplify, performatively establishes them as having already been authoritative for events within that domain. These systems are "fictional" in a sense comparable to Kukla's (2000) argument, following Sellars, that the normative authority of concepts must be established "mythologically."[36] Prior to the establishment of the experimental systems and conceptual models that provide standards for deploying and reasoning with those concepts, the world was unarticulated in those respects. There simply was no way to say, rightly or wrongly, how things are in those ways or to engage one's surroundings in ways that were appropriately or inappropriately responsive to those aspects of the world.[37] Once those practices and conceptual norms opened a scientific domain, however, they *became* already authoritative over what happens and had happened within that domain through our letting them govern what we say and do.[38]

36. See my extended discussion of Kukla's argument in chapter 5 and also in my previous book (Rouse 2002, 352–58).

37. One might argue against this claim that, for example, hygienic measures were appropriately responsive to the microbial causes of disease even though there were no experimental practices that could articulate that connection between "microscopically" invisible organisms and human morbidity. I would respond that hand washing may well have been *effective* against microbial causes of disease, but it could not have been *appropriately* responsive *to* that aspect of disease phenomena. There was no intelligible normative accountability that could allow those relations to be manifest as authoritative for anyone until the experimental work of Pasteur, Koch, and others developed bacteriology as a domain.

38. I argued in chapter 5 that such a retrospectively "mythological" or "fictional" constitution of normative authority as already (having been) in place only marks one side of the temporal constitution of conceptual content, authority, and force. The retrospective determination of conceptual authority functions together with the prospective determination of content by what was at issue in how one used and responded to those conceptual possibilities and what was at stake in the differences among intelligible responses (such that those responses could be right or wrong, appropriate or inappropriate, revealing or concealing, just or unjust, and so forth, with respect to how those different possibilities matter to who we are and will become in our dependence upon our

To talk about laboratory "fictions" as helping constitute conceptual normativity does not challenge or compromise a scientific commitment to truth. On the contrary, these fictional constructions help establish norms according to which new truths can be expressed and, correspondingly, erroneous understanding recognized as such. Such laboratory operations thereby help establish a domain of possible judgments and comportments as intelligibly "true-or-false" (Hacking 2002, ch. 11) in ways that also allow further reasoning about them. Conceptual articulation through experimental practice is itself vulnerable to empirical failure, but such failures show up differently than does the falsity of specific judgments within a domain. Empirical failure in opening or sustaining a conceptual domain is manifest in conceptual confusion or incoherence distributed across the domain rather than in the falsehood of particular judgments within it. To recognize a mistake, by settling upon various judgments as correct or incorrect for intelligible reasons, is to rely upon and implicitly endorse the relevant conceptual skills and standards as appropriately authoritative in that context (perhaps via revision and repair occasioned by more localizable incoherence). Moreover, as I have been arguing, such fictional constitution of normative authority is not stipulative or voluntaristic but is instead itself accountable to the world as it shows up intelligibly within our conceptually articulated environmental niche. Haugeland (1998, ch. 12) has rightly emphasized that normative accountability in science requires both resilient and reliable skill (allowing scientists to cope with apparent violations of the resulting conceptual norms by showing how they are *merely* apparent violations). The entire domain of scientific practices and concepts may even collapse into conceptual confusion if the interplay of empirical conflicts and conceptual confusions cannot be adequately resolved. Haugeland thus also insisted upon the need for scientists' resolute openness to acknowledging and accepting such collapse, if it were to occur, as partially constitutive of empirical contentfulness. Yet these skills and attitudes could not come into play unless and until there is a conceptually articulated domain of phenomena and experimental and theoretical practices toward which scientists could be resilient and resolute. The discursive function of experimental systems as "laboratory fictions" helps open

environmental niche). This complementary, futural aspect of the ongoing performative reconstitution of a conceptually articulated space of reasons is the subject of the next chapter on "Scientific Significance."

and sustain conceptual domains and thereby enables the attainment of scientific truth and the recognition of error.

Chapter 10, the final chapter of part 2, considers how such determinations of truth or falsity matter scientifically as significant in ways that bear on their conceptual content. Recognition of what is at stake in opening and sustaining a conceptually articulated domain—and thus how and why it matters to get right what is at issue in judgments and comportments bearing on that domain—is the futural counterpart to the retrospective, fictional constitution and ongoing reconstitution of conceptual authority.

TEN

Scientific Significance

Traditional conceptions of the scientific image identify scientific understanding with a comprehensive, internally consistent representation of the world. That comprehensive understanding has many gaps and promissory notes to fill in—the sciences are far from complete—but it already provides a conceptual structure within which those gaps can be located. The most general theories in the sciences—quantum mechanics and general relativity, thermodynamics, and evolution in biology—are comprehensive in scope, incorporating within their domain even the many phenomena that no one yet knows how to characterize in their terms. In some cases, such as quantum mechanics and general relativity, conflicting structural features of these comprehensive theories have so far seemed difficult to reconcile with one another in a single conception, but their mutual recalcitrance defines a central theoretical issue. In other cases, higher-order "emergent" entities or phenomena resist adequate characterization in terms of the lower-level entities and processes described in apparently more comprehensive theories. The comprehensiveness of the scientific image as traditionally understood is thus distinct from the aspiration to reduce more complex events and structures into patterns articulable in a single theoretical vocabulary. A comprehensive scientific image may be irreducibly multilayered, but correct descriptions of actual events in its terms must in the end be mutually consistent. A comprehensive conception of one world, even if it ultimately comprises a multiverse (Rubenstein 2014), must maintain an internal consistency that accommodates the evident compossibility of the events and structures it would describe.

I—The Partiality and Selectivity of Scientific Research

This familiar sense of the comprehensiveness of the scientific image may sit uneasily with the recognition that in many other respects, scientific understanding is highly selective and partial. Many scientific fields attend first and foremost to simplified or purified materials, processes, and interactions, which have been carefully isolated or shielded from "external" impingements, to constitute relevantly closed and simplified systems. Nancy Cartwright once remarked that "physics within its various branches works in pockets, primarily inside walls: the walls of a laboratory or the casing of a common battery or deep in a large thermos" (1999, 2). Physics is not alone in making isolated and simplified settings the proximate target of understanding. The conceptualization and experimental realization of relatively controlled, analyzed, and protected "microworlds" (Rouse 1987, ch. 4) or "experimental systems" (Rheinberger 1997) is a widespread feature of scientific practice, and I have argued that this work plays a central role in constituting scientific conceptual understanding as authoritative. Moreover, as Latour (1983, [1984] 1988) long ago noted, and Cartwright (1999) and others also remind us, such understanding is often extended beyond the laboratory less by adapting it to very different circumstances than by revising the circumstances elsewhere to resemble sufficiently the familiar and well-understood settings of laboratories or theoretical models.

Scientific understanding is selectively focused in other ways, as well. In many of the biological subfields that concern organismic function (e.g., physiology, genetics, genomics, development, or cell biology), a limited number of model organisms have been the massively disproportionate locus for experimental work and its conceptual articulation. Biologists now understand in remarkable detail many of the functional life processes of *Drosophila melanogaster, Caenorhabditis elegans, Escherichia coli K-12, Saccharomyces cerevisiae, Xenopus laevis, Mus musculus, Arabadopsis thaliana, Danio rerio,* and a relatively small number of other widely deployed model organisms. This extensive body of background knowledge in turn encourages further reliance upon these organisms in subsequent research, especially now that complete genomic sequences have been worked out for so many of the common model organisms (Ankeny 2001). Detailed knowledge of functional systems in the model organisms then stands in for the possibly quite different details of those processes in other species.

Selective scientific attention to a limited number of model organisms occurs mostly by choice, even if those choices were originally somewhat fortuitous and later were constrained by the comparative absence of relevant background knowledge or sources of supply for other organisms. To some extent, the narrow focus of detailed biological knowledge upon a handful of model systems reflects a combination of those organisms' distinctive advantages for certain kinds of experimental work (generation time, cost, ease of care, developmental canalization, etc.) and the judgment that they can appropriately stand in as "representative" of the same functional processes in larger taxonomic groups.[1] Other forms of selective scientific focus are more constrained. The standard theoretical models in fields such as classical mechanics, fluid mechanics, or basic quantum mechanics play a central role in those fields in significant part due to their analytical and computational tractability. Other systems are then understood via their deviations from the more analytically tractable models and often only to the pragmatically requisite precision. Scientific theory often yields small pockets of precise understanding surrounded by large areas where the same concepts provide at best a rough-and-ready conceptualization. In some cases, however, theoretical models are central as much for their conceptual role in articulating the domain as for their ease of analysis.[2] The various forms of harmonic motion, for example, are not merely readily analyzable; they also have a conceptual significance for classical mechanics that is likely disproportionate to the de facto propensity toward periodicity among macroscopic motions. In any case, as we saw earlier in chapter 7 in considering Cartwright's proposed neoempiricist restrictions on conceptual scope, there typically

1. The representational role of model organisms often does not depend upon a high degree of similarity to the features of other organisms for which they serve as models. *Drosophila melanogaster* has been one of the premier model organisms in developmental genetics and developmental biology more generally, despite its highly uncharacteristic form of syncytial development. Bolker (1995) argues that the standard model organisms in developmental biology embody a systematic bias toward highly canalized, ecologically insensitive patterns of development that are uncharacteristic of development in many other eukaryotes. In any case, as I argued in chapters 8–9, the conceptually articulative role of experimental systems is what enables them to help constitute a space of reasoning and representation for an entire law-governed scientific domain rather than first thinking of their representational role as directly constituted by some identifiable similarities.

2. Recall Andrea Woody's (2004b) argument, discussed in chapter 8, that the ideal gas law, $PV = nRT$, plays a central conceptual role in chemistry only in part due to its mathematical simplicity. The gas law, in describing the behavior of separated molecules, illustrates especially clearly the conceptual role of molecules for understanding the structure of matter in other phases and the concept of temperature as implicated in the thermodynamic character of chemical reactions.

seems to be no principled way to distinguish the subset of systems that are well characterized by available models of a more general theory from those for which no empirically adequate models have been developed. The scientifically significant set of modeled systems is usually a conceptually gerrymandered subset of the larger theoretical domain.

A different kind of selectivity in scientific understanding arises from the development of mechanistic models across a diverse range of sciences (Bechtel 2006; Darden 2006; Craver 2007). Mechanistic models also typically characterize processes that occur within walls or otherwise shielded spaces; yet even within those confines, they are more narrowly focused. Mechanistic models normally characterize in considerable detail what happens when the system functions normally or "properly," along with some of the more decisive or prominent enabling conditions under which the mechanism can be expected to function in these ways. Mechanistic models are nevertheless typically reticent about what happens or would happen when those enabling conditions are not met or are initially met but then violated. That the system then no longer functions normally is clear, but what happens instead is left largely indeterminate.

Selectivity takes a different form through what Mark Wilson memorably describes as "tropospheric complacency." Wilson's concern in calling attention to this phenomenon was to object to an all-too-easy assumption about the application of familiar concepts under significantly different conditions: "[*Tropospheric complacency*] represents our native inclination to picture the distribution of properties everywhere across the multifarious universe as if they represented simple transfers of what we experience while roaming the comfortable confines of a temperate and pleasantly illuminated terrestrial crust" (2006, 55). While endorsing Wilson's point, I note that this phenomenon also highlights the extent to which scientific understanding is disproportionately focused upon physical processes, chemical structures, biological functioning, and even psychological responses within that relatively familiar domain. This aspect of the selectivity of scientific significance is easy to overstate: there has been extensive and central scientific attention to phenomena at temperatures near 0° K or at the extraordinary temperatures within microseconds after the Big Bang and the somewhat cooler temperatures in the interior of stars. The extreme life conditions of many of the *Archaea*, the formation and behavior of gaseous outer planets, the physics of plasmas, collisions between galaxies, and so many more phenomena occurring under conditions that stretch our imaginative capacities have been very important to the development of scientific understanding.

Even these concerns that press beyond our familiar terrestrial circumstances are nevertheless selectively generated by issues within scientific practice and the conceptual concerns that have animated specific disciplinary trajectories. The sciences have not been inclined toward a disinterested comprehensiveness that would eschew distinctions between scientifically significant and insignificant issues and domains. Not surprisingly, the gravitation, atmosphere, temperature, and illumination familiar from our terrestrial habitat figures more centrally in scientific understanding than in the universe as a whole, and interest in other conditions can often be traced back to concerns with significant bearing on our familiar way of life. The sciences notoriously reveal the universe as not having been made for us, and as according us no special significance or standing, yet the centrality of that very *issue* for sciences from astronomy and cosmology to microbiology and evolutionary biology suggests that these disciplines still move along more broadly anthropocentric and terrestricentric trajectories.

The sense in which scientific understanding is highly selective is further clarified by the recognition that many situations whose complexity is largely overlooked or circumvented within the sciences have a different significance within engineering, medical, or other practically significant contexts. We need not seek any clear or sharp distinction between "basic" and "applied" or engineering sciences to recognize that some phenomena introduce analytical complexities or conceptual slippages that do not affect other scientific considerations but do need to be analyzed more carefully or precisely for various practical purposes. The demands for further development, articulation, or precision placed upon scientific concepts for practical purposes need not always make any scientific difference to the conceptual or theoretical domain that is being stretched or refined. Wilson (2006) is especially instructive in highlighting various discontinuities ("property dragging," "lifts," or just the shifts from one "theory facade" to another) within engineering and materials science, which nevertheless do not seem to threaten the coherence or reliability of the theories and concepts whose application they place under empirical or mathematical pressure. These differences that make no scientifically far-reaching difference stand out sharply in contrast to those occasions when seemingly small discontinuities or empirical approximations turn out to be conceptually revealing: the precession of the perihelion of Mercury, the cross-generational shifts in the color patterns of maize kernels that were first recognized as indicating genetic transpositions, the absence of interference patterns in the Michelson-Morley experiments, or apparent discrepancies in the energy balance of beta decay are salient examples of small but conceptually transformative effects.

II—Scientific Significance and Human Interests

Any adequate conception of scientific understanding as actually realized within the sciences must take account of these variations in significance. What *matters* scientifically is central to a scientific conception of the world. Not only would most truths about nature be of little or no scientific significance even if they were known, but the reasons *why* other truths are of greater importance are integral to scientific understanding. Moreover, scientific significance is neither static nor retrospective. Scientific research is organized around research *projects*, which in turn are responsive to *issues* and *opportunities* within a disciplinary or interdisciplinary field. Research is always directed ahead of its current understanding, toward what is not yet known but not altogether beyond our current ken. Rheinberger (1997) characterized experimental systems and research fields as directed toward an "epistemic thing" they seek to understand, which is indicated by but not yet fully or adequately articulable within current technical and conceptual capacities. Scientific understanding of "epistemic things" is directed toward neither the utter darkness of the unknown nor the brightly illuminated space of an already-articulated conceptualization but within the penumbra of the not-quite-grasped. We saw earlier that even the retrospective compilation of what has been achieved within scientific fields is regularly reconfigured in the review literature and textbooks in light of its bearing on where the field is going and what problems and opportunities lie ahead as not-fully-definite possibilities. The openness of conceptual understanding to further intensive and extensive articulation and possibly to revision or repair is thus not just a qualification that should be added to any compilation of scientific truth claims. The shape of scientific understanding *is* a configuration of possibilities for further development in ways that would matter conceptually and epistemically.

Scientific significance has often been regarded as shaped by two vector components. From one direction, contingent human interests or concerns pull inquiry toward practically useful or culturally meaningful achievements. In the other direction, scientific significance has an objective dimension. Even where human interests play a decisive role, it is not up to us which conceptualizations and judgments will actually serve those interests. Moreover, some aspects of the world may be so epistemically salient once encountered that their significance could not be greatly diminished by any contingent shifts in human interests and concerns. Some theorists of science have tried to reduce these two components to one. Objective theories of explanation, whether deductive-nomological, unificationist, structural-realist, or causal, have vested the epistemic significance of the sciences in relations that are largely insensitive to contingent human interests. In the other direction, social constructivist accounts of science have argued for the supremacy of human interests: as Shapin and Schaffer forcefully concluded, "It is ourselves and not reality that is responsible for what we know" (1985, 344). I propose a different kind of challenge to this two-component conception, however. I want to suggest that these two supposed components cannot be so readily disentangled.

The importance of practical considerations in the sciences may clearly seem to reflect human interests. A very high proportion of scientific work can be directly traced to a relatively small number of central human concerns: medicine, agriculture, energy sources and their uses, engineering and materials science, military capacities, or environmental protection and remediation. Yet those interests alone are not sufficient to determine the relevance or significance of the *sciences* for any of these endeavors. All these concerns were informed by relatively autonomous craft and technological traditions prior to the development of any related scientific research programs, and the eventual success of scientific research in making and sustaining inroads into those traditions could not be presumed. Moreover, the epistemic centrality of some of these concerns, such as medicine or agriculture, is not entirely up to us, even if their specific forms and histories are culturally variable. Human vulnerability to disease or starvation is hardly a merely contingent interest. Agriculture has not always or everywhere been central to human life, but it exemplifies the importance of niche construction in human evolution. We have become the organisms we are in significant part through a heritable transformation of our environment via land clearance and cultivation and our coevolution with a relatively small number of domesticated plant and animal species (Diamond 1998; Crosby 2004). Agriculture is also but one significant niche constructive transformation of human interests. What were once contingent human interests in many topics of scientific understanding have become less contingent as scientific capacities were built into human environments and ways of life. The extraction and burning of fossil fuels, and their consequences, are no longer merely optional concerns for human understanding and assessment. Electrical power grids, communications networks, new patterns of disease transmission, the computational intractability of factoring large numbers, the population dynamics of ocean fish, an extraordinary range of synthetic chemicals and genetically modified organisms, and much more have become integral parts of the environments in which we live and develop biologically.

Alongside the practical concerns that motivated scientific interest in topics of central relevance to human life, the sciences have also disclosed patterns within the world that have instead transformed human interests. As one prominent example, the concept of "climate" has shifted from referring to a relatively invariant feature of local habitats to indicating a global, dynamic system in ways that compel us to think differently about our terrestrial habitat. Recognition of the influence of anthropogenic CO_2 emissions on climate and the climate system's openness to relatively sudden and dramatic shifts in equilibrium even on time scales commensurate with human life spans have conferred new significance upon the historical patterns of climate change and the causal factors that can drive these changes at different temporal scales. These developments and their significance for the diversity of life on earth give a rather different and more portentous significance to Wilson's coinage, "tropospheric complacency."

A different example of the scientific reconstitution of human interests arises from recent developments in microbiology. Zoology has long taken a primary position among taxonomically defined fields within the life sciences, since it encompasses both human organisms and most other species of long-recognized primary interest to us. The so-called biological species concept (Mayr 1942) best fits animal speciation, for example, and the majority of model organisms in biology are animal species. Multiple developments in microbiology in recent decades have not only given new scientific prominence to microbes, however, but also begun to shift conceptions of the individuation of animal organisms and their evolution. Growing recognition of the symbiotic interdependence among animals (including humans) and their coevolved microbiomes is now fundamentally transforming scientific understanding of animals (Gilbert, Sapp, and Tauber 2012; Gordon 2012; McFall-Ngai et al. 2013). The role of microbes in eukaryotic development, immune responsiveness, neural function, and digestion challenges treatments of eukaryotic and prokaryotic organisms as fundamentally distinct. Understanding the extent of lateral gene transfer among microorganisms more radically undercuts the biological species concept and threatens familiar cladistic conceptions of evolutionary patterns (Dupre 2012). The use of mass genomic sequencing to identify bacteria that cannot be cultured by traditional methods highlights the extent to which the vast majority of bacterial species have been hitherto altogether unknown and that traditional characterizations of evolution in terms of increasing complexity of life forms fail to recognize that bacteria have been and remain the dominant terrestrial life forms (DeLong and Karl 2005; Helmreich 2009; Dupre 2012). These developments in microbiology not only further decenter the world from our anthropocentric conceptions but begin to decenter our very self-understanding as "anthropic." As I noted above, however, recognition of and interest in the considerations that would displace us from an imagined centrality within the universe itself remain indirectly but resolutely anthropocentric—these considerations transform the very sense of the "*anthropos*."

II—The Partial Autonomy of Disciplinary Domains

Perhaps the most fundamental ways in which changes in scientific significance have revised human interests, rather than being determined by them, arise from the opening or reconfiguration of conceptually articulated domains (including the above examples of climate science and microbial genomics). Recognition of the coherence of scientific domains has long been mediated by the establishment of scientific disciplines, and disciplines in turn were long identified primarily by their theoretical conceptualization (Shapere [1984] exemplifies philosophical attention to scientific domains as *theoretically* constituted). Rejection of a theory-centric conception of scientific disciplines has now rightly become widespread in science studies, however. The role of instrumentation and technical skills in establishing novel disciplinary formations has been highlighted in fields from biology to high-energy physics (Bechtel 1993; Kohler 1994; Rabinow 1996; Galison 1997; Rheinberger 1997). Disciplines also involve institution building as much as conceptualization (Kohler 1991; Lenoir 1997): the resources needed to train students and secure their employment, to establish journals and university departments, or to provide reliable funding for research programs rarely come from narrowly scientific interest alone. Lenoir (1997) in particular locates a distinction between disciplines and research programs along just those lines: research programs are defined by scientific problems, but discipline formation must be understood in terms of political economy.

More careful reading of these arguments nevertheless suggests an intermediate role for the conceptual constitution of scientific domains.

Lenoir's informative characterization of the role of political economy in shaping the choice of Carl Ludwig for a new professorial chair at Leipzig in the 1860s is a good example. He argues that Ludwig's appointment helped secure the conditions for a new disciplinary formation in physiology and the subsequent orientation of that discipline toward its contribution to medicine, drawing upon a wide range of extradisciplinary resources and concerns. Lenoir nevertheless also shows how emerging conceptual and technical interdependencies within that scientific domain gave specific shape to the organization of the new discipline: "Ludwig drew an important conclusion . . . : to advance his own research interests and to defend the validity of his explanatory strategy, he needed to incorporate the skills and techniques (but not the disciplinary perspective) of microscopic anatomy into his program. . . . Similar experiences arising out of his work on respiratory and urinary physiology led Ludwig to appreciate the importance of introducing a chemical element into his explanatory strategy and of enlisting the skills of a trained physiological chemist" (1997, 66).

Bechtel (1993) makes similar claims about the emerging field of cell biology from the 1930s to 1950s. The new cell biologists came from a variety of fields, but their concern with the biological function of cellular structures drew on distinct aspects of cellular morphology, biochemistry, and physiology. What marked their disciplinary interest as cell biologists were precisely the conceptual relations among elements of these extant fields of study: not biochemistry for its own sake but for its contribution to biological function through localization within cellular structures. As Bechtel notes, very few biochemists changed fields to become cell biologists, but relevant aspects of biochemistry became central to the training and research of most cell biologists. The point to emphasize is that the interdependence of specific issues and skills in constituting a domain reconfigures their scientific significance. These aspects of cellular structure, cellular physiology, and the biochemical reaction pathways that link structure and function matter scientifically precisely because together they demarcate a conceptually coherent domain with characteristic patterns of counterfactual stability—that is, laws.

We saw in chapters 8 and 9 that laws only maintain their counterfactual stability holistically as members of domain-constitutive sets of laws. Which features of the world bear upon one another in these robust ways is a mostly empirical matter. Yet that empirical determination cannot be readily disentangled from what is at stake in its outcome. The degree of accuracy and precision with which the laws must remain stable, and the range of relevant counterfactual considerations under which they must hold nontrivially, depend upon the broader patterns of practical and conceptual significance within which that domain is situated. All these factors are bound up with the practical and technical capacities for articulating that domain practically as well as theoretically. Yet as we shall see, what is at stake in understanding a scientific domain can also depend upon its relations to other domains and other concerns.

We can see such interdependence exemplified with broad strokes in a capsule history of genetics as a domain, and the shifting conceptualization of genes implicated within that history. The prehistory of genetics as a conceptual domain took shape in the latter parts of the nineteenth century as conceptions of biological heredity began to incorporate patterns of both stability and change at multiple levels. Darwinian gemmules, Mendelian factors, Weismannian germ plasm, and other material elements postulated as transferred from organisms to their descendants emerged at the confluence of multiple stakes: the clearer boundary between living and nonliving things established by the rejection of spontaneous generation, taxonomic continuity and change within species evolution, genealogical relations among individual organisms, and attempts to secure and justify boundaries and hierarchies among socially salient but increasingly troubled classifications of people by sex, race, ethnicity, and socioeconomic class. As earlier chapters noted, genetics emerged from this larger, overdetermined conceptual space of "heredity" and "blood." It became a distinct scientific domain only through the material demarcation of a more constrained pattern of correlations among relatively discrete phenotypic traits, statistical "linkages" among those traits in standardized breeding experiments with pure and hybrid lines, and visible cytogenetic changes in cellular chromosomes in a relatively small number of model organisms. Genes were thereby transformed from theoretically posited elements or factors to spatially localizable objects with trackable patterns of interaction.

Ruthlessly simplifying a complex and contested history, we can note multiple shifts in the conceptual and practical configuration of that domain over the ensuing century of intensely focused research. In one direction, work on eye pigmentation and auxotrophic mutations further articulated the primary mode of postulated "gene action" as biochemical (Kohler 1994, ch. 7; Keller 1995). In a different direction, the mathematical modeling of Mendelian factors in population genetics and its correlations with genetic diversity in natural populations forged suggestive links between cytogenetic linkage patterns and evolutionary systematics (Provine 1971; Mayr and Provine 1980; Smocovitis 1996). The identification of genes with base sequences in nucleic acids simultaneously accomplished a more articulated materialization of genes and their abstract reformulation as "information" that could be understood as stable, despite repeated changes of its material embodiment from base sequences in DNA and RNA, to amino acid sequences in proteins, and to cascading reaction pathways (Olby 1974; Morange 1998; Kay 2000). The identification of DNA in endosymbiotic organelles consolidated hypothesized differences between nuclear and cytoplasmic inheritance patterns (Sapp 1987). The identification of the 'lac' operon initiated and expanded the functional role of genes from structural, information-bearing elements to incorporate dynamic regulation of gene expression (Morange 1998). The tracking of genetic pathways in organismic development, and their evolutionary significance, initiated a partial and contested reconciliation between the domains of genetics and embryology/developmental biology, which had significantly diverged in the earlier partition of genetics from the larger conceptual field of heredity (Robert 2004; Amundson 2005; Laubichler and Maienschein 2007; Pigliucci and Müller 2010). The development of technical capacities for the insertion, knockout, or amplification of identifiable DNA sequences made genes into technologically constructed and manipulable objects as well as in vivo functioning units (Rabinow 1996; Morange 1998; Beurton, Falk, and Rheinberger 2000). The establishment of large-scale, rapid base-pair sequencing, a more detailed articulation of the editing and splicing of genetic sequences in transcription and translation, recognition of the dynamic stability of the genome via mismatch repair mechanisms, and the identification of genetic homologs such as Hox and Pax genes across broad taxonomic and functional differences began to reorganize the entire domain. Genetics and the identification of individual genes began to give way to genomics as a dynamic, holistic conceptual domain (Beurton, Falk, and Rheinberger 2000; Keller 2000). The history of genetics reminds us that, despite considerable historical continuity through this convoluted, overlapping series of conceptual reformulations, the significance relations brought forth by the conceptual organization of systematically intraconnected scientific domains can shift dramatically over relatively short periods of time. What genes are; how they are involved in biological function and its evolutionary reconstitution; whether genes are discrete entities or only nodes in more holistic functional patterns; and which phenomena disclose genetic patterns, mechanisms, and functions have repeatedly been reconfigured as patterns of prospective conceptual significance.

Genetics shows in turn that a more intensive articulation of concepts within a scientific domain, strikingly exemplified by the history of the gene concept, also displays a close engagement with the exploration and reconfiguration of its conceptual relationships to other domains. Genetics became a distinct scientific domain within the larger field of heredity due to the establishment of a small number of experimental systems that developed "pure" plant or animal lines, which could then yield hybridized recombinations (Müller-Wille and Rheinberger 2012, ch. 6). These systems quickly displayed a relatively self-contained field of conceptual relationships among discrete, heritable phenotypic traits, their statistical linkage relationships in breeding experiments, the increasingly densely spatialized mapping of those statistical relations, and the correlation of the spatial linkages with visual transformations of chromosomes. The prospects for constructive exploration and articulation of that emergent domain dramatically enhanced the scientific significance of its domain-opening experimental systems and conceptual relations: they enabled a much more finely grained understanding and assessment of the aspects of biological function and evolutionary continuity and change falling within that conceptual space. Yet we also saw in chapter 7 that the significance of those intradomain relations was dependent upon their not being entirely self-contained. Chromosomal trait-linkage maps were more than just a trivial curiosity because of the promise (initially a vague promise in many respects) that these intradomain relations were also informative about a broader pattern of biological relationships. Even this brief, drastically simplified summary of key transitions in the history of genetics reminds us of the extent to which the significance of genetics has been bound up with possibilities for how the finely articulated relations among genetic loci, sequences, codes, regulatory processes, and biochemical pathways could be recognized as *intentional* patterns in the sense of being informative *about* other aspects of organismic function and evolution. Thus in one direction, enhanced conceptual understanding requires the intensifying articulation of inferential relations and experimental skills and practices within relatively self-enclosed domains, whose stability and coherence enable inductive confirmation and counterfactual reasoning.[3] From a different direction, conceptual understanding requires some grasp of how and why it matters to work out a more fine-grained articulation of this conceptual

3. I am here using the term 'inferential' in a very broad way to incorporate the wide range of ways in which the sciences have discerned and articulated informative patterns in the world. These include inferential relations among sentences as paradigmatic cases, but they also include the wide range of ways in which scientists have learned to track and interconnect conceptual relationships: mathematically, visually, diagrammatically, and spatially, as well as the construction and refinement of experimental and theoretical models.

domain and to get it right. There must be something further at stake in the articulation and assessment of those inferences for them to be genuinely conceptual. In this way, the articulation and refinement of conceptual domains in the sciences exemplifies the partial autonomy of the conceptual that enables its characteristic two-dimensional normativity.

This partial autonomy of conceptually articulated domains in the sciences highlights the importance of conceptual relationships *among* scientific domains as well as within them. Philosophical interest in the conceptual relations between disciplinary domains has long been dominated by the issue of whether the domains of the "special sciences" could be incorporated via reduction or supervenience within what was thereby understood to be the "fundamental" domains of physics. Recognition of serious problems with reduction and even supervenience nevertheless posed the issue of how then to understand both the practical and the metaphysical significance of the seemingly recalcitrant autonomy of different disciplinary domains. This concern can even be found at the heart of the "unity-of-science" movement. As Nancy Cartwright and Thomas Uebel (Cartwright et al. 1996) have prominently argued, Otto Neurath's vision of unity was the antithesis of a conceptual imperialism: he conceived the unity of science as encyclopedic rather than systematic, with the achievements of multiple sciences brought together at the point of application to practical concerns that exceed the competence of any one domain of expertise.

IV—Conceptual Articulation as Both Homonomic and Heteronomic

The issue nevertheless looks somewhat different once one takes the domains in question to have their own internal systematicity or even necessity. Donald Davidson influentially posed this issue as constituting one of two fundamentally different kinds of conceptual relations:

> On the one hand, there are generalizations whose positive instances give us reason to believe the generalization itself could be improved by adding further provisos and conditions stated in the same general vocabulary as the original generalization. Such a generalization points to the form and vocabulary of the finished law: we may say that it is a *homonomic* generalization. On the other hand, there are generalizations which when instantiated may give us reason to believe there is a precise law at work, but one that can be stated only by shifting to a different vocabulary. We may call such generalizations *heteronomic*. (1980, 219)

Davidson acknowledged that much of our ordinary talk and understanding is heteronomic but thought that such inquiries, while often important, resist systematization within a single law-governed domain. In the mental and physical domains, whose unity or diversity was his central concern, there is an important place for heteronomic inquiry, since Davidson took these distinct conceptual domains to cross-classify many of the same events, with our interests often engaged by the causal relations between mental and physical events in both directions. Yet those causal relations would not be sufficient to permit systematic inquiry conjoining the two domains, he argued, because "causality and identity are relations between individual events no matter how described[,] but laws are linguistic; and so events can instantiate laws, and hence be explained and predicted in the light of laws, only as those events are described in one or another way" (1980, 215). Davidson then insisted that "there cannot be tight connections between these realms if each is to retain allegiance to its proper source of evidence" (1980, 222), since they are constitutively governed by the quite different ideals of rationality and exceptionless causal law.

The account of conceptual understanding put forward in this book differs from Davidson's conception of the relation between homonomic and heteronomic inquiries in three far-reaching ways, despite considerable common ground.[4] First, Davidson takes the realm constituted by "strict physical law" to extend throughout the natural sciences as a single homonomic domain. I have instead argued for a nomological pluralism in which the patterns that display holistic counterfactual stability constitute distinct domains of natural scientific inquiry governed by different constitutive issues. Heteronomic inquiry is thus pervasive within the sciences wherever research explores the boundaries of domains or draws upon concepts or results from one domain to refine the conceptualization of another. A second difference arises in my resistance to Davidson's attempt to distinguish the causal realm of events from the linguistic realm of laws. As I argued in chapter 8, laws are patterns in the world that incorporate the practices and skills that let those patterns be recognizable, so their lawful systematicity and their discursive articulation in scientific practice cannot be kept distinct.[5] Finally, Davidson

4. Davidson also restricts himself for philosophical purposes to the use of first-order extensional locutions, whereas I have argued that alethic and normative modalities are indispensable in understanding our conceptual capacities, but that difference does not play a direct role in my discussion below.

5. The irreducible entanglement of causal relations and their discursive articulation showed up more extensively in my previous book, *How Scientific Practices Matter* (Rouse 2002). There I argued

takes the boundaries and the constitutive issues and stakes that govern the physical and psychological domains to have already been settled, along with their heteronomic irreducibility. I have instead argued that scientific research fields are appropriately taken to be holistically law-governed domains, even though the configuration of those fields and even ultimately their status as intelligible domains are at issue in ongoing practice.

These considerations give a very different sense to the Davidsonian difference between homonomic and heteronomic inquiry and understanding. The sciences articulate and reconfigure the world as a "space of reasons" first and foremost by opening and further developing many distinct, holistic conceptual domains, within which we can talk and reason about and act toward aspects of the world that would otherwise be concealed or opaque. Such "homonomic" fields of material-discursive practice, to adapt Davidson's term, have a distinctive significance for scientific understanding due to the "expressive freedom" they enable (Brandom 1979, sec. 3). The systematic development of experimental or other empirically disclosive systems or operations works in concert with systematically interconnected concepts whose inferential relations are accountable to those research practices.[6] Opening and refining such domains thereby further extends the characteristic two-dimensional normativity of conceptually articulated understanding. It becomes possible within such domains to track the *appropriateness* of theoretical calculations, experimental manipulations, "data" recordings, and their concep-

both that Davidsonian radical interpretation (or Brandom's parallel model of discursive scorekeeping) was itself a causal phenomenon and that determinate causal structure must incorporate its conceptual articulation as structured in that way (see especially Rouse 2002, 284–93). This recognition also challenges Davidson's formulation of the distinction between homonomic and heteronomic conceptual domains. Davidson (1980, essay 11) argues against Goodman (1954) that the problem with the inductive projectability of "grue" is not a problematic feature of that concept by itself but rather that "grue" is heteronomically related to other familiar concepts, such as "emerald." "All emeralds are grue" is heteronomic, whereas "all emerires are grue" would be a well-formed, projectably homonomic generalization. Davidson's point would be well taken if concepts were only linguistically interrelated. The recognition that empirical concepts are inextricably bound up with practical and perceptual skills that constitute the recognizability of the relevant patterns, argued for in chapters 7–8, shows why Davidson's claim is mistaken. "Grue" is not and cannot be a genuine empirical concept (as opposed to a philosophical concept entirely parasitic upon its defined relation to familiar color concepts) without its integration into an appropriate form of skillful recognition. The difficulty of being able to tell when a present object was first observed suggests that the needed recognitive skills are unattainable (for a related argument against the intelligibility of "grue" as an empirical concept, see Haugeland [2013, 243–48]).

6. In chapter 7, I referred to this interconnection of experimental systems, models, and the concepts thereby projected inductively as the "double mediation" of the relation between theories and data.

tual expression and to do so with some degree of independence from the empirical *correctness* of the outcome. This partial independence enables us to *say* contentful things about some aspect of the world, which *matter* in that they may or may not be correct, in ways that make a further difference to what we (should) say and do. There is a sui generis or "fictional" character to these conceptual domains in that there is no domain at all without the establishment of experimental systems or other fields of phenomena within which what is conceptually appropriate and what is empirically correct mostly coincide. Typically, constructing that space requires developing an experimental system whose controlled operations produce salient patterns that exemplify conceptual relationships, alongside theoretical models and inferential norms that extend those patterns beyond their exemplifying instances. Davidson (1984) argues that a background of massive success as a standard for the assessment of error is a constitutive norm for the interpretation of utterances and beliefs more generally. Scientific practices nevertheless also constitute such patterns more locally by establishing a regimented field of conceptual operations whose possibilities for further defeasible extension open a scientific domain.

The significance of the difference between homonomic and heteronomic inquiries arises from this indispensable combination of systematicity and open-endedness that characterizes conceptually articulated domains. Concepts are only inductively projectable and inferentially regulated within systematically interconnected conceptual spaces that are appropriately taken to be demarcated by laws with the requisite counterfactual stability. Yet my criticisms of Hacking and Cartwright in chapter 7 showed why these domains cannot be self-enclosed; it is only through their openness to further intensive or extensive articulation that those domains are *conceptually* articulated as having a content—that is, as making a difference that matters. The open-ended inferential and articulative possibilities that secure conceptual significance can go in multiple directions. The obvious cases are research fields whose concepts undergo further articulation to work out just what should be said about situations not yet fully articulated in these terms. Yet we also saw cases such as geometrical optics (briefly discussed in chapter 7) whose internal refinement seems complete and yet whose limited significance is still secured by its place within a broader discursive context that highlights its empirical limitations, which must consequently be supplemented by a different set of laws. Only because these models are *about optics*, and hence open to empirical correction and theoretical extension in other terms, are they conceptually contentful at all as a counterfactually stable set that can be intelligibly understood ceteris paribus.[7]

The significance of heteronomic inquiry in its multiple forms in turn arises from this constitutive open-endedness of conceptual domains. Scientific inquiry is directed toward and across the boundaries of its articulated domains, including efforts to build connections between domains or place them with respect to one another. Understanding a scientific domain is in part recognizing where and how it confronts open questions and further possibilities for inquiry, even though it is a separate issue whether those boundaries need actually be explored or extended in subsequent research. Sometimes research that draws upon resources developed in different domains takes up a limited project, in which concepts, materials, skills, or results from one field suggest or enable interesting possibilities to explore within another area of research. Sometimes it addresses broader concerns that extend the resources of several research domains to constitute a heteronomic research "field" that does not yet seem to constitute an autonomous research domain. From Darden and Maull's (1977) classic paper on interfield theories to Galison's (1997) analogy between interdisciplinary "trading zones" and linguistic pidgins and creoles, philosophers of science have rightly recognized the existence of more tentative or fragile fields and modes of research that draw upon the resources and concerns of more than one scientific discipline. At the outset, it is often unclear whether such heteronomic explorations will turn out to be a limited common effort, a persistent interdisciplinary trading zone, or the locus for the emergence of a new disciplinary domain, perhaps as the conceptual reconstitution of one or more of its predecessors. The conceptual open-endedness of research domains reflects a practical *commitment* to taking one's concepts as inductively projectable as part of a counterfactually stable set but one whose full contours are not yet determined.

The gradual emergence of the interdisciplinary field variously known as evolutionary developmental biology ("evo-devo") or developmental evolution ("devo-evo") provides a telling example of how the configuration, direction, and significance of a heteronomic conceptual field can itself be at issue in its ongoing exploration. This field has a complex prehistory. On any account, that prehistory goes back at least to the different configurations of genetics in Germany and the United States during the early years of the field (Harwood 1993; Rheinberger 2010,

7. For an extended discussion of ceteris paribus laws, and how the implicit provisos invoked are not trivially permissive, see Lange (2004).

pt. 2), as German genetics maintained a commitment to the integration of genetics and embryology/development within a more unified conception of heredity. The disintegration and subsequent eclipse of the German genetic tradition during the National Socialist period was reinforced by the emergence of the modern evolutionary synthesis, which notoriously excluded embryology from its purview. For some contributors, however, the historical origins of the field trace back to Geoffroy St. Hilaire, Richard Owen, and the nineteenth-century "structuralist" tradition in morphology and embryology (Amundson 2005; Laubichler 2007), and the conceptual priority they accord to development suggests its alternative characterization as "developmental evolution" or "devo-evo." The emergence of a new interdisciplinary research field at the borders of evolutionary genetics and developmental biology (which may also be the reemergence of this structuralist tradition) was made possible for all concerned by striking results at the molecular level, notably the discovery of a homeotic gene complex (Lewis 1978) and developmental signaling cascades (Nüsslein-Vollhard and Wieschaus 1980) in *Drosophila*. Yet the significance, direction, and even evidential norms of this new field remain very much at issue. From one direction, evo-devo looks to be an extension of the modern synthesis, which attends to the evolution of regulatory, developmental genes into a "developmental tool-kit" that is then highly conserved across taxa (Carroll, Grenier, and Weatherbee 2001). From other perspectives, developmental evolution contributes to a revisionist project in evolutionary theory that emphasizes genes as relatively plastic resources for epigenetic processes in development that indicate the close interconnections of ontogeny and phylogeny (Wagner 2000, 2001; Wagner and Larsson 2003; Laubichler and Maienschein 2007). The point I want to emphasize is that recognition of the centrality of many of the same results, in an interdisciplinary field that by widespread agreement draws upon evolutionary genetics, comparative morphology, developmental biology, and systematics among other fields, is nevertheless consistent with quite different visions of where the field is going, what are its most central concerns and directions of research, and how its achievements and prospects matter to biology generally.

A different kind of heteronomic inquiry occurs at the boundaries of classical and quantum mechanics, where a variety of models and techniques have been developed to account for phenomena that cannot be readily modeled in terms of either theory alone, even though these models are explicitly limited to understanding these border phenomena. Quantum mechanics purports to be a thoroughly inclusive field, with classical mechanics invoked only as approximately accurate for those events whose scale is large relative to Planck's constant, often as a first approximation that treats other terms as if they were negligible. Yet the two fields are taught and practiced in their own terms, applying distinct families of theoretical models, despite their conceptual and methodological inconsistency, and we have seen Cartwright (1999) defend the legitimate autonomy of the laws constitutive of each field. At multiple points, however, physicists have had to develop "semiclassical" models that draw upon elements of both sets of laws, for example, for making sense of quantum mechanical analogues to classical deterministically chaotic systems or modeling the behavior of quantum mechanical particles in classically understood electromagnetic fields. Winsberg (2010, ch. 5) calls attention to a philosophically remarkable example of heteronomic inquiry across these borders in which incompatible models from continuum mechanics, classical molecular dynamics, and quantum mechanics were developed to account for the nanomechanical behavior of the same materials at different scales. Since the phenomena modeled at one scale have nonnegligible effects at other scales, scientists then had to develop more localized "hand-shaking" models to bring the results of theoretically incompatible models to bear upon one another. Here is a case where heteronomic reasoning was highly localized yet in ways that could accommodate significant conceptual incompatibilities.[8]

While heteronomic inquiry is thus sometimes quite local, Bechtel (1993) and Rheinberger (1995) provide a telling example of the unanticipated emergence of a novel conceptual domain at the borders of familiar disciplinary domains.[9] They describe the constitution of what became modern cell biology through the introduction of the ultracentrifuge and later the electron microscope into research programs originating within more traditionally constituted medical or biological domains. Albert Claude's work with chicken sarcoma cells, for instance, shifted the conceptual significance of the boundary between the normal and the pathological in cytological research as the precipitated fractions of sarcoma cells came to indicate cellular components with a better defined *normal* functional significance. Correlations between the fractionated materials and the visible structures revealed by light microscopy enabled the biochemical investigation of those structural components of cells. These

8. Readers of Wilson (2006) will not be surprised, as he has shown how the use of conceptually incompatible models to account for similar phenomena at different scales, different energies, or different levels of requisite precision is ubiquitous in applied mechanics and materials science.

9. My presentation of this example is especially indebted to Rheinberger (1995).

developments pointed beyond the static structures discerned by classical cytology toward a dynamic functional organization of cells. Biochemical processes previously analyzed in vitro could then be situated within a spatially organized field that enabled a more detailed decomposition and tracking of biologically significant processes within a cell.

The introduction of new instruments or the discovery of new phenomena can thus disclose novel conceptual relationships that reconfigure the patterns of inductive projectability, counterfactual reasoning, and experimental maneuvering that constitute conceptually articulated domains of inquiry. James Bono (1990) has argued for a comparable and more pervasive role for "metaphorical exchange" across scientific and other discursive domains. Metaphor and other tropes are more commonly discussed within philosophy of science in terms of Max Black's (1962) interactive conception of metaphors that emphasizes a transfer of meaning from one context to another (e.g., Boyd 1979; Arbib and Hesse 1986). In speaking instead of metaphorical "exchange," Bono shares my recognition that scientific concepts function holistically as part of a larger conceptual domain, even though his account does not take into account the modal significance of the collective counterfactual stability of a set of laws. The modal relations discussed in chapter 8 further reinforce the holistic interdependence of concepts in "homonomic" domains. Invoking a term that is normally employed in one context in order to extend or adapt work in another domain typically requires recognizing a broader set of inferential relations that implicitly accompany the term from its original context, including their holistically homonomic counterfactual stability. The new uses of the term will usually draw upon some but not all of those inferential roles elsewhere. These inferential roles are also usually not all compatible with other concepts and conceptual roles within the new setting, thus requiring mutual adjustment. These novel uses and inferential adjustments may in turn reverberate back into their original context as the inferential adjustments in the new setting suggest limitations or conflicts that call for further adjustment in turn.

The conceptual dynamics that result from such metaphoric exchanges become more significant on Bono's account because he sees them not just occurring between scientific domains but drawing upon and contributing to other discursive domains. He also argues that such dynamics result not merely from explicit theoretical inferences across conceptual domains but from the implicit heteronomic conceptual relations invoked in acquiring and maintaining a conceptual repertoire: "Metaphors and tropes may be transmitted over time, but their meaning must always be reconstituted synchronically. That is to say, such meanings are socially and culturally situated, carrying resonances that speak forcefully to individual members of specific communities. But this very process of reconstituting the meaning of metaphors subjects them to the interference of other discourses. . . . The metaphors and tropological features of extrascientific discourses—whether religious, political, social, economic, or 'literary'—through individual acts of interference and interaction come to constitute a synchronically coherent, if now metaphorically reordered and situated language" (Bono 1990, 77). The role of such metaphorical exchanges in shaping the configuration of disciplinary domains and their significance has been recognized in multiple cases. Paul Edwards (1996), for example, has extensively tracked a broad range of scientific and extrascientific metaphorical exchanges that played constitutive roles in the development of computers and the emergence of cognitive psychology as a disciplinary domain. Mary-Jane Rubenstein (2014) has recently displayed the multiple philosophical and theological entanglements that inform contemporary theorizing in physics about a possible "multiverse"; multiverse cosmologies are often explicitly put forward as challenges to anthropocentric or theological interpretations of the "fine tuning" of physical constants that enables human life. Vassiliki Betty Smocovitis's (1996) cultural history of the modern evolutionary synthesis has similarly shown the rich and diverse issues and stakes that can attach to the very project of scientific unification across disciplinary domains.

The history of genetics once again provides especially clear illustration of these issues. Central themes in that history, such as the positing of genes as stable elements "underlying" developmental changes in the body, the Weismannian distinction between somatic and germ plasm, quasi-homuncular conceptions of "gene action" or "gene activity" (Keller 1995), and the productive conception of genes as material embodiments of immaterial information that constitutes a developmental program, have an extensive history of conceptual entanglement with other discursive formations. Genetics has engaged theological and psychological discourses about souls and personal identity, kinship as grounded in "blood" relations, powerfully gendered conceptions of reproductive roles, and newly emergent talk of information, programming, computing, and coding (Sapp 1987; Martin 1991; Keller 1992, 1995; Nelkin and Lindee 1996; Haraway 1997, ch. 4; Kay 2000; Goodman, Heath, and Lindee 2003). These heteronomic inferential entanglements help explicate not only some of the productive conceptual developments in the history of genetics but also the distinctive significance accorded to genes, genetics, or genetic manipulation within the life sciences, by scientists and others alike, often expressed in quasi-theological language invoking "secrets of life," "the Holy Grail," or "tampering." These invocations should not be dismissed or diminished as "external" impingements on internal scientific developments. Scientists bring a more extensive cultural-conceptual heritage into their education within and understanding of their disciplinary domains, and that discursive familiarity with broadly expressed issues and stakes does not go away in the course of a scientific career. Nor are those issues irrelevant to the detailed development of scientific fields themselves, alongside the reciprocal influence of scientific work on the conceptualization of broader patterns of cultural understanding. Sciences matter in significant part through the ways in which their concepts and achievements draw from and bear upon issues that are already recognized as significant within broader conceptual and practical fields.

These broader heteronomic entanglements of genetics and other scientific domains arose directly from their emergence as intelligible, scientifically significant fields of research. Genetics became a distinct scientific domain, and an especially central and significant domain, in part because it was not entirely self-enclosed. Genetics was, to be sure, a circumscribed material and conceptual practice that enabled carefully regulated experimentation and inference within a limited range of laboratory-based systems. The rapid recognition of its scientific significance resulted to a great extent from its practitioners' achievement of extraordinary empirical productivity by means of rigorous experimental and conceptual control. Yet from the outset, those localized patterns of reasoning mattered scientifically in significant part because they also promised to be informative about other biological domains, with stakes that were also not limited to biology. Indeed, at some points in its history, the promissory relevance of genetics to organismic function, reproduction, and evolution more generally, and of human genetics to broader aspects of human life, has encouraged a synecdoche that identifies biology as a whole with the domain of genetic influence or determination. Genetics is one of a small group of scientific subfields that have acquired such promissory generality at various historical junctures. Quantum field theory and general relativity in physics, or the many efforts to build more systematic heteronomic connections between them, may now seem to hold a similarly distinctive place within the sciences more generally. Some practitioners make comparable claims for the unifying role of economics or psychology within the human sciences or for the physical chemistry of intramolecular bonding and intermolecular forces as constitutive of what we might call the "material sciences."[10] The prospective generality of these disciplinary domains, as promising a more comprehensive and complete scientific understanding, has at the very least given these fields an especially central place in *philosophical* reflection upon the sciences.

This recognition of relations between generality and scientific significance brings us back to the issue that initiated this chapter as well as this part of the book: the familiar conception of the scientific image as a comprehensive, internally consistent representation of the world as a whole. The selective foci of scientific research, and its directedness toward historically contingent and contested issues and stakes, suggest that the sciences themselves are not the source for this aspiration to a comprehensive representation of the world, even though some scientific subfields gain promissory significance from their own aspiration to and promise for greater generality.[11] This counterindication is reinforced by recognition of scientific research as itself a consequential form of niche construction. The sciences remake parts of our surroundings through the construction of experimental systems, theoretical models, and other conceptually articulative activities that allow some aspects of the world to be intelligible in new ways. The sciences open and continually refine and reconfigure conceptual spaces, which we can recognize as aspects of the Sellarsian "space of reasons," now reconceived as part of our biological heritage and way of life. They produce not a systematic representation of the world but a space of possible—that is, intelligible—conceptualization, which is oriented by contested issues and a sense of what is at stake in their possible resolution. Understanding scientific research in this way as sustaining a temporally extended conceptual field

10. The conceptually unifying significance of chemistry for the "material sciences" becomes clear from the typical undergraduate science curriculum, for which introductory inorganic and organic chemistry provide what Latour (1986) calls an "obligatory passage point" for all the other sciences except physics in a way that physics does not. Indeed, Lange's (2007) claim that only the "gross features" of physical laws are relevant to the conceptual norms for reasoning in other disciplines is reflected in the typical construction of an introductory sequence of "physics for the life sciences," which only aims to provide biologists, neuroscientists, and premedical students with an understanding of the "gross features" of physical concepts and laws, often utilizing only minimal mathematical articulation of the domain.

11. When I say that the demand does not come from the sciences themselves, I mean that it is not part of the ways in which scientists frame research questions, assess their significance, pursue those questions in their research, and take account of the results of previous research in the course of doing so. Philosophical views about the aims and methods of the sciences are also part of general intellectual culture, and many scientists hold broader philosophical views that are not mandated by their own research practices.

suggests that the aspiration to a comprehensive scientific representation may be a philosophical demand, shaped by the discipline's historical commitment to understanding and assessing the use of "knowledge" as a general philosophical category. Naturalists should always be suspicious when confronting such philosophically imposed demands for what the sciences must aspire to.

Yet we should recognize that this challenge to a comprehensive, representational scientific image also resists any opposing conception of scientific disunity that postulates an irreducibly fragmented conceptual space. Aspirations to systematicity and generality do mark an important dimension of conceptual normativity. Those considerations continue to shape scientific understanding, even though conceptual unification has been pursued in different ways (Morrison 2000) and is continually challenged by the proliferation of more localized research fields and their multiple heteronomic interconnections. The open-endedness and broader accountability of conceptual domains both encourages and relies upon the exploration of such heteronomic conceptual relations among diverse scientific and "extrascientific" domains, without demanding or aiming toward a comprehensive unification among them. The patchiness and heterogeneity of scientific conceptual development should not be surprising given its character as systematically directed but also open-ended and contested. We can now recognize that conceptual understanding depends upon open-endedness of this sort, which enables its constitutive two-dimensional normative accountability. The conception of scientific understanding developed here in part 2 recognizes this ongoing, productive tension between systematicity and proliferation as a central feature of the "space of reasons" articulated in scientific practice. "The scientific image," understood as a composite grasp of the world drawn from a multifarious research enterprise, is not a single comprehensive representation in either its unificationist-reductionist or its pluralist versions. It is not an image or representation at all. The sciences continually reconfigure our involvement in the world as an open-ended field of conceptual possibilities, fraught with productive tensions, focused upon shifting issues, and oriented toward working out and reconfiguring how those possibilities make a difference to our lives and the world we inhabit.

Conclusion

ELEVEN

Naturalism Articulated

This book has attempted to work out a more adequately naturalistic understanding of conceptual capacities, especially capacities for scientific understanding. To understand those capacities naturalistically is to show how they are a natural phenomenon, intelligibly part of the natural world as scientifically understood. "Understanding" is normative; not just any account purporting to situate scientific understanding of nature within nature will do. Such explications must answer to what is at issue in claiming to *understand*, to understand the *sciences*, and to do so *naturalistically*. Among those issues, four stand out. A naturalistic account of scientific understanding must accomplish the following:

1. Answer to the continuing work of the sciences, taking into account the best available research that bears upon its concerns;
2. Account for scientific understanding as actually embodied and achieved in scientific research rather than a projection of what the sciences must be like to fulfill our philosophical expectations or preconceptions;
3. Explicate scientific understanding in ways that would not undercut its authority as conceptually contentful, empirically accountable, and truthful; and
4. Appeal to nothing supernatural—that is, magical, transcendent to nature, or inexplicable.

These four issues do not stand out arbitrarily by philosophical fiat. They are philosophically salient because they express the principal respects in which other contemporary approaches to a naturalistic understanding have failed.

Defending that critical assessment has not been among the argumentative tasks undertaken in this book. That assessment forms the background for my arguments, a lesson to be learned jointly from the Sellarsian, Quinean-Davidsonian, and Heideggerian traditions in philosophy; from recent work in interdisciplinary studies of scientific practice; and from important developments in the biological and other sciences. Many philosophers will undoubtedly take issue with that assessment. A different book would have been needed to engage critically with other contemporary attempts at a naturalistic account of scientific understanding as conceptual, and had I written that book, the important constructive work in response to those criticisms would still remain to be done. As it stands, my efforts to respond constructively to these four issues should provide ample basis for readers to assess for themselves whether they could do better with their preferred approach to a naturalistic philosophical orientation or by trying to make sense of scientific understanding without commitment to naturalism.

I do think that most contemporary advocates of naturalism in philosophy have failed to varying degrees on all four counts. Philosophical naturalists have not yet adequately come to terms with the "extended synthesis" in evolutionary biology and genomics. Many have promoted philosophical conceptions of science and the scientific image that diverge from scientific understanding in practice, to their detriment: some have overlooked or idealized away the sciences' disciplinary divisions, patchy modeling, experimental idealizations, and selective orientation toward scientifically significant concerns; others have taken such disunity to heart in ways that would diminish the sciences' conceptual scope and significance. Many supposedly naturalistic accounts of knowledge and mind would, if correct, yield impoverished capacities that do not do justice to how the sciences have refined and articulated our conceptual capacities while remaining empirically accountable. Often these latter shortcomings have been concealed by ascribing to the sciences an impossible "god's-eye view" of the world and ourselves from "sideways on," a conception of scientific language or theoretical models as magically contentful apart from the extensive scientific work needed to bring those concepts to bear on the world in definite ways, or a philosophical conception of scientific laws and their necessity disconnected from scientific concepts and practices.

The inference I draw from these failures is not that we should therefore abandon any stringent commitment to naturalism, as do many philosophers today (including many whose work has importantly contributed to my project). The commitment guiding this book is instead that, as aspiring naturalists, we can do better. I have argued that doing better requires two mutually supporting groups of revisions to our most familiar and widely accepted philosophical views about conceptual capacities and about scientific understanding. My account of the normativity of conceptual articulation is exemplified by how scientific research refines and reconfigures our discursive environment as a space of reasons. That account in turn shows how contemporary scientific research enables a more adequate grasp of our own capacities and prospects as environmentally dependent organisms and as rationally, empirically accountable agents. The central aim of the book has been to develop those revisions and their rationale and to show how their mutual support provides a more adequate, and more adequately naturalistic, understanding of our capacities and achievements as a scientific culture. This concluding chapter summarizes the principal claims of the book and the resulting conception of the sciences, of ourselves, and of naturalism as its guiding philosophical orientation. The epilogue suggests one way in which this conception illuminates some of the most pressing contemporary issues confronting us politically, scientifically, and biologically.

I—Conceptual Understanding

Any naturalistic understanding of our capacities to understand and reason about the world we live in must now start with our lives as organisms, making a living through ongoing intra-action with our environment.[1] We thereby sustain and differentially reproduce the biological lineage of which we are a part. Organisms and their lineages are patterns of ongoing environmental intra-action that sustain and reproduce that very pattern as a way of life. As thermodynamically open systems, organisms extract energy and their constituent materials from their surroundings, export entropy and waste, and maintain the integrity of that ongoing process against internally or environmentally induced disruptions. Not all organisms and lineages succeed in doing so, but those life patterns

1. Naturalism must begin with biology not because of some eternal, invariant conceptual or preconceptual organization of the world but because we inherit an intellectual tradition that has made our physiology, development, and evolution as organisms in a lineage salient and conceptually indispensable. As we shall see, that intellectual tradition is itself part of the ongoing process of behavioral niche construction that can now be understood in biological terms. We find ourselves in the midst of that intellectual tradition, both drawing upon its resources and responsive to its challenges, in the same way that as organisms we find ourselves in the midst of our environment, and having to sustain our way of life within it. These two aspects of our lives work "in the same way" because the former is part of the latter via behavioral niche construction.

that fail to sustain and reproduce themselves disappear. Those patterns of exchange articulate the world by configuring two boundaries within it. The primary boundary is between the organism itself as a goal-directed process of self-maintenance and the environmental conditions that characteristically enable or affect that process. The secondary boundary is between that organism-configured environment—comprising the interconnected aspects of the world that the organism's physiology, development, and differential reproduction collectively pick out as enabling, engaging, and/or threatening its ongoing way of life—and the larger physical setting, opaque to the organism's life processes, against which that environment stands out.

These boundaries are continuously reconfigured by two-way, nonlinear, boundary-sustaining intra-action across them. Organismic patterns reproduce differentially in response to selection pressures imposed by environmental circumstances, while organisms' developmental and selective environments are reconfigured by the cumulative, niche-constructive effects of their life processes. Those changes also mark out shifting "outer boundaries" of organismal environments. Some aspects of the larger setting become newly salient for the organism's way of life, while formerly salient aspects of its environment recede into opacity. The dynamics of this process are driven by the complex ways in which different life patterns intra-act in turn. Each organism encounters other organisms and lineages as integral components of its environment, and the nonlinear dynamics of each organism/environment intra-action reverberate through this complexly articulated web of life. The resulting complexity and diversity of life on earth, replete with symbiotic mutuality and hierarchy, divergent lineages, complexly intra-active physiology and behavior, and linked population dynamics, should not surprise us in hindsight. The dynamic equilibrium of solar input and the earth's radiative output has, in the fullness of deep time, provided the energetic resources needed to fuel those entangled processes and nonlinear intra-actions of diversifying differential reproduction.

For all its entangled diversity and complexity, however, organismic differentiation from and consequent articulation of its earthly environment is one-dimensional. The dimensional metaphor is apropos here due to the opacity of larger dimensionality to smaller that Edwin Abbott's (2010) *Flatland* long ago made salient. Only from within our two-dimensionally articulated way of life can the "limitation" to one dimension of other articulative life processes manifest itself. The scare quotes indicate that, from within those ways of life, one-dimensionality is no limitation, and the two-dimensionality of conceptually articulated understanding is not encountered as absent.[2] Yet for all their complex articulation in that dimension, other organisms' life processes do articulate the world only one-dimensionally. *We* can distinguish, within our conceptual repertoire, multiple aspects of organismic life: cellular and organismic physiology, development, behavior, ecology, reproduction, and evolution, for example. We can correlatively distinguish multiple aspects of the organism's developmental and selective environment, since different aspects of the organism's surroundings affect its ongoing way of life in different ways as nutrients, camouflage, shade, potential mates, threats, or perceptual cues. The organism itself nevertheless enacts these aspects of its life pattern only holistically in response to these features of its environment. Everything it does is simultaneously physiological, behavioral, ecologically situated, perceptually responsive, reproductively relevant, and so forth and must "balance" those diverse demands and constraints. It does not, however, discriminate and then recombine these distinct factors in appropriate balance. Organisms are instead tightly and holistically coupled with their environments. What seem to us distinct components of their way of life instead integrate the whole at every point: every perceptual processing (and its enabling neurological wiring and connection) is a physiological process burning nutrients, every movement is a perceptual shift, every developmental stage is under selection pressures, and every nutritional fulfillment is ecologically situated amid predators, parasites, potential mates, and so forth.[3] Organisms are able to be finely attuned and responsive to their environments in ways that are instrumentally rational in context and multifactorial from our perspective, precisely because they need not discriminate and reintegrate discrete "factors" but can flexibly respond to those environmental configurations as a whole.

2. The fact that the conceptually articulated content and reasoning embedded in our discursive performances is opaque to other organisms is what is right about McDowell's (1994) claim that conceptual normativity is sui generis. Conceptual normativity cannot be explicated in terms of the kind of one-dimensional responsiveness to their surroundings characteristic of other organisms (or simply in causal or information-theoretic terms, in a different variety of "bald" naturalism). It nevertheless can be understood as a biological phenomenon, once one recognizes the ways that discursive niche construction belongs to a larger pattern of organismic life. McDowell follows this line of argument only partway, recognizing *that* we are rational *animals* whose rationality arises within our habituated "second nature" while insisting that this recognition leaves the sui generis character of rational normativity unaffected.

3. The inclination to identify organismic behavior with its primary significance within our biological explanations (e.g., as hunting, eating, mating, predator avoidance, etc.), in abstraction from the holistic physiological and behavioral complex within the organism's way of life, parallels the ways in which we overlook the material apparatus that enables the display of conceptually salient patterns in the laboratory (see chapter 7). In each case, we erase the materially enabling conditions for the manifestation of a conceptually significant pattern.

We, too, are organisms, and everything just said is true of us. Our conceptual capacities are part of a holistically integrated way of life, albeit as extraordinarily demanding components. Neurological tissue and neural processing are energy-intensive and time-consuming; our niche-constructive off-loading of cognitive processes onto symbolic expressions, equipmental complexes, and social institutions magnifies those costs even while mostly compounding their effectiveness; and the developmental complexity of reproducing that way of life introduces relations of neotenous dependence that extend throughout most of our life span on one side or the other of those relations. One result is that, as an element of other organisms' environments, we are an extraordinarily voracious and destructive species, and our way of life depends upon and actively seeks out an ever-expanding and diversifying range of environmental resources. The boundary between the human environment and its biologically opaque background has thereby dramatically shifted outward over the history of our lineage.

From within our biologically articulative way of life, however, we have also opened a conceptually articulated space of reasons as partially autonomous from the broader way of life to which it belongs. The underlying capacity is the ability to track and produce a wide range of behaviors and environmental features in two dimensions simultaneously: as integral parts of our biological way of life but also as proximally responsive to their interconnections within a relatively autonomous field of conversational, intralinguistic, social-discursive relations. We saw in chapters 3 and 4 that this capacity and the environment within which it is reproduced first emerged in rudimentary form under unusual selection pressures and opportunities. It was then sustained, reproduced, and further articulated over many generations through extensive and pervasive forms of behavioral and material niche construction. Language, other symbolically expressive activities, and extensive social-institutional-equipmental complexes have thereby become salient aspects of our normal developmental environments. Discursive niche construction now articulates the world along diverse lines, distinguishing conceptual contents, institutions, occupations, rituals, art, games, equipment,[4] and social, legal, moral, sacred, or other statuses, all of which remain almost completely

4. Tool use has a long and contested history among proposed demarcations of human from other organismic ways of life. Distinguishing one- and two-dimensional normativity displaces that issue because instrumental use of tools as means to ends is only one-dimensional. Equipment, which involves interrelated tools with distinct roles and purposes, appropriate and inappropriate uses, and associated skills and occupations for its users, is two-dimensional. Instrumental effectiveness for a task at hand is one-dimensional, but equipment is also open to a second dimension in

opaque within other organisms' ways of life. Language and other public expressive repertoires within our behaviorally reconstructed environment are integral to these capacities to track and differentially reproduce these boundaries, but the intralinguistic relations never stand alone. New patterns of talk and other expression are sustained by reconfigured circumstances and changed ways of life.[5] The importance of experimental systems, museum collections, field sites, and models in mediating theoretical understanding in the sciences exemplifies this larger pattern of discursive niche construction, which conjoins expressive behavior with material reconstruction.[6]

Discursive practices are patterns of interdependent performance. No one can speak a sentence of English, teach a class in philosophy, compete in basketball, worship a triune god, or use functional magnetic resonance imaging to correlate neural activity with cognitive processes except as part of a much larger pattern of performances. Just what one is doing in undertaking any of these activities depends upon the larger pattern of performances and circumstances to which each belongs. The possibility of engaging in any conceptually articulated practice thus depends upon mutually supporting performances and enabling circumstances. Such patterns of practice are vulnerable to two kinds of failure, which blend together at the margins. Practices can "die out" if they stop being reproduced. They can also be reproduced in sufficiently divergent ways that it becomes unclear *how* to continue to reproduce them or just what is the pattern to which they belong. Discursive practices do tolerate divergences, as various performers extend past patterns of practice in somewhat different ways or in different circumstances. They can differentiate into distinct traditions or separate practices. Such divergences nevertheless generate countervailing pressures for realignments to sustain the mutual interdependence of performances within a more or less shared way of life: participants adjust their own performances to accord with others, challenge other participants to change their ways, or rearrange the circumstances to reduce divergence.

---

which one assesses the appropriateness of the use and the skillfulness and occupation (or other statuses) of the user.

5. This interdependence of linguistic and other expressive repertoires with the forms of life and worldly circumstances in which they are used is not merely contingent. They can be only partially autonomous from their circumstantially situated patterns of use, because only in relation to that broader pattern of involvements can there be anything at stake in the use of that repertoire or the choice of one expression over another.

6. Latour (1987, ch. 6) provides a useful indication of the extent of the mobilizations, tracings, juxtapositions, changes of scale, calculative operations, and their institutional supports that are integral to scientific conceptual articulation.

Our social-discursive way of life thereby articulates the world two-dimensionally. We have seen that in "making a living," all organisms introduce meaningful boundaries between organism and environment, which also distinguish the entire pattern of organism-cum-environment from the larger material setting that is effectively opaque to that organism's way of life.[7] The closely coupled, finely tuned response of organism to environment *takes up* those surrounding affordances in often flexible, instrumentally rational, but holistically interdependent ways, which consequently cannot differentiate environmental uptake from further-articulated "takings-as" that could be *mis*taken. Other organisms can succeed or fail in maintaining themselves and their environmental support, but they cannot mistake any aspects of their environment *as* anything other than whatever they normally engage with as a whole. The holistic pattern of their way of life cannot differentiate any takings-as within its overall environmental uptake.[8] We, however, do articulate and sustain further significant divisions within the world by tracking and assessing how to situate events and performances within various social-discursive practices and how those performances respond to what is at issue and at stake in the ongoing reproduction of our conceptually articulated organismic way of life.

Both one- and two-dimensional normativity arise as temporally extended patterns. Organisms and their lineages are patterns of life activity that are identifiable as such by their ongoing, active self-differentiation from what thereby becomes their environment. Death or extinction is the cessation of that activity of continuing, self-maintaining differentiation. As Okrent summarized, "To describe an entity as alive is to attribute to it both a *structure* that has been and is maintained and a *pattern* of events in the entity, a metabolism, or process of interchange with its environment, through which the structure is maintained. . . . To be alive is to do something such that those patterns continue. But, . . . at any

7. Strictly speaking, the organism's activities only configure an environment as a space of affordances for its way of life; the environment "stands out" intra-actively with its life patterns, but the opacity of the "background" to that way of life is nearly complete, except indirectly as part of the environments of those other organisms that are integral to its environment. Along with the internal differentiation of other organisms' holistic responsiveness to their environment, the broader physical background for organism/environment couplings thus only emerges within and for conceptually articulated understanding.

8. We, as professional biologists or common-sense interpreters, can attribute conceptually articulated differentiations within other organisms' way of life, for our own explanatory and predictive purposes, and do so correctly or incorrectly and informatively or not. Those assessments are nevertheless accountable to what is at issue and at stake *in* the organism's way of life *for* our conceptually articulated understanding. We do not thereby isolate the organism's own relation to the world from "sideways on."

moment, the life process *might* cease. . . . From [a] future-oriented perspective, the pattern of the entity's life functions emerges as a standard or norm—though not necessarily a realized one—against which what actually does happen can be measured" (2007, 71–72). Neither the structure nor the pattern in question can be defined in isolation from its temporal extension, however. Organisms develop and their lineages evolve; the resulting changes nevertheless count as successfully maintaining the structure and continuing the pattern. Their goal-directed normativity can only be specified anaphorically, referring back to some prior structure and pattern *as* aiming to sustain over time a structural organization and temporal continuity that *would be* identifiable as *its* continuation. What is *at issue* for an organism is whatever threatens[9] to end its continuation as an identifiable, goal-directed pattern, and what is *at stake* in its response to those issues is whether it succeeds in maintaining its continuity over time.

Conceptually articulated practices are also constitutively temporally extended patterns that are only identifiable anaphorically. Like functioning organisms, they are holistically interconnected patterns, such that one cannot engage in one conceptually articulated performance without having the capacity to undertake many. Similarly, they are also temporally extended in ways that constitute norms for the assessment of the actual pattern. The partial autonomy of various conceptually articulated repertoires, from language and iconic expressions to equipmental complexes, introduces a new dimension to our environmental intra-actions, however. It allows a differentiation of how we take things to be (their significance within and for the partially autonomous repertoire) from whether and how to understand that repertoire and its local demands within a broader pattern of human life. What is at stake in conceptually articulated practices is then not just whether the practice continues but how it continues and how the specific performance fits into the larger pattern. The most widely recognized form of two-dimensionality distinguishes meaning from truth within language, as an exemplary case of distinguishing appropriateness within a conceptual repertoire from "correctness" overall. Truth and meaning are systematically interrelated. As Davidson long ago noted, one can only determine whether an utterance is true if one has fixed its meaning for the purposes of that assessment,

9. I am using "threatens" in a very general sense to refer not only to other entities whose effect on the organism might prevent the continuation of its life pattern but also to the absence of conditions or resources needed to sustain that continuation and to internal malfunctions or absences among its constituent processes and their components. Threats are anything the organism has to "work" to overcome in order to maintain itself.

but there is no nonarbitrary basis for interpreting an utterance except by maximizing the truth and rationality of the larger pattern of utterances and other performances within which it is situated.[10]

In everyday discursive engagements, people often tack back and forth fairly effortlessly between these two dimensions of responsiveness. In conversations or other social interactions, for example, we "make sense" of others' performances (and make our performances sensibly) against a background of other performances and circumstances, *as* having responded to that background in an intelligible way—that is, in ways opening onto other possible performances as intelligible in turn.[11] We may in turn "stand back" from such performances and the sense they make in order to consider whether and how that performance was something "to be done": whether an utterance is true or warranted, whether a joke was funny, and more generally whether what was said or done was something to build upon, repudiate, or circumvent, and so forth. More complex combinations are also possible, as critical assessment of whether another performance was "to be done" might lead to more charitable reassessment of the initial sense making. The key point is to recognize the two-dimensionality of understanding that is involved: interpreting various events or performances in two distinct registers, both as part of a more localized pattern of situated social-discursive practice and as part of our organismic response to our developmental and selective environment (although we need not *think* of our pattern of response to our surroundings in social or biological terms, or indeed, *think* of it at all; much of what we do as agents and organisms is below the threshold of explicit reflection, even though taken up and worked out in conceptually intelligible ways).

The difference between linguistic expressions and the conceptual contents they help articulate becomes clearer when we recognize the significance of semantic ascent. Natural languages play a central role in opening and sustaining conceptually articulated understanding. Introducing and using a term, whose patterns of appropriate use in various circumstances mark out and help track the conceptual relations in question,

10. Davidson proposes a truth-theoretical model for the interpretation of speakers and agents, however, using only the resources of first-order, extensional language. For reasons indicated below, and in chapter 8, I regard the two-dimensional normativity of language and other conceptually articulated repertoire to involve modal relations that are ineliminable.

11. "Intelligible" here marks out an issue within discursive practice rather than an overarching and determinate norm that is constitutive of it. The term arises within ongoing discursive practice to articulate and track a difference in *how* performances and circumstances interact in ways that affect subsequent responses.

is among the most common and readily understood means of conceptual articulation.[12] Conceptual contents are nevertheless not identifiable with the uses of a term because of the temporal extension of the practices to which their use belongs. Conceptual contents are instead anaphorically identifiable issues within ongoing discursive practices.[13] Learning how to employ new terms and to distinguish some of the circumstances in which they are appropriately applied enables one to talk about aspects of the world that were previously inaccessible. In the experimental practices that allowed the differentiation of genetics and development within the field of heredity (discussed in chapter 9), for example, the term 'gene' was deployed to mark out stable correlations among relative chromosomal locations, observable cytological features of the chromosomes, and phenotypic patterns in anatomy and physiology that incorporate or presuppose the normal embryological development of those phenotypes. Having learned one's way around this conceptual domain, one can then use semantic ascent to ask whether these terms and their established patterns of recognitive and inferential use are themselves adequate to express the conceptual differences and interrelations that only became manifest through their use. The linguistic relations are integral to, but not identical with, the conceptual relations they help articulate. Such iterated two-dimensional assessments now often proceed in tandem almost from the outset of domain-opening conceptual articulation, tacking back and forth between efforts to formulate what various performances are "getting at" conceptually, in words or other expressive productions, and consideration of whether those formulations are the best ways to indicate and articulate those same conceptual relations.

The two-dimensionality of conceptually articulated practices stands out more clearly in those domains in which the further elaboration of conceptual relations is subject to careful critical assessment. Scientific inquiry is prominent among such practices. We saw in chapter 8 that understanding conceptually articulated scientific domains involved

12. Understanding a new linguistic expression normally involves both a grasp of aspects of its inferential and grammatical relations to other expressions and some understanding of the circumstances in which its application would be appropriate. Not all new expressions can be employed noninferentially in reporting those circumstances directly, but there must at least be some inferential relationship to worldly circumstances for linguistic expressions to have a conceptual role.

13. The issue of whether and how the semantic significance of uses of a term at one time might be dependent upon the pattern of future uses of that term or its descendants has been extensively discussed in debates over "temporal externalism" in the philosophy of language, extending the more familiar material and social externalisms proposed by Hilary Putnam (1975) and Tyler Burge (1979). The possibility of a temporal externalism was first raised by Wilson (1982) and Donnellan (1983). Ebbs (2000), Tanesini (2014), and Rouse (2014) provide critical overviews of the issues.

maintaining the holistic stability of the domain's conceptual relationships under relevant actual or counterfactual perturbation. It is not enough to be able to take its recognized conceptual relations into account (e.g., in designing an experiment or planning a research program) or to use its linguistic expressions and models appropriately. In research, scientific skills include refining and extending established conceptual relations to new circumstances in potentially significant ways while also maintaining their holistic stability throughout the domain.[14] Such abilities are ineliminably modal. Recognizing when a measurement or other experimental outcome *cannot* be right, or some inferential developments of its concepts are *incoherent* together, must be able to lead to correcting one's performances, revising the relevant skills, or repairing their more recurrent failings while maintaining the holistic stability of the domain.

The sciences are not unique in this respect, however. Law is another domain in which concepts are continually extended or refined self-critically. Sometimes new cases can reveal unforeseen incoherence among concepts or their applications, much in the way that thought experiments do for scientific concepts. Sometimes developments in other domains, such as technological innovations or new strategies in business or political campaigns, can put pressure on established legal precedents and practices. Such contested extensions can highlight assumptions that were implicitly part of the systematic articulation of the relevant conceptual relations and that may need to be adjusted or revised. Such interconnected domains project and regulate the application of conceptual norms and the skills for their recognition and application in ways that allow for incompatibilities. In the face of those incompatibilities, one must either revise some projected applications or adjust the conceptual relations to accommodate them. Because the legal system's role is to regulate conflicts of interest or performance, it regularly confronts new issues that require further adjudication according to what is at stake in relevant parts of the domain. Those stakes are typically contested, but they are also responsive to ongoing adjudication that shifts the configuration of the conflicts.

14. We saw in chapter 10 that grasping which novel applications or inferential relations *matter* to ongoing scientific work is integral to understanding in a scientific field. Scientific domains are always open to more intensive or extensive conceptual articulation, but not all such developments are scientifically significant. Such defeasible judgments of significance shape what is at issue in ongoing research and answer to what is at stake in further articulation of the domain, whether homonomically or heteronomically.

The sciences and law are domains in which conceptual relations are often under continuing, incompatibility-generating pressures and in which those relations are often subject to explicit, authoritative explication. Other conceptually articulated domains also implicitly invoke modal relations. I have previously called attention to the difference between merely instrumental uses of tools and the two-dimensional conceptual normativity of equipmental complexes of interrelated equipment, skills, roles, and purposes. Equipmental complexes open conceptual domains such as plumbing or auto repair that are less explicitly demarcated than most scientific domains. They still each mark out an open-ended range of "possible" situations in their domain, for which different kinds of equipment and skills are appropriate, with resilient skill needed to handle less straightforward problems. The conceptual relations in such everyday domains are usually looser than in the sciences or law. Recalcitrant problems can more often be circumvented rather than resolved, as defective components are replaced rather than diagnosed and repaired, for example, so that the sources of failure remain more opaque. Heteronomic relations to economic, legal, and social considerations are also usually more pervasive in shaping the conceptual configuration of these fields.

Conceptual domains need not be sharply demarcated from one another for their constitutive holistic relations of invariance to do significant work. Language is itself a conceptually articulated domain, with its own characteristic but shifting relations of possibility and impossibility, which also have normative significance both within language as a social practice and for its heteronomic relations to other practices. Language's versatility as a partially autonomous domain aids the tracking of conceptual relations in other domains of social practice, and it is pervasive throughout almost every aspect of human life. Conceptual articulation nevertheless extends beyond language, both because there are nonlinguistic conceptual repertoires (images, symbols, music, equipmental complexes, etc.) and because "language" itself is not self-contained but incorporates its circumstantial settings and a wide array of vocative and recognitive social relations.

Goodman's (1954) and Sellars's (1997) discussions of the normative and modal character of ordinary perceptual concepts indicates the ubiquity of systematic interrelations within conceptual domains. Just as experimental systems establish exemplary conditions for conceptual norms in the sciences, so there are paradigmatic conditions of lighting, viewing position, color juxtapositions, and conditions of embodiment that demarcate the partially constitutive relations between "is green" and "looks green." Color concepts must also remain invariant across other actual or counterfactual changes: appropriate ascriptions of 'green' are unaffected by when an object was first discovered, whether it is sacred, which day of the week it is observed, or whether one is in the company of strangers. Colors must be spatially extended and can blend but not overlap. Color concepts are also open to more intensive and extensive articulation: they can vary in shade, hue, saturation, chroma, or "colorfulness"; color samples can play exemplifying symbolic roles within the domain (Elgin 1991); and indexical and demonstrative locutions allow the discrimination and tracking of color variations that are below the threshold of any canonical color vocabulary (McDowell 1994, 56–60). As in other conceptual domains, color concepts' heteronomic relations to and within related domains, such as painting, interior decorating, computer graphics, or fashion, can shift the salience and significance of variations in color and enable further extensive articulation along such lines as compatibility or complementarity, affective response, and social significance. The characteristic two-dimensional normativity of conceptually articulated involvement in the world conjoins alethic-modal relations of systematic invariance within and across conceptual domains with normative directedness toward issues and stakes posed by involvement with those aspects of the world.

Conceptual articulation thereby stands out from the often subtle and complex differentiations that are integral to the environmental responsiveness of many other organisms, from primates to social insects. Dominance hierarchies, communicative indications of hidden or distant circumstances, and myriad forms of mimicry or deception exemplify the kinds of complex behavior that can be produced by organisms that are capable of flexible response to multiple conflicting indications.[15] Even the comparatively rigid behavioral responses of "detection agents" can produce complex behavioral cascades when those response patterns are sequentially chained, such that the normal output of one such response

15. Deferential behavior that establishes and sustains dominance relations in access to food, mates, or collective protection is widespread among social organisms. Vervet monkey predator calls and bee dances show how organisms' flexible responsiveness to the configuration of current circumstances can extend beyond what is perceptually accessible individually, although such indirect responsiveness must typically draw upon saliently fitness-relevant concerns such as food access or predator avoidance (see chapter 4 for discussion of why such limited symbolic displacement, difficult though it is for most organisms, is not yet two-dimensionally normative; Bickerton [2009] emphasizes that bees and ants also exhibit such limited forms of symbolic displacement). Sterelny (2003) argues persuasively that behavior that mimics or blocks other organisms' salient perceptual cues tends to arise wherever other organisms become influential components of one another's selective and developmental environments and often produces selection pressures for more flexible, multiply cued patterns of behavioral response.

is an initiating cue for its successor (Sterelny 2003, ch. 2). More flexible response patterns allow still greater differentiation when the outcomes of an organism's own prior behavioral and physiological responses are among the considerations affecting subsequent responsiveness. Such complex patterns are nevertheless still one-dimensional in the sense that there is no way to distinguish what environmental pattern the organism is responding to except through its holistic correlation with the organism's overall physiological and behavioral response. Social animals only track and respond to dominance relations, genetic relatedness, or the communicative expressions of other organisms as part of larger patterns of life activity, which are flexibly responsive to multiple aspects of their circumstances, with overlapping and sometimes countervailing significance for the sustenance and reproduction of life and lineage. Organisms with partially autonomous expressive repertoires could track distinct aspects of their environment and with respect to those repertoires could thereby distinguish signal from noise within different aspects of their life patterns. Absent such repertoires, everything the organisms "take up" perceptually and practically is relevantly signaled.

Local, domain-specific differences between two dimensions of normative assessment are writ large in the overall biological significance of the two-dimensional, conceptually articulated normativity of human life and the human lineage. Most living organisms and their lineages are only open to nonarbitrary normative assessment along one dimension: does its physiology and behavior succeed in sustaining and reproducing its lineage and way of life? An organism's way of life may change in the course of sustaining and reproducing itself as circumstances change. Those changes may or may not enhance its prospects for continued reproductive success in changing environments. Yet the goal-directed normativity of biological life provides no further basis for assessing what the organism *ought* to be doing, or what its anatomical and physiological organization *ought* to be, apart from what it normally does, successfully or unsuccessfully. Other organisms' goal-directedness does not even provide a basis for concluding that they *ought* to succeed in sustaining their lineage but only that they aim to do so, thereby differentiating success from failure. Human lives and the human lineage are also goal-directed patterns that differentiate organism from environment and strive to sustain and reproduce that differentiation. Our way of life nevertheless also generates a secondary dimension of normative assessment concerning what that way of life is and will be. Other organisms' ways of life work to sustain their boundary-defining pattern of intra-action with their environment, whatever that pattern actually turns out to be; in contrast, the *possibilities* for what human life is and will be are also at issue in its ongoing maintenance and reproduction. Many of the possibilities that matter within human life are themselves only manifest within linguistic or other conceptual repertoires, but they thereby acquire a bearing upon whether and how that way of life unfolds.

The two-dimensional normativity of human life is a thoroughly natural, biological phenomenon. Securing the physiological, developmental, and reproductive needs of human bodies pervades every aspect of our lives. Those needs are not a biological substratum for a social and cultural superstructure, however. Our neural capacities and organization, their physiological and developmental realization, and the environment that sustains them have coevolved within the ongoing reconstruction of our discursively articulated environmental niche. These coevolutionary patterns of natural selection and niche construction have also radically changed our physical environment and the consequent selection pressures on our lineage. Human life now integrally involves agricultural cultivation and domestication, multifaceted resource extraction, built habitats, prosthetic embodiments (clothing, transportation, equipment), medical and sanitary interventions, and waste export (nowadays most notably of atmospheric and oceanic CO_2). These material transformations are thoroughly entangled with the linguistic and other social-discursive interactions that pervade normal human developmental environments. We live our lives in discursively articulated ways, picking up on extant patterns of talk and interaction and extending them in ways we can "make sense" of. These discursive patterns are part of our material surroundings, and they both respond to and guide further reconstruction of those surroundings. Human organisms are tightly coupled to this environmental complex, just as other organisms are to their environmental circumstances, but our environment incorporates causally efficacious patterns of public performance and perceptual-practical uptake that are virtually opaque to all other organisms.

Naturalists who might otherwise endorse this incorporation of all aspects of human life within human biology are nevertheless often suspicious of the normative idiom (Turner 2010; Roth forthcoming). Naturalists were once similarly suspicious of teleological locutions. Those suspicions were mostly dissipated by the widespread acknowledgment that teleological language, and the functional or adaptive normativity that it licenses, are indispensable for biological understanding (Dennett 1995; Davies 2001; Lewens 2004; Okrent 2007). Their residual import is that teleological appeals to goals or functions can still never be an unexplained explainer. Biological purposiveness can be irreducible, but nature has no intrinsic purposes. Functionality or goal-directedness instead arises as part of a temporally extended, self-sustaining pattern of organization within the world, responsive to changes in the conditions that constitute an organism's developmental and selective environment.

Conceptual normativity is a complex extension of such biological goal-directed maintenance and reproduction of our way of life. Human beings now inhabit a discursively articulated environment, replete with linguistic expressions, equipmental complexes, socially differentiated roles, and other expressive and articulative patterns. We develop into and as mature human beings in part by learning how to recognize and respond to those patterns in ways that others around us can recognize and respond to in turn. Our capacities for recognition and response are honed by the discursive practices amid which they develop: we learn to speak extant languages, occupy social roles that fit in with others, use available equipment for tasks that contribute to and are supported by others' activities, and so forth. Those patterns of past and continuing practice are nevertheless open to further articulation, refinement, and contestation. Just as biological teleology is a purposiveness without a fixed purpose, so conceptual understanding opens a space of normativity without determinate norms. Norms of conceptual contentfulness and justification are only specifiable anaphorically from within the conceptually articulative practices they govern. They express what is at issue in how those practices develop from their antecedent performances. What is at stake in the resolution of those issues is whether *and* how the practice and its organismic lineage can be sustained over time.

The normative authority that those issues hold over a practice's constitutive performances and performers arises from the latter's mutual interdependence. We can see what this means by comparison to one-dimensional organismic life. The normativity of organismic functioning also arises from the interdependence of (what we can recognize as) its constituent components. Hearts "ought" to pump blood only in the sense that if they do not do so, other cells in the body will not receive nutrients and oxygen and the organs they compose will not enable perception, motility, nutrient intake, or waste excretion, all with the consequence that those hearts will then no longer be able to pump blood or do anything else "as hearts."[16] The scare quotes indicate that hearts

16. Such patterns of goal-directed normativity can also cross canonical boundaries between organisms. The recognition that eukaryotic organisms are symbiotically intra-active with an extensive microbiome (Gilbert, Sapp, and Tauber 2012; McFall-Ngai et al. 2013) highlights the one-dimensionality of life patterns, since the functionally significant microorganisms that inhabit animal or plant bodies "ought" to function as part of the larger whole in the same sense, and with

can only express and fulfill their goal in their actual functioning—that is, within the holistic one-dimensional repertoire of environmentally responsive bodily activity. What is at stake in the ongoing interdependence of that bodily pattern is nothing more nor less than its own continuation and reproduction in whatever form it can continue. Our conceptual capacities are also part of our one-dimensional functioning as individual organisms and as a lineage. The ability to produce and respond to conceptually articulated performances is as much a part of normal human bodily functioning as any other functionally significant capacity, and those abilities contribute significantly to how we make a living in our multifarious life circumstances.

Conceptual capacities also function interdependently in another, partially autonomous register, however, alongside their contribution to organismic functioning. The sentences we utter, the images we make, the equipment we use, and the social roles we take up also belong to other patterns in the world, and such performances can only continue if they function together with other such performances in ways so as to sustain the larger patterns of performance to which they belong. Such performances depend upon others' uptake and response to have an effect. Those effects matter in two distinct dimensions: as part of our organismic functioning that sustains our lives and lineage and as part of ongoing patterns of conceptually articulated practice. They are "conceptually articulated" in the sense that they conjoin worldly patterns with capacities for pattern tracking, recognition, expression, and assessment, which together make a difference within our biological way of life as a whole. The normativity that allows us to speak of pattern "recognition" rather than just response arises from the holistic counterfactual stability of conceptual domains and the heteronomic relations to other domains through which there is something at stake in recognition or misrecognition of the relationships within those domains.[17] Conceptually articulative practices establish and maintain expressive repertoires within which meaningful boundaries emerge as part of a larger pattern of practical and perceptual involvement in the world. There is no way to express those boundaries or how they matter except within such repertoires. In this respect, my account parallels and expands upon Davidson's insistence

the same force, as do the animal's or plant's "own" organs and cells (those symbiotic organisms that are not obligate symbionts answer to multiple functional interdependences).

17. See chapter 8 for more extended discussion of the holistic counterfactual stability of laws and their intertwining with skills for pattern recognition and chapter 10 for the importance of both homonomic conceptual articulation and its heteronomic relation to other conceptual domains.

that the truth conditions for linguistic utterances are only identifiable "in language."

Three fundamental differences from Davidson's view dissipate any concern that conceptual articulation, so understood, would also "degenerate into a frictionless spinning in a void" (McDowell 1994, 66).[18] First, conceptually articulated domains of human life are not just intralinguistic relations but incorporate the skillful recognition and application of conceptual differences as part of our perceptual and practical involvement with our surroundings as living organisms.[19] Second, conceptual relationships are not just synchronically articulated as a self-contained, coherent system. They are temporally extended patterns of involvement in the world, accountable to what is at issue and at stake in sustaining and developing that pattern in its changing circumstances.[20] The conceptual distinctions articulated in scientific and other practices must be learnable, applicable, and extendable to novel cases and heteronomically accountable under the changing conditions and shifting issues of human interaction with one another and the world. This temporally extended accountability is also two-dimensional, to the "intelligibility" (the holistic, counterfactual stability) of a partially autonomous conceptual domain and to the heteronomic significance of that domain within our broader, organismic involvement with our environment (which also incorporates other conceptually articulated domains). The possible incompatibilities among homonomic conceptual relations within a domain and their heteronomic relations to other aspects of human life supply the needed empirical "friction."

18. These three differences are distinct from the differences between Davidson's and my uses of the homonomic/heteronomic distinction that were highlighted in chapter 10, although Davidson's insistence that there cannot be rational relations between the rational and causal domains figures prominently in shaping both lists.

19. Davidson would also argue that the rational relations constitutive of the mental domain are not just intralinguistic, because they are token-identical with causal relations. Token-identity is a merely notional relation, however, since Davidson insists that there are no rationally explicable relations between beliefs or assertions and their causal context. That is part of why it has been so important to acknowledge that the learnable/reproducible skills of pattern recognition are part of the constitutive pattern of any conceptually articulated domain.

20. Davidson (2005b, essay 7; 1984, essay 17) acknowledges an indispensable role—within the holistic relations of thought, utterance, and action that constitute personhood—for temporally extended patterns of utterance and action that outrun what can be interpreted synchronically within the truth theories he takes as models for interpreters' activity. He nevertheless treats metaphor and other extensions of concepts to new circumstances as merely causal occasions for conceptual change rather than as integral to conceptual understanding as a dynamic process. His effort to accommodate such temporally extended patterns of performance does not diminish this second difference between our accounts because it falls back into the problems with his dualism of the rational and causal domains that marked the first difference between us.

The third difference between Davidson's view and mine concerns the conceptually constitutive norms of rationality, intelligibility, or "sense making." For Davidson, both the rational normativity of the mental and the homonomic systematicity of natural laws are constitutive norms for the mental and physical domains, with no further explanation of their authority.[21] I take such higher-level normative considerations also to be at issue within the ongoing development of a conceptually articulated way of life.[22] Goodman's classic account of how to adjudicate inferences and their governing norms is writ large in the ongoing adjustment of conceptual relationships with the forms of intelligibility they both exemplify and answer to:

> I have said that deductive inferences are justified by their conformity to valid general rules, and that general rules are justified by their conformity to valid inferences. But this circle is a virtuous one. The point is that rules and particular inferences alike are justified by being brought into agreement with each other. *A rule is amended if it yields an inference we are unwilling to accept; an inference is rejected if it violates a rule we are unwilling to amend.* The process of justification is the delicate one of making mutual adjustments between rules and accepted inferences; and in the agreement achieved lies the only justification needed for either. (Goodman 1954, 64)

As with Davidson's account of truth- and rationality-preserving interpretation, Goodman's conception of a reflective equilibrium between deductive inferences and inference rules *would* spin in a vacuous circle if that equilibrium were synchronically self-contained. Logic, empirical sciences, natural languages, and other conceptual domains are not closed systems, however, and never reach equilibrium. Their conceptual systematicity extends to encompass the relevant skills for producing and assessing performances in each domain and the issues and stakes that arise accordingly within an ongoing, vulnerably reproducible way of life. The resulting temporal open-endedness, material involvement, and two-dimensional accountability let them meaningfully articulate the world as a "space of reasons."

21. That inexplicably constitutive role for rationality and lawful systematicity is one indication of the ways in which Davidson's view is antinaturalistic, despite his anomalously monistic physicalism.

22. As one telling example, logical truth, which plays a role in defining other forms of relevant counterfactual stability (by setting the terms in which counterfactual hypotheses are compatible or incompatible with a set of putative laws) is itself another case of a conceptually articulated domain demarcated by its holistic relations of counterfactual stability. See chapter 8.

II—The Scientific Image

Naturalists seek to situate their philosophical work within a scientific conception of the world and not impose philosophical limitations or preconceptions upon scientific understanding. I have argued in part 2 that most philosophical conceptions of scientific understanding have not satisfied this critical desideratum. Philosophers have mostly located scientific understanding in bodies of scientific knowledge. Whether understood as a systematic theoretical representation of the world (e.g., Sellars), a systematic theoretical representation of what is observable by us (e.g., van Fraassen), or a disunified patchwork of models that partially and often inconsistently represent aspects of the world for different purposes, "the scientific image" has been conceived as a product of scientific research whose epistemic authority stems from retrospective assessment of its justification, reliability, truth, explanatory insight, empirical adequacy, instrumental success, or more pluralistic virtues. Longstanding philosophical commitments to epistemology make this orientation seem familiar and obvious. Closer attention to scientific understanding in practice shows why this familiar orientation is instead a philosophical imposition upon the sciences that naturalists should not countenance.

This book has developed a conception of scientific understanding as embedded in the scientific research enterprise rather than in bodies of knowledge extractable from it. Scientific understanding is not static but is always directed ahead toward possibilities that outrun current capacities for clear and precise expression. Scientific understanding also extends beyond the linguistic or mathematical expression of knowledge claims to incorporate the experimental, clinical, observational, or field settings, and the scientific skills and practices, through which scientific concepts engage the world meaningfully. A substantial part of scientific work is directed toward establishing and maintaining the research domain as materially and conceptually articulated. Properly prepared materials, introduced into properly regulated and shielded settings, and with appropriate instruments and the right skills to oversee what is happening are the indispensable background that allows verbally or mathematically expressed claims to be "about" what happens within a scientific domain. Adjustment and refinement of inferential relations among terms or the further development and application of mathematical operations are important components of this process, but they do not stand alone. As I argued in chapter 7, theoretical understanding in the sciences is doubly mediated by experimental or other material systems and the models that connect theoretically interrelated terms with those proximate circumstances of use.

Research is a temporally extended process that builds upon established domains of interrelated skills, concepts, and material systems, but its orientation is futural. This futural orientation is not just directed toward possible revisions in the assignment of truth values or degrees of justification to various claims. What those claims *say* about the world is always open to and oriented toward further adjustments. Those adjustments involve ongoing revisions in the experimental systems and scientific skills that allow for their applicability and the further inferential relations that spell out what such application says about anything else. The configuration of a research field is not well expressed in terms of which beliefs are justified because research always points beyond the current configuration of the field. We should instead think in terms of what is at issue in the field and how those issues matter. Efforts to formulate those issues will always be partially contested and open-ended, but research fields are marked out by the changing shape of those conflicts and the possibilities for further development that they point toward.

The ongoing reconfiguration of conceptually articulated domains in the sciences is partially masked by continuities in the linguistic terms and other expressions they deploy. Here the literature on temporal externalism in the philosophy of language captures an important feature of scientific understanding (Ebbs 2000; Rouse 2014; Tanesini 2014). Most philosophers are now familiar with other forms of semantic externalism (Putnam 1975; Burge 1979), which emphasize that the semantic significance of linguistic expressions depends upon the material and social circumstances in which they are used. Temporal externalism recognizes that the relevant circumstances change over time, while continuity in the terms expresses the mutual accountability of those changing uses. Scientific terms such as 'gene,' 'acid,' 'electron,' or 'magnetism' remain in place through changing patterns of their inferential and reportorial use because those patterns of use are anaphorically interrelated. When we use the term 'gene' today, we are talking about what Wilhelm Johannsen and T. H. Morgan were talking about early in the twentieth century and what others will be talking about in similar terms after us, *whatever that is*. The semantic significance of scientific concepts is at issue over time, but their conceptual significance incorporates the changing configurations of the issue. Scientific realists have followed Putnam in trying to confine the relevant semantic significance to referential relations between words and (kinds of) things, but the point applies more generally to their inferential relations to other expressions and the skills and practices through which the use of those terms contributes to the articulation of conceptual domains. The temporal extendedness[23] of conceptual contents as issues within an ongoing practice of articulative engagement with the world reflects their characteristically two-dimensional normativity. Situating utterances and other performances within temporally extended practices allows for the differentiation of what they are "about" from how they take it be. Such two-dimensionality is not just a matter of word use but extends to any conceptually articulative performance. Morgan's use of the term 'gene' was about genes with the same sort of temporally extended significance with which the entire *Drosophila* experimental system was about genetics.[24]

The justification of scientific claims is also situated within this temporally extended two-dimensionality. When scientists address issues of justification, for example in the review literature or in textbooks, they are not asking about the acontextual justification of a timeless proposition but are instead asking about the reliability and significance of what is *expressed* by a claim in the context of ongoing research that seeks to improve that claim's formulation of the issue. The same is true even when an issue is taken up in other contexts outside of ongoing scientific research. When one asks about the justification and reliability of scientific claims for other purposes, such as for public policy analysis or corporate product development, assessment is keyed to what is at issue and at stake in the context of evaluation. That context is also temporally extended, even when those issues and stakes are expressed at a point in time, for different assessments over time are anaphorically directed toward the same issues. Shifting philosophical attention away from bodies of scientific knowledge toward the temporally extended practice of research does not diminish the importance of empirical and inferential justification in the sciences, however. This shift instead recognizes that justification is only one aspect of a fundamentally

23. I introduce this word because 'extension' already has a distinct, familiar use in linguistic contexts. The temporal extendedness of semantic contentfulness has to do with the iterative relations between uses of the "same" expression, which anaphorically constitute their normative interdependence as accountable to the same issues and stakes over time, whatever those turn out to be.

24. See the discussion in chapter 9 of how Morgan's experimental system played an integral role in the opening of genetics as a conceptual domain distinct from the larger field of heredity that incorporated development and now ecology. Just what Morgan's system was "about" continues to shift in light of the recognition of niche construction and "extended phenotypes" (Dawkins 1982; Griffiths and Gray 1994, 2001) as part of biological heredity and of the placement of genetics within larger patterns of genomics and epigenetic gene regulation.

two-dimensional normativity. Often the justification of a claim is preserved and strengthened in part by reinterpreting what the claim says or the settings in which it is appropriately applied. Charles Darwin's claim that natural selection played an important role in biological evolution is much better justified empirically than it was when he made it in 1859, in part because our understanding of just what Darwin's claim means has also changed. The sense and significance of claims and their justification shift together as part of a single process of mutual adjudication in changing circumstances.

Situating scientific understanding within its "natural habitat" of the ongoing research enterprise thereby offers a fundamentally different sense of "the scientific image" if we take that term to apply to the composite conceptual capacities of the sciences. Traditional conceptions of the scientific image as a more or less unified body of knowledge locate scientific understanding *within* the Sellarsian space of reasons, as a set of claims that are or should be endorsed on scientific grounds. The alternative developed in this book treats "the scientific image" not as a set of claims taken as true (or empirically adequate, instrumentally reliable, etc.) but as the ongoing reconfiguration of the space of reasons itself. It incorporates not just which claims "ought" to be endorsed but the space of alternatives to those claims, reasons for and against them, and the material and social practices through which those claims stand out as intelligible and significant.[25] "Intelligibility" here indicates the possibility of reasoning for and against those claims, acting upon them, revising them, registering perceptual and practical uptake in their terms, and so forth. The sciences have not merely produced a justifiable body of knowledge; they have contributed to the expansion, refinement, and reconfiguration of the space of reasons we inhabit in our conceptually responsive way of life.

The configuration of that space is ineliminably modal and normative. Conceptual relations constitute a partially autonomous repertoire whose performances can be assessed both "internally" and as contributions to a broader organismic way of life. The holistic relations of counterfactual stability that mark out patterns of relative invariance under changing circumstances are what allow for the internal articulation of

25. 'Ought' is in scare quotes in this passage because what ought to be endorsed scientifically is itself at issue within "the scientific image" in the more encompassing sense I am advocating. The various philosophical interpretations of the scientific image as a body of knowledge are *part of* the scientific image in my sense—that is, they are claims whose intelligibility and justifiability depends upon the larger field of scientific practice and culture within which they are situated.

such conceptual repertoires. The canonical cases of such stability are the relations of logical, physical, biological, and other forms of necessity expressed by sets of scientific laws. The inductive projectability of conceptual relations to new cases is sustained by their invariant stability under counterfactual perturbation consistent with the laws. Such stability need not be expressed explicitly but instead typically reflects a kind of practical competence in holding those relations stable, "ceteris paribus" (Lange 2002). Such mitigating circumstances can be open-ended, partially indeterminate, and contestable, and are indexed to what is at issue and at stake in the domain.

Such flexibility is possible because the counterfactual stability of conceptually articulated domains is not a matter of synchronic fact ("at time *t*, the set of statements *L* taken as laws within domain *S* remains subnomically stable under any actual or counterfactual hypothesis consistent with *L*") but is an issue in a temporally extended practice. Scientific laws only acquire their modal significance as part of an ongoing practice of research that sustains their holistic counterfactual stability as a distinctive kind of necessity. The laws, the conceptual relations they express, and the skills to recognize them are open to ongoing correction, revision, and repair to maintain the coherence and stability of the domain. The reliability and resilience of those capacities for adjustment of conceptual relations are an integral part of that stability. The resulting ability to sustain and reproduce the practice over time is sufficient for conceptually articulated intelligibility, allowing for mutual reliance among its performances in response to what is at stake in maintaining the relevant conceptual relations under changing circumstances.

Such revisability might suggest the worry that the defeasibility conditions for apparent violations of counterfactual stability might become so tolerant and inclusive as to render vacuous the supposed conceptual relations involved. Avoiding such collapses of conceptual content is not a constitutive criterion for conceptual domains, however, but a live issue within them. The projectable stability of the inferential and reportorial relations that mark out a conceptual domain is what allows it to articulate patterns that make a difference to the ways of life within which they are situated. That is another reason conceptual domains must incorporate the skills for recognizing and applying the relevant conceptual differentiations: a failure in the resilience or reliability of those skills is just as destructive for an ongoing discursive practice as would be the inability to maintain the inferential and reportorial stability of its concepts due to empirical failures or conceptual incompatibilities or to a failure to sustain their significance because of overly tolerant standards.[26] The moral is that what matters to conceptual articulation is the combination of the alethic-modal invariance of putative conceptual relations within a domain with their two-dimensional normative accountability. The latter is two-dimensional in tracking the maintenance of its "internal" conceptual coherence against the broader practical and biological effects of maintaining and upholding those holistically interconnected conceptual relations.

The broader human environment within which conceptual relations might make a difference to our way of life is now discursively articulated in multiple ways. For that reason, heteronomic relations among conceptually articulated domains play a very significant part in the two-dimensional normativity of any one domain. Conceptual domains are always open to more intensive, homonomic development as their concepts, skills, practices, and claims are revised and extended in response to issues arising in their ongoing reproduction. Scientific research fields are actively oriented toward such further development of their own conceptual capacities. We saw in chapters 7 and 10 that the issues within a scientific domain must also bear on matters outside the domain as a condition for their contentfulness and significance. An entirely self-contained domain of conceptual relations and skills would have nothing at stake in getting them right and hence no basis for adjudicating any issues they confront. We might be tempted to describe such self-enclosed practices as "just a game" whose internal conceptual relations are unconstrained apart from what the players are willing to accept.[27]

26. Although Lange himself officially identifies disciplinary domains in the sciences with the holistic counterfactual stability of their laws, his defense of open-ended ceteris paribus clauses in the laws makes clear that the scientific practices for maintaining the relevant conceptual distinctions are integral to the stability of the laws. Thus he argues against one prominent criticism of ceteris paribus provisos that

> Earman and Roberts (1999) worry that if "in advance of testing" there is no statement of "what the content of a law is, without recourse to vague escape clauses," then there is no way to "guarantee that the tests are honest" because "the scientific community as a whole" could "capriciously and tacitly change what counts as an 'interfering factor' in order to accommodate the new data as they come in" (p. 451). But how does a claim *without* a "vague" *ceteris-paribus* clause supply the "guarantee" that Earman and Roberts crave? Even if the hypothesis is "explicit," there is never a *guarantee* that the scientific community will exercise good faith rather than tacitly reinterpret its hypothesis. (Lange 2002, 410)

27. Haugeland (1998, ch. 13) argues for just such a distinction between sciences and games, which he claims accounts for why the former are empirically refutable in ways that the latter are not. Even Haugeland grants that some games are empirically unplayable, though. If baseball required pitchers to throw balls that paused momentarily en route to the batter before resuming their ballistic trajectory, it would not be playable. I argue elsewhere (Rouse 2002, 244–46) that even games like chess can be refuted, because what is at stake in chess play may be unattainable with current

Even games, however, raise issues and stakes to which the standards and skills of game play are accountable. Games can be *relatively* self-contained, but whether and how they continue to be played depends in significant part on their place in larger patterns of human life. That broader, heteronomic significance provides a reference point for reasoning about their constitutive possibilities, standards, and skills. We might indeed sometimes say that some scientific field has become "just a game," but that is a claim about *what* is at stake in its success or failure and not about *whether* there is any basis for differentiating success from failure there.

The importance of heteronomic issues and stakes in scientific practice has been my primary point of divergence from many recent advocates of scientific disunity. Disunifiers have rightly argued that the conceptual relations within the plurality of scientific domains cannot be homonomically united within a single conceptual domain, despite long-standing scientific and philosophical aspirations to conceptual unity. Different sciences, and different conceptual domains within those sciences, establish their own systematic but open-ended relations of holistic conceptual stability, conjoined with the skills and practices that both sustain and adjudicate the recognition of those conceptual patterns. Scientific domains often differ irreducibly in the experimental or other material settings in which their conceptual relations are developed and displayed; in the scientific skills and practices that sustain and display such conceptual relations; in the accuracy, precision, and noise tolerance with which those relations are sustainable; and in the scope and structure of their constitutive counterfactual invariance.[28]

Scientific domains are nevertheless only partially autonomous. Apparent incompatibilities and other infelicities in their relationships

practices, such that the constitutive rules themselves need revision. In terms of my argument in this book, the point is that even games are heteronomically accountable.

28. That many scientific domains are mutually irreducible in these respects does not entail that previously distinct domains are *never* united homonomically. Chapter 9 discussed one such example: inquiry into "cytoplasmic inheritance" as a supposedly distinct mode of biological heredity, with its own characteristic experimental systems (e.g., the *Kappa* particle in *Paramecium*), was assimilated within a unified field of genetics during the 1950s and 1960s through the discovery of protein-coding DNA in nonnuclear subcellular organelles. Heteronomic relations between distinct domains of inquiry are often at issue in ways that reconfigure those domains, but typically homonomic consolidation reconfigures the conceptual relations in both fields rather than reducing one to another as imagined in philosophical models of reduction. Historically, the proliferation of relatively autonomous domains of inquiry and heteronomic "interfield theories" (Darden 2006) and "trading zones" (Galison 1997) at the borders of established domains has nevertheless significantly outpaced their homonomic consolidation.

with other conceptual domains, or possibilities for their constructive interrelations, often play important roles in the internal reconfiguration of a scientific domain's concepts, skills, and practices. Often, only relatively gross features of the more fine-grained conceptual relations developed within one scientific domain matter within another (Lange 2007), but they are not thereby irrelevant.[29] New tools, skills, or conceptual issues in other domains can also have fruitful or disruptive significance for some scientific practices. The political economy of the sciences has often played a prominent role in shaping the directions of their development and the significance of those developments, but significance-constituting heteronomic conceptual relations extend beyond the sciences in other ways also.[30] Nor are they limited to technological efforts to apply scientific concepts and skills for other purposes. Scientific concepts and practices develop and take shape within a conceptually articulated environmental niche, and as naturalists, we need to acknowledge and account for the myriad ways in which such relationships contribute to "the scientific image." Philosophically motivated distinctions between what is "internal" to scientific conceptual domains and what are external influences upon the sciences have no place within a naturalistic account of scientific understanding, except as further contributions to the heteronomic shaping of scientific significance.

Philosophers do not usually think of conceptual and material relations among scientific practices and between scientific and other conceptual domains in evolutionary-biological terms. The historical emergence of the sciences has been far too rapid to register within neo-Darwinist conceptions of human evolution as changing gene frequencies in human populations. That recognition is not diminished by the extent to which the extraordinarily rapid growth of the human population in the

29. Comparison of the uses of the quantum mechanical formalism in physics and physical chemistry provides a useful illustration of this point. Quantum theoretical considerations governing chemical bonds in complex molecules cannot be expressed or calculated with the rigor or precision characteristic of many canonical uses of the theory in physics. Physical chemistry also takes into account concerns, such as the chemically effective geometry of molecules or the qualitative ordering of chemical properties, that do not have the same relevance for physics (Woody 2004a, 2004b, 2014). Physical theory is not thereby irrelevant to physical chemistry, even if the connections between them are much more localized, sporadic, and imprecise than philosophical accounts of theory reduction would suggest.

30. Anthropologists and historians of science have extensively explored some of the ways in which heteronomic conceptual relations between scientific practices and other domains have been mutually influential. Illustrative examples include Martin (1994, 2007), Helmreich (1998, 2009), Edwards (1996, 2010), Rabinow (1996, 1999), Kay (2000), Traweek (1992), Dumit (2004, 2012), Franklin (2007), Smocovitis (1996), Myers (2015), Kevles (1985), Downey and Dumit (1997), Reid and Traweek (2000), Franklin and Lock (2003), and Goodman, Heath, and Lindee (2003).

past few centuries has been aided and abetted by scientifically mediated capacities for agricultural production, energy extraction and use, and medical/sanitary intervention.[31] The selection pressures on human populations had already been substantially mitigated by other changes in human ways of life, and such human cognitive and discursive capacities seem largely disconnected from any recently or currently operative evolutionary processes.[32] As a result, those scientists and philosophers who have sought to understand human cognitive capacities in evolutionary terms have looked for guidance to the selection pressures on human populations in the late Pleistocene rather than to conceptual developments within historically accessible time frames (e.g., Barkow, Cosmides, and Tooby 1992). Even if current human ways of life were to turn out in retrospect to be a pathway to rapid extinction rather than adaptive success, the relevant considerations would be larger patterns of habitat destruction and unsustainable population growth rather than any fine-grained conceptual relationships.

Recognition of niche construction as a central mechanism of evolution fundamentally changes that situation. Cumulative niche construction can lead to significant, heritable changes in patterns of organismic development and in the selection pressures that affect those organisms and their lineages. The nonlinear relations between organismic ways of life and their developmental and selective environments can take place on rather more rapid time scales than those tracked by evolutionary population genetics. To be sure, the emergence of languages and other conceptually articulative practices was a decisively transformative development in human evolution that took place relatively early, along with the neurological reorganization needed to track and produce intelligible word sequences (Bickerton 2014, esp. ch. 5). The coevolution of languages and human neurological capacities nevertheless leaves language structure significantly underdetermined and to be worked out variously in the plurality of linguistic practices. Moreover, languages only function within conceptually articulated ways of life, and evolutionary reconstruction of human environments continues apace. The sciences

31. There are good reasons to think that the sciences have played only a minuscule role in enabling rapid human population growth. The growth and expansion of urban communities and their agricultural support networks, and the shifting patterns of morbidity and mortality from disease in human populations, not to mention the effective abolition of predation by other species, largely predate the emergence of any effective scientific understanding of the underlying processes.

32. Bickerton (2014) highlights the grossly "excessive" development of human linguistic and cognitive capacities, beyond any apparent role for natural selection in driving those changes, as "Wallace's Problem," in honor of Alfred Russell Wallace's early recognition of and response to the issue.

exemplify this interdependence of language as a public phenomenon with other transformations of our developmental environment. Scientific talk is only intelligible amid more extensive transformations of our discursively articulated environment, from laboratory materials and practices to institutional structures and pedagogical routines.

Niche construction theory thus situates conceptual normativity centrally within the evolutionary process in scientifically intelligible ways. It can account for not only the continuities between our conceptual capacities and the flexible, instrumentally rational responsiveness of many other organisms to their developmental, physiological, and selective environments but also for the crucial discontinuities between them. We are adaptively and reconstructively responsive to a very different environment, which has coevolved with our conceptual capacities. The key transformation was the development of partially autonomous performative and recognitive repertoires through the ability to track and assess them in two dimensions. We are responsive to a dual significance of various performances and circumstances, both for appropriateness within their proximate domains and for their broader significance for our lives and ways of life. At first, this broader significance was directly selective, as the emergence of protolanguage and its subsequent conceptual articulation contributed to the survival and growth of the human lineage. That developmental and selective context allowed conceptually articulated practices to proliferate: languages with growing expressive power, visual depiction and other nonverbal symbolic practices, equipmental complexes and their associated skills, and the social practices and institutions thereby established and sustained. As these practices became integral to normal human developmental environments, heteronomic relations among conceptual domains partially buffered immediate selection pressures while also allowing for more complex assessments of our conceptually articulated performances. The physiological and selective aspects of our organismic life have not gone away, however, and heteronomic relations among conceptually articulated domains are permeated by considerations of human vulnerability and the nutritional, affective, neotenous-developmental, and other aspects of our organismic physiology. The interplay between homonomic conceptual articulation and its complex heteronomic significance nevertheless complicates the considerations that govern the ongoing reproduction of our ways of life: we struggle to not only maintain and reproduce that way of life but shape its conceptually articulated unfolding.

The sciences are exemplary cases of discursive niche construction. Scientific practices are conjoined material-behavioral patterns in the world that must be continually maintained and reproduced. As practices that aim at their own continual expansion and refinement, their differential reproduction is more actively transformative than most other aspects of human ways of life.[33] While the growth, refinement, and heteronomic influence of the sciences have had substantial material effects on our environment, the most striking outcomes of their ongoing differential reproduction have been articulative. Scientific practices have allowed aspects of the world to show themselves in newly differentiated ways, such that we can talk and reason about them, act responsively to them, and assess their significance for other conceptual domains. Realists and antirealists have debated whether the conceptual patterns disclosed in scientific practice are already there in the world or are instead dependent upon our perceptual capacities, practical concerns, or social practices and norms. Both are mistaken.[34] Scientific practices *let* the world show itself in a conceptually articulated way, in patterns that encompass both the emergent conceptual differentiations and their skillful, discursive recognition. Neither aspect of these emergent, articulated patterns would be intelligible apart from the other.

In bringing the sciences and other conceptually articulative practices into evolutionary purview, however, this account significantly transforms our criteria for naturalistic understanding. Naturalistic philosophical accounts of the scientific image have typically taken the language and scientific skills through which that image is expressed as a transparent depiction of the world from "sideways on" or a "god's-eye view."[35] I have argued instead that any discursive practice, including scientific practices, can only articulate the world from within. Scientific practices and scientific language are part of the world we seek to understand, and

33. Several influential intellectual traditions emphasize that modern economies also practice continual creative destruction and suggest that the research imperative of the modern sciences is merely a component of that larger process. I do not here take a position within those debates about the place of scientific practices within modernity or postmodernity.

34. From another perspective, one might say both are partially correct. They are mistaken in the question they are trying to answer, but there are important constructive elements retrievable from each response to it. The mistaken question concerns how to understand the apparent conformity between scientific conceptualization and its objects (the world, experience, social practices, etc.). The mistake is in treating conceptual contentfulness as a relation between words and things or their manifestations rather than an articulation *of* the world that incorporates scientific practices as reconstructive from within.

35. I take these two expressions to be roughly equivalent characterizations of a way of understanding ourselves in the world that is conceivable but not possible (we can understand and express the conditions that would enable such a conception if there were one while also recognizing that those conditions do not and cannot obtain). Haugeland (1998, ch. 13) gives a clear account of how and why the scope of what is articulable and recognizable ("conceivable") within any conceptual domain must exceed the range of what is nomologically possible.

they transform the world and our way of life within it in ways that allow it to be intelligible. "Intelligibility" and "naturalistic" are terms that indicate contested issues within our discursive practices rather than serving as impossibly independent criteria for ahistorical assessment. In chapter 1, I characterized naturalism as a historical project whose conceptualization has developed dialectically. We can now see why its constitutive temporal extendedness is futural as well as historical. To advocate a naturalistic understanding is not merely to describe the world and our place within it in a way that countenances no appeals to anything supernatural, transcendent, or mysterious. Naturalism is a project to remake our way of life in ways that would block off such appeals and develop our conceptual resources and capacities accordingly. That project is anything but arbitrary; it is a reasoned response to our history, current situation, and prospects for maintaining our conceptually articulated way of life and flourishing within it. Naturalism is not neutral, however. As a naturalist, I am proposing that we develop our scientific and other conceptual practices, and the way of life to which they belong, in some ways rather than others. That proposal also has far-reaching implications for how conceptual domains outside of the sciences ought to be conceived and pursued, even though those implications have not been foregrounded in the book.[36] Even if my proposal were successful in its own terms, however, that would not settle the scientific, philosophical, and broadly political issues involved but would instead pose new issues and stakes in whether and how our way of life unfolds from there.

36. As one illustrative example, Elise Springer (2013) proposes a revisionist approach to moral theory that is complementary to the naturalistic account of conceptually articulated domains that I have been advocating. Springer argues that moral theory should aim not to determine the moral norms that anyone ought to take as authoritative but instead to understand and assess the practices through which we articulate, communicate, and seek to resolve moral concerns arising in our lives together. Morality is conceived as a practice of engaging critically with one another concerning how our actions and ways of life affect one another and other agents and circumstances. So conceived, moral evaluation addresses what is at issue and at stake in our current situation and our ongoing way of life as a moral domain and thereby exemplifies a broadly naturalistic account of moral normativity without determinate moral norms.

Epilogue: Naturalism and the Contingency of the Space of Reasons

> Once upon a time, in some out of the way corner of that universe which is dispersed into numberless twinkling solar systems, there was a star upon which clever beasts invented knowing. That was the most arrogant and mendacious minute of "world history," but nevertheless, it was only a minute. After nature had drawn a few breaths, the star cooled and congealed, and the clever beasts had to die. One might invent such a fable, and yet he still would not have adequately illustrated how miserable, how shadowy and transient, how aimless and arbitrary the human intellect looks within nature. There were eternities during which it did not exist. And when it is all over with the human intellect, nothing will have happened. —FRIEDRICH NIETZSCHE (1979, 79)

Nietzsche's bleak image of our peripheral, vulnerable, and ephemeral place in astronomical space and deep time relies upon the very capacities for scientific understanding that it denigrates as miserable, transient, and aimless. The spatial and temporal scales and benchmarks that make intelligible the duration and variety of life on earth or the expansion and materialization of the universe were instituted by a tradition of imaginative and painstaking scientific research. Without the transformative efforts that built and sustained laboratories, observatories, and the discursive practices around them, these and other scientifically disclosed domains could never have escaped "the darkest of all prisons, the prison of utter obscurity" (Haugeland 2013, 173).

Nietzsche's irony thereby suggests a somewhat more disturbing view of the project of reconciling Sellars's manifest and scientific images. Sellars's project arose from the recognition that the comprehensiveness of the scientific image seemed to undercut its own epistemic authority and semantic content: our best scientific understanding of ourselves and the world seemed unable to accommodate its own rational authority. The underlying issue was continuous with Frege's and Husserl's reasons for rejecting a psychologism, naturalism, or historicism that would ground the normative authority of logic or transcendental consciousness in nature or history. The widespread philosophical turn toward naturalism in the latter half of the twentieth century has only partially departed from Husserl's or Frege's stance: the normative force of rationality has been sought in human biology or social life, but its content and authority have usually been taken to have a universal scope and applicability unconstrained by contingencies of nature or history.[1]

Nietzsche extends that shared concern for the coherence of naturalism a step further. The issue he raises is not merely a variant upon an all-too-familiar epistemological skepticism. His fable, after all, casts no doubt on the *truth* of scientific claims about the universe and our inconsequential place within it but rather presumes their truth. Nietzsche instead confronts us with the possibility that the difference between truth and falsity makes no difference that matters. Copernican astronomy, multiverse cosmologies, Darwinian evolution, geological time, and now microbial symbiosis have notoriously displaced humanity from a presumed centrality to the cosmos, the universe, life, earth, or even our own bodies. Naturalists endorse that displacement, with perhaps one salient exception. The last bastion of a defiant anthropocentrism may be the significance of human capacities to understand the natural world and acknowledge our ever-more-diminished place within it. Nietzsche himself understood this defiance all too well:

> Science also rests on a faith. . . . The question whether *truth* is needed must not only have been affirmed in advance, but affirmed to such a degree that the principle, the faith, the conviction finds expression: "*Nothing* is needed *more* than truth." . . . Those who are truthful in that audacious and ultimate sense that is presupposed by the faith

1. Thus, for example, in their consideration of the normative force of rational conceptual authority, Quine (1965), Dennett (1987, 1995), or Millikan (1984) appeal to the selective advantage accrued by any organism that could reliably predict its environment, Davidson (1980, essay 11) argues for the causal force of a constitutive norm of rationality that arises from the token identity of mental and physical states, and Sellars (2007, ch. 14) takes over the "perennial" conception of persons as rational agents as gaining its force from a collective intention.

> in science *thus affirm another world* than the world of life, nature, and history. . . . It is still a *metaphysical faith* upon which our faith in science rests—that even we seekers after knowledge today, we godless anti-metaphysicians still take our fire, too, from the flame lit by a faith that is thousands of years old, that Christian faith which was also the faith of Plato, that God is the truth, that truth is divine. (Nietzsche 1974, 282–83)

A thoroughly consistent naturalism that refuses to "affirm another world than the world of life, nature, and history" at least to this extent may seem to threaten self-immolation of the human intellect.

The conception of language and conceptual understanding advanced in this book may seem to reanimate Nietzsche's worry in a new guise. Language and conceptual normativity, including scientific understanding, are forms of behavioral niche construction and hence a historically particular phenomenon arising within our biological lineage. There are, of course, many natural languages, living and dead. Yet these belong together as homologous biological traits, parts of a single lineage rather than instances of a general kind. Moreover, there is no alternative to internal comparative connection for establishing the homology (in this respect, language and conceptual normativity are no different from more traditionally recognized homologies). Functionally similar traits may not be homologous, and homologous traits need not display readily apparent morphological, functional, or behavioral similarities: quadruped forelimbs, bipedal arms, bird wings, and whale fins are all homologous. Although the ancestral forms of human language left no direct evidence for comparison, recognition of the niche constructive coevolution of language and human neural capacities strongly suggests that earlier homologs of living languages were structurally and functionally rudimentary by comparison. This point also strengthens the positive case for recognizing the symbolic and syntactic acquisitions of the chimpanzee Kanzi (Savage-Rumbaugh, Shanker, and Taylor 1998; Lloyd 2004; chapter 3 of this book) as homologous with the neural developmental patterns and capacities that enable language acquisition. Objections that Kanzi's limited syntactic combinatory capacities and expressive repertoire are not really analogous to human conceptual capacities are irrelevant to a claim for homology. Moreover, in that case, the common ancestral neural capacities that enable language acquisition were not adapted for anything like language as we know it.

Recognizing language and conceptual understanding as the product of extended behavioral niche construction in a single lineage does not altogether rule out the possibility that analogous traits may independently evolve and hence the possibility of situating our linguistic and conceptual capacities as instances of a more general functional similarity. If language and conceptual understanding are unique to our species, the difficulties of achieving the requisite neural complexity, energetic balance, and fine-grained perceptual and expressive repertoire might have explained why that is so, without challenging the functional value of a conceptually articulated grasp of the world. Yet we can now see why conceptually articulated understanding may be an isolated peculiarity of human evolution rather than a unique, crowning achievement. Apart from our own idiosyncratic evolutionary history, symbolic displacement and conceptual normativity may well be dysfunctional for any organisms with a sufficiently flexible behavioral repertoire that might otherwise make conceptual understanding achievable. Language and conceptual normativity, not to mention the extraordinary articulation of scientific understanding, begin to look like hypertrophic oddities within the human lineage rather than a general capacity for reason with its own constitutive normative authority. The sense that conceptually articulated scientific knowledge has an unquestioned and unquestionable normative authority for "anyone" with the capacity to recognize that authority has nevertheless long animated much of the naturalistic tradition with which this book makes common cause.

Vassiliki Betty Smocovitis's cultural history of the evolutionary synthesis succinctly summarizes the apparent stakes in recognizing the natural and historical contingency of science and the space of reasons more generally:

> The Enlightenment ideals of the proper systematic study of "Man," culminating in evolutionary humanism, liberalism, progress, and the unity of science and of all knowledge, would hold sway by the early 1950s. Knowledge would be unified by reduction to the physical sciences, whereas the diversity and variety of knowledge would emerge from the social sciences above. Evolution, partly reducible *to* the physical world, but also emergent *from* the physical world, would lead ultimately to the progressive divergence of knowledge. The "growth" of scientific knowledge was thus to take the trajectory of a progressively diverging path. The ever-branching, ever-ramifying "tree of life" began to map a one-to-one correspondence with the ever-branching, every ramifying "tree of knowledge." Bearing special signification for religious systems of thought, this metaphor herein represented an end to conventional Judeo-Christian thought: a secular, yet meaningful evolutionary humanism had thus emerged. (1996, 150–51)

The 1950s versions of this grand unifying narrative, institutionally symbolized by the appointment of evolutionary biologist Julian Huxley as the first director-general of a then-Western-dominated United Nations

Educational, Scientific, and Cultural Organization (UNESCO) may now seem naive. Yet the persistence of a more sophisticated progressivist sense of the rational authority of the sciences within the naturalist tradition may nevertheless seem disjunctively challenged by the particularity and contingency of the sciences and scientific understanding.

One side of the disjunction comes from Nietzsche's recognition of a residual commitment to vestiges of natural theology within the naturalistic tradition in philosophy and the sciences. Laplace's successors within that tradition rightly deny any need to postulate God as an explanatory hypothesis within a scientific conception of the world and often militantly oppose efforts to reintroduce "intelligent design" via appeals to biological complexity or the anthropic principle (Rubenstein 2014). Many naturalists have nevertheless been more reluctant to forswear the intelligibility of God's postulated epistemic or semantic position and have often implicitly conceived of scientific understanding as aspiring to a similar positioning. Such vestiges of the theological origins of a scientific conception of the world are not limited to those who "dream of a final theory" (Weinberg 1992) or aspire to the completion of scientific knowledge. Most contemporary naturalists would endorse a thoroughgoing fallibilism and a finite computationalism (Cherniak 1986) that deny even the possibility of omniscience. Yet recognition that the *intelligibility* of a scientific conception of the world depends upon a history of material and behavioral niche construction calls attention to the extent to which naturalists have presumed that the conceptual authority of *the* scientific image transcends such particularity. The unity, comprehensiveness, and explanatory power of the emerging scientific image have long seemed to confer upon it a significance that does transcend the contingencies of its location within a particular historical and biological trajectory.

In the absence of such transcendent authority for scientific understanding, however, Nietzsche's fable would represent a complementary challenge. Enormous effort, ingenuity, and resources have been and are devoted to achieving an articulated scientific understanding of the world. As a form of behavioral niche construction, that achievement only persists through its ongoing reproduction in each generation. Moreover, what needs to be reproduced is not merely a body of accumulated knowledge and a widespread competence to understand, appreciate, and utilize it but a tradition of ongoing research that aims to revise and surpass its own heritage. There may once have been a time when sustaining the scientific enterprise and a rationalist or naturalistic culture seemed both inevitable and inevitably "progressive." If so, that complacent sense of a progressive, secular, rational scientific modernity has since been swept away as yet one more indefensible illusion. Without some pretension to the transcendent significance of scientific knowledge, however, why maintain the enormous collective efforts to sustain the scientific research enterprise, and why commit ourselves to a naturalistic self-understanding?

Nietzsche's fable may even have been overly sanguine about the prospects for scientific and naturalistic understanding. His imagined "clever beasts" presumably remained clever until the death of their life-sustaining star ended their quest for knowledge and self-understanding. We can recognize vulnerabilities in our scientific cultures and naturalistic self-understanding that are much closer at hand. Religious or other cultural hostility to science and naturalistic secularism, commercial corruption of research (Krimsky 2003; Greenberg 2007), or ideological opposition to inconvenient scientific authority (Shulman 2006; Oreskes and Conway 2010) are familiar, recurrent, but still localized threats to scientific and naturalistic understanding. A different sense of the global vulnerability of scientific cultures perhaps first arose with the construction of nuclear weapons and their subsequent proliferation, in cultural resonance with Promethean myths of self-destructive hubris. Subsequent recognition of the scale and significance of CO_2-driven anthropogenic climate change, coupled with a way of life dependent upon voracious consumption of finite energy resources, suggests the possibility that scientific-technological cultures might already be self-destructive rather than just potentially so. Subtler threats to our scientific, naturalistic heritage arise from the realization that the political, economic, or cultural regimes that might suffice to deflect or contain these global threats to the continuation of a scientifically sophisticated way of life could nevertheless undermine that way of life conceptually. No one yet knows whether the material and political conditions for the continuation of scientific-technological niche construction will compromise its cultural, psychological, and pedagogical reproduction or compel difficult choices between scientific-technological and moral-political commitments that might lead us *rightly* to repudiate a naturalistic, scientific way of life as morally unacceptable.

If, or when, these threats to the continuation of scientific niche construction are realized, the loss will not merely be the disappearance of one instance of a perennial possibility for conceptual understanding, which might then be realized elsewhere in the universe. Advocates of protecting biodiversity remind us that extinction is forever. Should we then follow Nietzsche in thinking that, when the geologically and cosmologically brief opening of a two-dimensional space of reasons once again collapses into that dark prison of utter obscurity, nothing will have happened?

Nietzsche's fable provides a useful frame for recapitulating the project of this book. The book contributes to ongoing efforts to work out a consistent, coherent philosophical naturalism. Naturalism has often seemed self-defeating because situating our self-understanding within a scientific conception of the world seems to challenge the very authority and significance of scientific understanding. In response, philosophical understanding of science has often projected its transcendence of the local and the human to account for its authority. I take aspirations to escape our historical and ecological embeddedness to have been erroneous and in any case at odds with a continuing commitment to naturalism. As naturalists, we must understand the normativity of scientific practice from within the natural world as disclosed through scientific research.

I have argued that we should do so by recognizing conceptual understanding in the sciences and elsewhere as exemplifying biological niche construction. Our discursive practices have effected a material transformation of the world and our way of life, which lets the world show itself and affect us in new ways. Our understanding of nature does not and cannot occupy an imaginary standpoint outside nature that would let us represent it as a whole in an intralinguistically articulated "image." Scientific understanding is intraworldly, partial, historically situated, and unable to transcend its own worldly involvements. Yet those involvements extend outward from scientific practices in the narrowest sense to encompass the place of scientific understanding within human life more generally. Conceptually articulated niche construction extends throughout human life. The sciences are important to us *because* of their integration within those broader issues, not as separate and relatively self-contained. In this respect, scientific understanding belongs within the contingencies of human history and culture.

In disagreement with many fellow naturalists, I thus take naturalism to require the rejection of any essentialist conception of science or scientific understanding. Scientific understanding specifically and conceptually articulated understanding more generally are not perennial possibilities always available in human history or to rational or intelligent beings of different biological species or planetary ecologies. Sciences are historically specific practices that emerged within human history, with significance and justificatory standards that continue to change. This recognition ought to broaden the scope of philosophical reflection upon the sciences. A central theme of this book has been that the naturalist tradition in metaphysics and epistemology and the philosophy of mind and language has been insufficiently attentive to work in the philosophy of science that transforms our understanding of what the sciences do and how their achievements occur. Despite earlier, circumscribed efforts to integrate philosophical and historical understanding of the sciences, many philosophers of science remain insufficiently attentive to a rich body of historical, anthropological, and sociological scholarship that shows how the aims, sense, and significance of scientific work are mutually engaged with broader cultural concerns.

A brief consideration of the range of that historical reconsideration is instructive. Mario Biagioli (1993), Steven Shapin and Simon Schaffer (1985), James Bono (1995), and others have shown how the very sense and significance of describing the natural world and reasoning about those descriptions have been at issue in the emergence of new scientific institutions and practices. They do not merely show that Galileo, Boyle, or Harvey understood their own aspirations and achievements differently from our sense of what it is to describe the world and get it right. They also show how subsequent self-understandings emerged in part in response to both the difficulties and the opportunities salient within these different discursive, institutional, and political contexts. Different conceptions of the grounds and authority of natural philosophy have also not surprisingly yielded different senses of natural order and its significance. A generation of historical scholarship on Newton (Westfall 1980; Dobbs 1991; Dobbs and Jacob 1995) has brought attention to the divergence between Newton's and later Newtonian conceptions of the natural world, but scholars from Foucault (1971) to recent historians of chemistry and natural history (Golinski 1992; Roberts 1992) have also defamiliarized eighteenth-century senses of natural order and its significance. The mutual entanglement of natural science or natural philosophy with conceptions of political order has been a recurrent and productive theme: Shapin and Schaffer (1985) on the Royal Society and the restoration, Desmond (1989) on the political controversies preceding and surrounding Darwin's work on evolution, Forman's (2011) classic but controversial work on Weimar physics, a wide range of reflections on science and race (e.g., Haraway 1989; Anderson 2003; Reardon 2005), and an equally wide range of work on science and the Cold War (such as Graham 1993; Hollinger 1995; Edwards 1996; Erickson et al. 2013; Cohen-Cole 2014) sketch some of the relevant territory. Recognition that conceptual understanding is a form of niche construction gives new significance to the many studies linking the sciences with the material infrastructure of the modern world (Smith and Wise 1989; Galison 2003;

Hughes 2005; Edwards 2010). Such work need not be solely retrospective, as anthropologists increasingly attend to how scientific practices engage contemporary cultural and political issues, from shifting conceptions of ourselves as persons (Franklin and Lock 2003; Dumit 2004; Thompson 2005; Martin 2007) to the transformative significance of reconfigured scientific domains (Martin 1994; Rabinow 1999; Helmreich 2009; Edwards 2010).

We should not think of such work as indicating a "contextual" imposition of cultural meaning upon an autonomous scientific enterprise from the "outside." Framing the historiography of science in that way not only misunderstands how scientific work acquires conceptual significance and authority but also contributes to a truncated conception of culture and society that too often overlooks the integral role of scientific work in shaping our conceptually articulated way of life. Recognizing scientific work as continuous with our ongoing biological history of niche construction reminds us that human society or culture always involves an understanding of the place of our way of life within nature. The historical specificity of scientific understanding is integral to the biological specificity of language and conceptual understanding more generally.

The specter of epistemic or conceptual relativism has often haunted any philosophical acknowledgment of the historical specificity and contingency of scientific understanding. Such concerns dissipate with the recognition that what is historically specific is the truth-or-falsity and the significance of scientific claims rather their truth. The sciences matter, and make authoritative claims upon us, *because* of rather than *despite* their historical and cultural specificity, and truth is a concept that expresses that authority. Sciences are powerful but historically specific extensions of the conceptually articulated way of life that is our biological heritage. They do not instantiate an ideal possibility perennially available with sufficient intellect and social support. They likewise cannot transcend our historical contingency in order to take on a "god's-eye view" of ourselves and the world. Science is indeed a precarious and risky possibility that only emerged in specific circumstances, and could disappear.[2] Yet the contingency of conceptual understanding generally,

2. There is no *necessary* tension between essentialist conceptions of science as a perennial possibility within human life and a recognition of the vulnerability of a scientific ethos and the way of life it sustains. One might think of the conceptual and epistemic norms of the sciences as always making claims upon us as rational beings, even though recognition and uptake of those claims is at risk. I nevertheless am making a stronger claim: the normative authority of scientific practices, concepts, and claims only emerges within a historically and biologically specific context, such that

and of scientific understanding specifically, does not thereby undercut the sciences' authority or significance. Their contingency instead calls attention to what is at stake in whether and how those practices continue and develop.

The contingent historical emergence and open-ended future possibilities for scientific understanding are not "just" one historical possibility among many, whose fate would be a matter of arbitrary indifference from the standpoint of the universe. *Nothing* matters from the imagined standpoint of the universe (which is itself only conceivable from a specific location within it), but we do not and cannot actually occupy such a standpoint. In living in the midst of that history, these possibilities are the horizons for our lives and how they matter.[3] Indeed, living in terms of possibilities that matter is precisely what characterizes our distinctively two-dimensional normativity. Other organisms are goal-directed and often instrumentally rational in what *we* can recognize as their efforts to sustain and reproduce their way of life. What is at stake in our conceptually articulated way of life is not merely whether that way of life will continue, however. Who we are and shall be; what our world is like and how it might further reveal itself; and what possibilities it might thereby open to us and our descendants or close off: those are at issue and at stake in the ongoing development of our social-discursive way of life, including our scientific practices. Nothing could matter more, or be less arbitrary from a naturalistic standpoint, from *within* the world that we live in and seek to understand.

maintaining that authority requires also sustaining the way of life within which those practices, concepts, and claims could be authoritative for us. Recognizing the contingency of scientific practices and norms does not undercut their authority but instead intensifies the significance of what is at stake in sustaining a scientific way of life. There is nevertheless an important insight in essentialist conceptions of the normative authority of the sciences. They are best understood not at face value as descriptions of the ahistorical "nature" of science but instead as efforts to focus what is at issue in specific conflicts or tensions over the maintenance of the intelligibility of a scientific way of life and a scientific culture as we know it. They should thereby be understood as an important aspect of a scientific way of life. They are situated, reflective efforts to articulate who we are, how we live, and why it matters to sustain that way of life from within its horizons. In doing so, they help sustain and to some extent transform that way of life by bringing its normative claims and their authority to reflective attention.

3. Samuel Scheffler (2013) has recently argued that many of the projects and activities that we care about most deeply would not and could not matter in the same way without the expectation that the lives of other human beings would continue after our deaths in ways that sustain connections to those projects and activities. Scheffler's claim that "valuing is itself a diachronic or temporally extended phenomenon" (2013, 81) exemplifies the more general account of normativity advanced in this book, in which conceptual, epistemic and, moral normativity, as well as explicitly evaluative assessments, arise from what is at issue and at stake in ongoing, conceptually articulated practices. His lectures offer a thoughtful exploration of some important aspects of the issues raised in this epilogue.

References

Abbott, Edwin. 2010. *Flatland*. Cambridge: Cambridge University Press.

Akins, Kathleen. 1996. Of Sensory Systems and the "Aboutness" of Mental States. *Journal of Philosophy* 96: 337–72.

Althusser, Louis. 1971. Ideology and Ideological State Apparatuses. In *Lenin and Philosophy*, tr. B. Brewster, 127–86. New York: Monthly Review Press.

Amundson, Ronald. 2005. *The Changing Role of the Embryo in Evolutionary Thought*. Cambridge: Cambridge University Press.

Anderson, Warwick. 2003. *The Cultivation of Whiteness*. New York: Basic Books.

Ankeny, Rachel. 2001. Model Organisms as Models: Understanding the "Lingua Franca" of the Human Genome Project. *Philosophy of Science (Proceedings)* 68: S251–S261.

Arbib, Michael, and Mary Hesse. 1986. *The Construction of Reality*. Cambridge: Cambridge University Press.

Aristotle. 1941. *Metaphysics*. Translated by W. D. Ross. Oxford: Oxford University Press.

Armstrong, David. 1983. *What Is a Law of Nature?* Cambridge: Cambridge University Press.

Avital, Eytan, and Eva Jablonka. 2000. *Animal Traditions*. Cambridge: Cambridge University Press.

Baird, Davis. 2004. *Thing Knowledge*. Berkeley: University of California Press.

Barad, Karen. 2007. *Meeting the Universe Halfway*. Durham, NC: Duke University Press.

Barkow, Jerome, Leda Cosmides, and John Tooby, eds. 1992. *The Adapted Mind*. Oxford: Oxford University Press.

Beatty, John. 1995. The Evolutionary Contingency Thesis. In *Concepts, Theories, and Rationality in the Biological Sciences*, ed.

G. Wolters and J. Lenno, 45–81. Pittsburgh, PA: University of Pittsburgh Press.

Bechtel, William. 1993. Integrating Sciences by Creating New Disciplines. *Biology and Philosophy* 8: 277–99.

———. 2006. *Discovering Cell Mechanisms*. Cambridge: Cambridge University Press.

Bennett, Jonathan. 1985. Critical Notice of Davidson's *Inquiries into Truth and Interpretation*. *Mind* 94: 601–26.

Beurton, Peter, Raphael Falk, and Hans-Jörg Rheinberger, eds. 2000. *The Concept of the Gene in Development and Evolution*. Cambridge: Cambridge University Press.

Biagioli, Mario. 1993. *Galileo, Courtier*. Chicago: University of Chicago Press.

Bickerton, Derek. 1975. *Dynamics of a Creole System*. Cambridge: Cambridge University Press.

———. 2009. *Adam's Tongue*. New York: Hill and Wang.

———. 2014. *More than Nature Needs*. Cambridge, MA: Harvard University Press.

Black, Max. 1962. *Models and Metaphors*. Ithaca, NY: Cornell University Press.

Bogen, James, and James Woodward. 1988. Saving the Phenomena. *Philosophical Review* 98: 303–52.

Bolker, Jessica. 1995. Model Systems in Developmental Biology. *BioEssays* 17: 451–55.

Bono, James. 1990. Science, Discourse and Literature: The Role/Rule of Metaphor. In *Literature and Science*, ed. S. Peterfreund, 59–89. Boston: Northeastern University Press.

———. 1995. *The Word of God and the Languages of Man*. Madison: University of Wisconsin Press.

Boyd, Richard. 1973. Realism, Underdetermination and a Causal Theory of Evidence. *Noûs* 7: 1–12.

———. 1979. Metaphor and Theory Change: What Is "Metaphor" a Metaphor For? In *Metaphor and Thought*, ed. A. Ortony, 356–08. Cambridge: Cambridge University Press.

———. 1980. Scientific Realism and Naturalistic Epistemology. In *PSA 1980, Volume 2*, ed. P. Asquith and R. Giere, 613–62. East Lansing, MI: Philosophy of Science Association.

———. 1990. Realism, Approximate Truth, and Philosophical Method. In *Scientific Theories*, ed. W. Savage, 355–91. Minneapolis: University of Minnesota Press.

Brandom, Robert. 1979. Freedom and Constraint by Norms. *American Philosophical Quarterly* 16: 187–96.

———. 1994. *Making It Explicit*. Cambridge, MA: Harvard University Press.

———. 2000. *Articulating Reasons*. Cambridge, MA: Harvard University Press.

———. 2002. *Tales of the Mighty Dead*. Cambridge, MA: Harvard University Press.

———. 2008. *Between Saying and Doing*. Oxford: Oxford University Press.

———. 2011. Platforms, Patchworks, and Parking Garages: Wilson's Account of Conceptual Fine-Structure in *Wandering Significance*. *Philosophy and Phenomenological Research* 82: 183–201.

Brandon, Robert. 1990. *Adaptation and Environment.* Princeton, NJ: Princeton University Press.

———. 1997. Does Biology Have Laws? The Experimental Evidence. *Philosophy of Science* 64: S444–S457.

Brandon, Robert, and Janis Antonovics. 1996. The Coevolution of Organism and Environment. In *Concepts and Methods in Evolutionary Biology,* ed. R. Brandon, 161–78. Cambridge: Cambridge University Press.

Burge, Tyler. 1979. Individualism and the Mental. In *Studies in Metaphysics: Midwest Studies in Philosophy, Volume 4,* ed. P. French, T. Uehling, and H. Wettstein, 73–122. Minneapolis: University of Minnesota Press.

Carman, Taylor. 2008. *Merleau-Ponty.* New York: Routledge.

Carnap, Rudolf. 1950. Empiricism, Semantics and Ontology. *Revue Internationale de Philosophie* 4 (1950): 20–40.

———. 1967. *The Logical Construction of the World.* Berkeley: University of California Press.

Carr, David. 1974. *Time, Narrative and History.* Bloomington: Indiana University Press.

Carroll, Sean, Jennifer Grenier, and Scott Weatherbee. 2001. *From DNA to Diversity.* Malden, MA: Blackwell.

Cartwright, Nancy. 1983. *How the Laws of Physics Lie.* Oxford: Oxford University Press.

———. 1989. *Nature's Capacities and Their Measurement.* Oxford: Oxford University Press.

———. 1999. *The Dappled World.* Oxford: Oxford University Press.

———. 2003. From Causation to Explanation and Back. In *The Future for Philosophy,* ed. B. Leiter, 230–45. Oxford: Oxford University Press.

———. 2007. *Hunting Causes and Using Them.* Cambridge: Cambridge University Press.

———. 2008. Reply to Daniela Bailer-Jones. In *Nancy Cartwright's Philosophy of Science,* ed. S. Hartmann, C. Hoefer, and L. Bovens, 38–40. New York: Routledge.

Cartwright, Nancy, Jordi Cat, Lola Fleck, and Thomas Uebel. 1996. *Otto Neurath.* Cambridge: Cambridge University Press.

Chalmers, David. 1996. *The Conscious Mind.* Oxford: Oxford University Press.

Chang, Hasok. 2004. *Inventing Temperature.* Oxford: Oxford University Press.

Chemero, Anthony. 2009. *Radical Embodied Cognitive Science.* Cambridge, MA: MIT Press.

Cheney, Dorothy, and Robert Seyfarth. 1990. *How Monkeys See the World.* Chicago: University of Chicago Press.

Cherniak, Christopher. 1986. *Minimal Rationality.* Cambridge, MA: MIT Press.

Chomsky, Noam. 1995. *The Minimalist Program.* Cambridge, MA: MIT Press.

Clark, Andy. 2003. *Natural-Born Cyborgs.* Oxford: Oxford University Press.

———. 2008. *Supersizing the Mind.* Oxford: Oxford University Press.

Cohen-Cole, Jamie. 2014. *The Open Mind.* Chicago: University of Chicago Press.

Colwell, Robert. 1992. Niche: A Bifurcation in the Conceptual Lineage of the Term. In *Keywords in Evolutionary Biology*, ed. E. F. Keller and E. Lloyd, 241–48. Cambridge, MA: Harvard University Press.

Cowie, Fiona. 1999. *What's Within?* Cambridge: Cambridge University Press.

Craver, Carl. 2007. *Explaining the Brain*. Oxford: Oxford University Press.

Crosby, Alfred. 2004. *Ecological Imperialism*. Cambridge: Cambridge University Press.

Cummins, Robert. 1996. *Representations, Targets, and Attitudes*. Cambridge, MA: MIT Press.

Darden, Lindley. 2006. *Reasoning in Biological Discoveries*. Cambridge: Cambridge University Press.

Darden, Lindley, and Nancy Maull. 1977. Interfield Theories. *Philosophy of Science* 44: 43–64.

Daston, Lorraine, and Peter Galison. 2007. *Objectivity*. New York: Zone Books.

Davidson, Donald. 1980. *Essays on Actions and Events*. Oxford: Oxford University Press.

———. 1984. *Inquiries into Truth and Interpretation*. Oxford: Oxford University Press.

———. 2001. *Subjective, Intersubjective, Objective*. Oxford: Oxford University Press.

———. 2005a. *Truth and Predication*. Cambridge, MA: Harvard University Press.

———. 2005b. *Truth, Language and History*. Oxford: Oxford University Press.

Davies, Paul Sheldon. 2001. *Norms of Nature*. Cambridge, MA: MIT Press.

Davis, Natalie Zemon. 1983. *The Return of Martin Guerre*. Cambridge, MA: Harvard University Press.

Dawkins, Richard. 1982. *The Extended Phenotype*. Oxford: Freeman.

———. 1986. *The Blind Watchmaker*. New York: Norton.

Deacon, Terence. 1997. *The Symbolic Species*. New York: Norton.

DeLong, Edward, and David Karl. 2005. Genomic Perspectives in Microbial Oceanography. *Nature* 437: 336–42.

Dennett, Daniel. 1987. *The Intentional Stance*. Cambridge, MA: MIT Press.

———. 1991. Real Patterns. *Journal of Philosophy* 88: 27–51.

———. 1995. *Darwin's Dangerous Idea*. New York: Simon and Schuster.

Derrida, Jacques. 1967a. *De la Grammatologie*. Paris: Minuit. English translation. 1974. Translated by G. Spivak. *Of Grammatology*. Baltimore MD: Johns Hopkins University Press.

———. 1967b. *La Voix et la Phénomène*. Paris: Presses Universitaires de France. English translation. 1973. *Speech and Phenomena*. Translated by D. Allison. Evanston, IL: Northwestern University Press.

Desmond, Adrian. 1989. *The Politics of Evolution*. Chicago: University of Chicago Press.

Diamond, Jared. 1998. *Guns, Germs and Steel*. New York: Norton.

Dobbs, Betty Jo. 1991. *The Janus Faces of Genius*. Cambridge: Cambridge University Press.

Dobbs, Betty Jo, and Margaret Jacob. 1995. *Newton and the Culture of Newtonianism*. Atlantic Highlands, NJ: Humanities Press.

Donnellan, Keith. 1983. Kripke and Putnam on Natural Kind Terms. In *Knowledge and Mind*, ed. C. Ginet and S. Shoemaker, 84–104. Oxford: Oxford University Press.

Dor, Daniel. 1999. From Symbolic Forms to Lexical Semantics: Where Modern Linguistics and Cassirer's Philosophy Start to Converge. *Science in Context* 12: 493–511.

———. 2000. From the Autonoly of Syntax to the Autonomy of Linguistic Semantics: Notes on the Correspondence between the Transparency Problem and the Relationship Problem. *Pragmatics and Cognition* 8: 325–56.

Dor, Daniel, and Eva Jablonka. 2000. From Cultural Selection to Genetic Selection: A Framework for the Evolution of Language. *Selection* 1–3: 33–55.

———. 2001. How Language Changed the Genes. In *New Essays on the Origin of Language*, ed. J. Trabant and S. Ward, 149–75. Berlin: de Gruyter.

———. 2004. Culture and Genes in the Evolution of Human Language. In *Human Paleoecology in the Levantine Corridor*, ed. N. Goren-Inbar and J. Speth, 105–15. Oxford: Oxbow Press.

———. 2010. Plasticity and Canalization in the Evolution of Linguistic Communication: An Evolutionary-Developmental Approach. In *The Evolution of Human Language: Biolinguistic Perspectives*, ed. R. Larson, V. Déprez, and H. Yamakido, 135–47. Cambridge: Cambridge University Press.

Douglas, Heather. 2009. *Science Policy and the Value-Free Ideal*. Pittsburgh, PA: University of Pittsburgh Press.

Downey, Gary, and Joseph Dumit. 1997. *Cyborgs and Citadels*. Santa Fe, NM: School of American Research.

Dretske, Fred. 1977. Laws of Nature. *Philosophy of Science* 44: 248–68.

———. 1981. *Knowledge and the Flow of Information*. Cambridge, MA: MIT Press.

Dreyfus, Hubert. 1979. *What Computers Can't Do*. 2nd ed. New York: Harper and Row.

———. 1991. *Being-in-the-World*. Cambridge, MA: MIT Press.

———. 2000. Responses. In *Heidegger, Coping, and Cognitive Science*, ed. M. Wrathall and J. Malpas, 313–49. Cambridge, MA: MIT Press.

———. 2005. Overcoming the Myth of the Mental: How Philosophers Can Profit From the Phenomenology of Everyday Expertise. *Proceedings and Addresses of the American Philosophical Association* 79: 47–65.

———. 2007a. Response to McDowell. *Inquiry* 50: 371–77.

———. 2007b. The Return of the Myth of the Mental. *Inquiry* 50: 352–65.

———. 2013. The Myth of the Pervasiveness of the Mental. In *Mind, Reason, and Being-in-the-World*, ed. J. Schear, 15–40. New York: Routledge.

Dumit, Joseph. 2004. *Picturing Personhood*. Princeton, NJ: Princeton University Press.

———. 2012. *Drugs for Life*. Durham, NC: Duke University Press.

Dummett, Michael. 1975. *Frege: Philosophy of Language*. New York: Harper and Row.

Dunbar, Robin. 1996. *Grooming, Gossip, and the Evolution of Language*. London: Faber and Faber.

Dupre, John. 1993. *The Disorder of Things*. Cambridge, MA: Harvard University Press.

———. 2012. *Processes of Life*. Oxford: Oxford University Press.

Earman, John, and John Roberts. 1999. *Ceteris Paribus*, There Is No Problem of Provisos. *Synthese* 118: 439–78.

Ebbs, Gary. 2000. The Very Idea of Sameness of Extension over Time. *American Philosophical Quarterly* 37: 245–68.

———. 2009. *Truth and Words*. Oxford: Oxford University Press.

Edelman, Gerard. 1987. *Neural Darwinism*. New York: Basic Books.

———. 1992. *Bright Air, Brilliant Fire*. New York: Basic Books.

Edwards, Paul. 1996. *The Closed World*. Cambridge, MA: MIT Press.

———. 2010. *A Vast Machine*. Cambridge, MA: MIT Press.

Elgin, Catherine. 1991. Understanding in Art and Science. In *Philosophy and the Arts*, ed. P. French, T. Uehling Jr., and H. Wettstein. Vol. 16 of *Midwest Studies in Philosophy*, 196–208. Notre Dame, IN: University of Notre Dame Press.

———. 1996. *Considered Judgment*. Princeton, NJ: Princeton University Press.

———. 2009. Exemplification, Idealization, and Scientific Understanding. In *Fictions in Science: Philosophical Essays on Modeling and Idealization*, ed. M. Suárez, 77–90. New York: Routledge.

Elton, Charles. 1927. *Animal Ecology*. London: Sidgwick and Jackson.

Erickson, Paul, Judy Klein, Lorraine Daston, Rebecca Lemov, Thomas Sturm, and Michael Gordin. 2013. *How Reason Almost Lost Its Mind*. Chicago: University of Chicago Press.

Evans, Gareth. 1982. *Varieties of Reference*. Oxford: Oxford University Press.

Feyerabend, Paul. 1962. Explanation, Reduction and Empiricism. In *Scientific Explanation, Space and Time*, ed. H. Feigl and G. Maxwell. Vol. 3 of *Minnesota Studies in the Philosophy of Science*, 28–97. Minneapolis: University of Minnesota Press.

———. 1975. *Against Method*. London: New Left Books.

Fine, Arthur. 1986a. *The Shaky Game*. Chicago: University of Chicago Press.

———. 1986b. Unnatural Attitudes: Realist and Instrumentalist Attachments to Science. *Mind* 95: 149–79.

Fleck, Ludwik. 1979. *Genesis and Development of a Scientific Fact*. Translated by T. Trenn and R. Merton. Chicago: University of Chicago Press.

Fodor, Jerry. 1979. *The Language of Thought*. Cambridge, MA: Harvard University Press.

———. 1981. *Representations*. Cambridge, MA: MIT Press.

———. 1998. *Concepts*. Oxford: Oxford University Press.

Forman, Paul. 2011. *Weimar Culture and Quantum Mechanics*. London: Imperial College Press.

Foucault, Michel. 1971. *The Order of Things*. Translated by A. Sheridan. New York: Pantheon.

———. 1978. *The History of Sexuality, Volume 1*. Translated by R. Hurley. New York: Random House.

———. 1982. The Subject and Power. In *Michel Foucault*, ed. H. Dreyfus and P. Rabinow, 208–26. Chicago: University of Chicago Press.

Franklin, Allan. 1989. *The Neglect of Experiment*. Cambridge: Cambridge University Press.

Franklin, Sarah. 2007. *Dolly Mixtures*. Durham, NC: Duke University Press.

Franklin, Sarah, and Margaret Lock, eds. 2003. *Remaking Life and Death*. Santa Fe, NM: School of American Research.

Friedman, Michael. 1975. Explanation and Scientific Understanding. *Journal of Philosophy* 71: 5–19.

———. 1999. *Reconsidering Logical Positivism*. Cambridge: Cambridge University Press.

Galison, Peter. 1987. *How Experiments End*. Chicago: University of Chicago Press.

———. 1997. *Image and Logic*. Chicago: University of Chicago Press.

———. 2003. *Einstein's Clocks, Poincaré's Maps*. New York: Norton.

Galison, Peter, and David Stump. 1996. *The Disunity of Science*. Stanford, CA: Stanford University Press.

Gallagher, Shaun. 2005. *How the Body Shapes the Mind*. Oxford: Oxford University Press.

Gibson, James. 1979. *The Ecological Approach to Visual Perception*. Boston: Houghton Mifflin.

Giere, Ronald. 1985. Philosophy of Science Naturalized. *Philosophy of Science* 52: 331–56.

———. 1988. *Explaining Science*. Chicago: University of Chicago Press.

———. 1999. *Science without Laws*. Chicago: University of Chicago Press.

———. 2006. *Scientific Perspectivism*. Chicago: University of Chicago Press.

Gilbert, Scott, Jan Sapp, and Alfred Tauber. 2012. A Symbiotic View of Life: We Have Never Been Individuals. *Quarterly Review of Biology* 87 (4): 325–41.

Godfrey-Smith, Peter. 2002. On the Evolution of Representational and Interpretive Capacities. *The Monist* 85: 50–69.

Golinski, Jan. 1992. *Science as Public Culture*. Cambridge: Cambridge University Press.

Goodman, Alan, Deborah Heath, and Susan Lindee, eds. 2003. *Genetic Nature/Culture*. Berkeley: University of California Press.

Goodman, Nelson. 1954. *Fact, Fiction and Forecast*. Cambridge, MA: Harvard University Press.

Gordon, J. I. 2012. Honor Thy Symbionts Redux. *Science* 336: 1251–53.

Graham, Loren. 1993. *Science in Russia and the Soviet Union*. Cambridge: Cambridge University Press.

Greenberg, Daniel. 2007. *Science for Sale*. Chicago: University of Chicago Press.

Grice, H. Paul. 1989. *Studies in the Way of Words*. Cambridge, MA: Harvard University Press.

Griffiths, Paul, and Russell Gray. 1994. Developmental Systems and Evolutionary Explanation. *Journal of Philosophy* 91: 277–304.

———. 2001. Darwinism and Developmental Systems. In *Cycles of Contingency*, ed. S. Oyama, P. Griffiths, and R. Gray, 195–218. Cambridge, MA: MIT Press.

Grinnell, Joseph. 1924. Geography and Evolution. *Ecology* 5: 225–29.

Gu, Quanli, Carl Trindle, and J. L. Knee. 2012. Communication: Frequency Shifts of an Intramolecular Hydrogen Bond as a Measure of Intermolecular Hydrogen Bond Strengths. *Journal of Chemical Physics* 137: 091101.

Habermas, Jürgen. 1971a. *Knowledge and Human Interests*. Translated by J. Shapiro. Boston: Beacon Press.

———. 1971b. *Toward a Rational Society*. Translated by J. Shapiro. Boston: Beacon Press.

———. 2003. *The Future of Human Nature*. Cambridge: Polity Press.

Hacking, Ian. 1983. *Representing and Intervening*. Cambridge: Cambridge University Press.

———. 1992. The Self-Vindication of the Laboratory Sciences. In *Science as Practice and Culture*, ed. A. Pickering, 29–64. Chicago: University of Chicago Press.

———. 2002. *Historical Ontology*. Cambridge, MA: Harvard University Press.

———. 2009. *Scientific Reason*. Taipeh: National Taiwan University Press.

Hankinson-Nelson, Lynn. 1990. *Who Knows?* Philadelphia, PA: Temple University Press.

Hanna, Patricia, and Bernard Harrison. 2004. *Word and World*. Cambridge: Cambridge University Press.

Hanson, N. R. 1958. *Patterns of Discovery*. Cambridge: Cambridge University Press.

Haraway, Donna. 1989. *Primate Visions*. New York: Routledge.

———. 1991. *Simians, Cyborgs and Women*. New York: Routledge.

———. 1997. *Modest_Witness@Second_Millenium.FemaleMan©Meets_OncoMouse™*. New York: Routledge.

———. 2008. *When Species Meet*. New York: Routledge.

Harding, Sandra. 1991. *Whose Science? Whose Knowledge?* Ithaca, NY: Cornell University Press.

Harwood, Jonathan. 1993. *Styles of Scientific Thought*. Chicago: University of Chicago Press.

Haugeland, John. 1982. Heidegger on Being a Person. *Noûs* 16: 6–26.

———. 1998. *Having Thought*. Cambridge, MA: Harvard University Press.

———. 2000. Truth and Finitude. In *Heidegger, Authenticity and Modernity*, ed. M. Wrathall and J. Malpas, 43–78. Cambridge, MA: MIT Press.

———. 2007. Letting Be. In *Transcendental Heidegger*, ed. S. Crowell and J. Malpas, 93–103. Stanford, CA: Stanford University Press.

———. 2013. *Dasein Disclosed*. Cambridge: Cambridge University Press.

Hauser, Mark. 1996. *The Evolution of Communication*. Cambridge, MA: MIT Press.

Hauser, Mark, Noam Chomsky, and W. Tecumseh Fitch. 2002. The Language Faculty: Who Has It, What Is It, and How Did It Evolve? *Science* 298: 1569–79.

Hegel, Georg W. F. 1977. *Phenomenology of Spirit*. Translated by A. Miller. Oxford: Oxford University Press.

Heidegger, Martin. 1927. *Sein und Zeit*. Tübingen: Niemeyer. English translation. 1962. *Being and Time*. Translated by E. MacQuarrie and J. Robinson. New York: Harper and Row.

———. 1950. *Holzwege*. Frankfurt: Vittorio Klostermann. English translation. 2002. *Off the Beaten Track*. Translated by J. Young and K. Haynes. Cambridge: Cambridge University Press.

———. 1985. *History of the Concept of Time*. Translated by T. Kisiel. Bloomington: Indiana University Press.

Helmreich, Stefan. 1998. *Silicon Second Nature*. Berkeley: University of California Press.

———. 2009. *Alien Ocean*. Berkeley: University of California Press.

Hempel, Carl. 1965. *Aspects of Scientific Explanation*. New York: Free Press.

Hitchcock, Christopher. 2003. Of Humean Bondage. *British Journal for the Philosophy of Science* 54: 1–25.

Hollinger, David. 1995. Science as a Weapon in Kulturkämpfe in the United States during and after World War II. *Isis* 86: 440–54.

Horst, Steven. 1996. *Symbols, Computation and Intentionality*. Berkeley: University of California Press.

Hughes, R. I. G. 1999. The Ising Model, Computer Simulation, and Universal Physics. In *Models as Mediators*, ed. M. Morgan and M. Morrison, 97–145. Cambridge: Cambridge University Press.

Hughes, Thomas. 2005. *Human-Built World*. Chicago: University of Chicago Press.

Humphreys, Paul. 2004. *Extending Ourselves*. Oxford: Oxford University Press.

Husserl, Edmund. 1970a. *Cartesian Meditations*. Translated by D. Cairns. The Hague: Martinus Nijhoff.

———. 1970b. *Logical Investigations*. Translated by J. Findlay. London: Routledge and Kegan Paul.

———. 1982. *Ideas, Volume 1*. Translated by F. Kersten. The Hague: Martinus Nijhoff.

Hutchinson, G. E. 1957. Concluding Remarks. *Cold Spring Harbor Symposia on Quantitative Biology* 22: 415–27.

Intergovernmental Panel on Climate Change (IPCC). 1990. *First Assessment Report*, http://www.ipcc.ch/publications_and_data/publications_and_data_reports.shtml.

———. 1995. *IPCC Second Assessment: Climate Change 1995*, http://www.ipcc.ch/pdf/climate-changes-1995/ipcc-2nd-assessment/2nd-assessment-en.pdf.

———. 2001. *IPCC Third Assessment Report: Climate Change 2001*, http://www.grida.no/publications/other/ipcc_tar/.

———. 2007. *Fourth Assessment Report (AR4)*, http://www.ipcc.ch/report/ar4/.

———. 2014. *Climate Change 2014: Synthesis Report*, http://www.ipcc.ch/pdf/assessment-report/ar5/syr/SYR_AR5_LONGERREPORT.pdf.

Jablonka, Eva, and Marion Lamb. 2005. *Evolution in Four Dimensions*. Cambridge, MA: MIT Press.

Jackson, Frank. 1998. *From Metaphysics to Ethics*. Oxford: Oxford University Press.

Jasanoff, Sheila. 2005. *Designs on Nature*. Princeton, NJ: Princeton University Press.

Kant, Immanuel. 1987. *Critique of Judgment*. Translated by W. Pluhar. Indianapolis: Hackett.

———. 1998. *Critique of Pure Reason*. Translated by P. Guyer and A. Wood. Cambridge: Cambridge University Press.

Kaplan, David. 1990. Words. *Proceedings of the Aristotelian Society*. Supplement. 64: 93–119.

Kay, Lily. 2000. *Who Wrote the Book of Life?* Stanford, CA: Stanford University Press.

Keller, Evelyn Fox. 1992. *Secrets of Life, Secrets of Death*. New York: Routledge.

———. 1995. *Refiguring Life*. New York: Columbia University Press.

———. 2000. *Century of the Gene*. Cambridge, MA: Harvard University Press.

Kelly, Sean. 1998. What Makes Perceptual Content Non-Conceptual? *Electronic Journal of Analytic Philosophy* 6 (1998), http://ejap.louisiana.edu/EJAP/1998/kelly98.html.

———. 2001. The Non-Conceptual Content of Perceptual Experience: Situation Dependence and Fineness of Grain. *Philosophy and Phenomenological Research* 62: 601–8.

———. 2000. Phenomenology, Dynamical Neural Networks, and Brain Function. *Philosophical Psychology* 13: 213–28.

Kevles, Daniel. 1985. *In the Name of Eugenics*. Cambridge, MA: Harvard University Press.

Kierkegaard, Søren. 1954. *Fear and Trembling and the Sickness Unto Death*. Translated by W. Lowrie. Princeton, NJ: Princeton University Press.

Kirschner, Marc, and John Gerhart. 2005. *The Plausibility of Life*. New Haven, CT: Yale University Press.

Kitcher, Philip. 1999. Unification as a Regulative Ideal. *Perspectives on Science* 7: 337–48.

———. 2001. *Science, Truth and Democracy*. Oxford: Oxford University Press.

Klein, Ursula. 2003. *Experiments, Models, Paper Tools*. Stanford, CA: Stanford University Press.

Kohler, Robert. 1991. *Partners in Science*. Chicago: University of Chicago Press.

———. 1994. *Lords of the Fly*. Chicago: University of Chicago Press.

Kordig, Carl. 1971. *The Justification of Scientific Change*. Dordrecht: D. Reidel.

Krimsky, Sheldon. 2003. *Science in the Private Interest*. Lanham, MD: Rowman and Littlefield.

Kripke, Saul. 1980. *Naming and Necessity*. Cambridge, MA: Harvard University Press.

———. 1982. *Wittgenstein on Rules and Private Language*. Cambridge, MA: Harvard University Press.

Kuhn, Thomas. 1970. *The Structure of Scientific Revolutions*. 2nd ed. Chicago: University of Chicago Press.

———. 1977. *The Essential Tension*. Chicago: University of Chicago Press.

Kukla, Rebecca. 2000. Myth, Memory and Misrecognition in Sellars's "Empiricism and the Philosophy of Mind." *Philosophical Studies* 101: 161–211.

Kukla, Rebecca, and Mark Lance. 2009. *Yo! and Lo!* Cambridge, MA: Harvard University Press.

Lakatos, Imre. 1978. *The Methodology of Scientific Research Programmes*, ed. J. Worrall and G. Currie. Cambridge: Cambridge University Press.

Lance, Mark. 2000. The Word Made Flesh: Toward a Neo-Sellarsian View of Concepts and Their Analysis. *Acta Analytica* 15 (25): 117–35.

Lange, Marc. 2000a. *Natural Laws in Scientific Practice*. Oxford: Oxford University Press.

———. 2000b. Salience, Supervenience and Layer Cakes in Sellars's Scientific Realism, McDowell's Moral Realism, and the Philosophy of Mind. *Philosophical Studies* 101: 213–51.

———. 2002. Who's Afraid of Ceteris Paribus Laws? Or How I Learned to Stop Worrying and Love Them. *Erkenntnis* 57: 407–23.

———. 2007. Laws and Theories. In *A Companion to the Philosophy of Biology*, ed. S. Sarkar and A. Plutynski, 489–505. Oxford: Blackwell.

———. 2009. *Laws and Lawmakers*. Oxford: Oxford University Press.

Latour, Bruno. 1983. Give Me a Laboratory and I Will Raise the World. In *Science Observed*, ed. K. Knorr-Cetina and M. Mulkay, 141–70. London: Sage.

———. 1984. *Les Microbes suivi par Irréductions*. Paris: A. M. Métailié. English Translation. 1988. *The Pasteurization of France*. Translated by A. Sheridan and J. Law. Cambridge, MA: Harvard University Press.

———. 1987. *Science in Action*. Cambridge, MA: Harvard University Press.

Laubichler, Manfred. 2007. Does History Recapitulate Itself? Epistemological Reflections on the Origins of Evolutionary Developmental Biology. In *From Embryology to Evo-Devo*, ed. M. Laubichler and J. Maienschein, 13–34. Cambridge, MA: MIT Press.

Laubichler, Manfred, and Jane Maienschein. 2007. *From Embryology to Evo-Devo*. Cambridge, MA: MIT Press.

Laudan, Larry. 1977. *Progress and Its Problems*. Berkeley: University of California Press.

Layne, Linda, ed. 1998. Anthropological Approaches in Science and Technology Studies. Special issue, *Science, Technology and Human Values* 23: 1–128.

Lenoir, Timothy. 1997. *Instituting Science*. Stanford, CA: Stanford University Press.

Levinson, Stephen. 2000. *Presumptive Meanings*. Cambridge, MA: MIT Press.

Lewens, Tim. 2004. *Organisms and Artifacts*. Cambridge, MA: MIT Press.

Lewis, David. 1973. *Counterfactuals*. Cambridge, MA: Harvard University Press.

Lewis, Edward. 1978. A Gene Complex Controlling Segmentation in *Drosophila*. *Nature* 276: 565–70.

Lewontin, Richard. 2000. *The Triple Helix*. Cambridge, MA: Harvard University Press.

Lloyd, Elisabeth. 1988. *The Structure and Confirmation of Evolutionary Theory*. New York: Greenwood Press.

———. 1996. Science and Anti-Science: Objectivity and Its Real Enemies. In *Feminism, Science and the Philosophy of Science*, ed. L. H. Nelson and J. Nelson, 217–59. Dordrecht: Kluwer.

———. 2004. Kanzi, Evolution, and Language. *Biology and Philosophy* 19: 577–88.

Longino, Helen. 1990. *Science as Social Knowledge*. Princeton, NJ: Princeton University Press.

———. 2002. *The Fate of Knowledge*. Princeton, NJ: Princeton University Press.

MacIntyre, Alasdair. 1980. Epistemological Crises, Dramatic Narrative, and the Philosophy of Science. In *Paradigms and Revolutions*, ed. G. Gutting, 54–74. Notre Dame, IN: University of Notre Dame Press.

Mahan, Bruce H. 1975. *University Chemistry*. Reading, MA: Addison-Wesley.

Martin, Emily. 1991. The Egg and the Sperm. *Signs* 16: 485–501.

———. 1994. *Flexible Bodies*. Boston: Beacon Press.

———. 2007. *Bipolar Expeditions*. Princeton, NJ: Princeton University Press.

Martin, Wayne. 2006. *Theories of Judgment*. Cambridge: Cambridge University Press.

Mauss, Marcel. 1979. *Sociology and Psychology*. London: Routledge and Kegan Paul.

Mayr, Ernst. 1942. *Systematics and the Origin of Species*. New York: Columbia University Press.

Mayr, Ernst, and William Provine. 1980. *The Evolutionary Synthesis*. Cambridge, MA: Harvard University Press.

McDowell, John. 1984. Wittgenstein on Following a Rule. *Synthese* 58: 325–63.

———. 1994. *Mind and World*. Cambridge, MA: Harvard University Press.

———. 2007a. Response to Dreyfus. *Inquiry* 50: 366–70.

———. 2007b. What Myth? *Inquiry* 50: 338–51.

———. 2009. *Having the World in View*. Cambridge, MA: Harvard University Press.

McFall-Ngai, Margaret; Hadfield, Michael; Bosch, Thomas; Carey, Hannah; Domazet-Loso, Tomislav; Douglas, Angela; Dubilier, Nicole et al. 2013. Animals in a Bacterial World, a New Imperative for the Life Sciences. *Proceedings of the National Academy of Sciences* 110 (9): 3229–36.

McManus, Denis. 2012. *Heidegger and the Measure of Truth*. Oxford: Oxford University Press.

Merleau-Ponty, Maurice. 1962. *Phenomenology of Perception*. Translated by Colin Smith. London: Routledge and Kegan Paul.

Millikan, Ruth. 1984. *Language, Thought and Other Biological Categories*. Cambridge, MA: MIT Press.

———. 2005. *Language: A Biological Model*. Oxford: Clarendon Press.

Mitchell, Sandra. 2003. *Biological Complexity and Integrative Pluralism*. Cambridge: Cambridge University Press.

———. 2009. *Unsimple Truths*. Chicago: University of Chicago Press.

Morange, Michel. 1998. *A History of Molecular Biology*. Translated by M. Cobb. Cambridge, MA: Harvard University Press.

Morgan, Mary, and Margaret Morrison. 1999. *Models as Mediators*. Cambridge: Cambridge University Press.

Morrison, Margaret. 2000. *Unifying Scientific Theories*. Cambridge: Cambridge University Press.

Müller, Gerd, and Massimo Pigliucci. 2010. *Evolution: The Extended Synthesis*. Cambridge, MA: MIT Press.

Müller-Wille, Staffan, and Hans-Jörg Rheinberger. 2012. *A Cultural History of Heredity*. Chicago: University of Chicago Press.

Musgrave, Alan. 1989. NOA's Ark: Fine for Realism. *Philosophical Quarterly* 39: 383–98.

Myers, Natasha. 2015. *Rendering Life Molecular*. Durham, NC: Duke University Press.

Nanney, D. L. 1983. The Cytoplasm and the Ciliates. *Journal of Heredity* 74: 163–70.

Nelkin, Dorothy, and M. Susan Lindee. 1996. *DNA Mystique*. Ann Arbor: University of Michigan Press.

Neurath, Otto. 1973. The Vienna Circle, or the Scientific Conception of the World. In *Empiricism and Sociology*, ed. M. Neurath and R. S. Cohen, 301–18. Dordrecht: D. Reidel.

Nietzsche, Friedrich. 1954. *Twilight of the Idols*. In *The Portable Nietzsche*, tr. W. Kaufmann, 464–565. New York: Viking.

———. 1974. *The Gay Science*. Translated by W. Kaufmann. New York: Viking.

———. 1979. *Philosophy and Truth*. Translated by D. Breazeale. Atlantic Highlands, NJ: Humanities Press.

———. 1998. *On the Genealogy of Morality*. Translated by M. Clark and A. Swenson. Indianapolis, IN: Hackett.

Nöe, Alva. 2004. *Action in Perception*. Cambridge, MA: MIT Press.

———. 2009. *Out of Our Heads*. New York: Hill and Wang.

Norton, Stephen, and Frederick Suppe. 2001. Why Atmospheric Modeling Is Good Science. In *Changing the Atmosphere*, ed. C. Miller and P. Edwards, 68–105. Cambridge, MA: MIT Press.

Nüsslein-Vollhard, Christiane, and Eric Wieschaus. 1980. Mutations Affecting Segment Number and Polarity in *Drosophila*. *Nature* 287: 795–801.

Odling-Smee, F. John, and Keith Laland. 2009. Cultural Niche Construction: Evolution's Cradle of Language. In *The Prehistory of Language*, ed. R. Botha and C. Knight, 99–121. Oxford: Oxford University Press.

Odling-Smee, F. John, Keith Laland, and Marcus Feldman. 2003. *Niche Construction*. Princeton, NJ: Princeton University Press.

Okrent, Mark. 2007. *Rational Animals*. Athens: Ohio University Press.

———. 2013. Heidegger's Pragmatism Redux. In *Cambridge Companion to Pragmatism*, ed. A. Malachowski, 124–58. Cambridge: Cambridge University Press.

Olby, Robert. 1974. *Path to the Double Helix*. Seattle: University of Washington Press.

Oreskes, Naomi, and Eric Conway. 2010. *Merchants of Doubt*. New York: Bloomsbury Press.

Oyama, Susan. 2001. Terms in Tension: What Do You Do When All the Good Words Are Taken? In *Cycles of Contingency*, ed. S. Oyama, P. Griffiths, and R. Gray, 177–93. Cambridge, MA: MIT Press.

Oyama, Susan, Paul Griffiths, and Russell Gray, eds. 2001. *Cycles of Contingency*. Cambridge, MA: MIT Press.

Pearl, Judea. 2000. *Causality*. Cambridge: Cambridge University Press.

Perry, John. 1979. The Problem of the Essential Indexical. *Noûs* 13: 3–21.

Peschard, Isabelle. 2010. Modeling and Experimenting. In *Models, Simulations, and Representation*, ed. P. Humphreys and C. Imbert, 42–61, New York: Routledge.

———. 2011. Making Sense of Modeling: Beyond Representation. *European Journal of Philosophy of Science* 1: 335–52.

———. 2012. Forging Model/World Relations: Relevance and Reliability. *Philosophy of Science* 79: 749–60.

Pickering, Andrew. 1984. *Constructing Quarks*. Chicago: University of Chicago Press.

Piggliucci, Massimo, and Gerd Müller. 2010. *The Extended Synthesis*. Cambridge, MA: MIT Press.

Pinker, Steven. 1994. *The Language Instinct*. New York: Harper.

Porter, Theodore. 1995. *Trust in Numbers*. Princeton, NJ: Princeton University Press.

Price, Huw. 2011. *Naturalism without Mirrors*. Oxford: Oxford University Press.

Provine, William. 1971. *Origins of Theoretical Population Genetics*. Chicago: University of Chicago Press.

Putnam, Hilary. 1975. The Meaning of "Meaning." In *Mind, Language and Reality*, 215–71. Cambridge: Cambridge University Press.

Quine, Willard v. O. 1953. *From a Logical Point of View*. Cambridge, MA: Harvard University Press.

———. 1960. *Word and Object*. Cambridge, MA: MIT Press.

———. 1965. Epistemology Naturalized. In *Ontological Relativity and Other Essays*, 69–90. New York: Columbia University Press.

———. 1981. *Theories and Things*. Cambridge, MA: Harvard University Press.

Rabinow, Paul. 1996. *Making PCR*. Chicago: University of Chicago Press.

———. 1999. *French DNA*. Chicago: University of Chicago Press.

Radder, Hans. 1996. *In and About the World*. Albany: State University of New York Press.

Radick, Gregory. 2007. *The Simian Tongue*. Chicago: University of Chicago Press.

Rawls, John. 1957. Two Concepts of Rules. *Philosophical Review* 64: 3–32.

Reardon, Jenny. 2005. *Race to the Finish*. Princeton, NJ: Princeton University Press.

Reid, Roddy, and Sharon Traweek. 2000. *Doing Science + Culture*. New York: Routledge.

Rheinberger, Hans-Jörg. 1995. From Microsomes to Ribosomes: "Strategies" of Representation, 1935–55. *Journal of the History of Biology* 48: 49–89.

———. 1997. *Toward a History of Epistemic Things*. Stanford, CA: Stanford University Press.

———. 2010. *An Epistemology of the Concrete*. Durham, NC: Duke University Press.

Richardson, Alan. 1998. *Carnap's Construction of the World*. Cambridge: Cambridge University Press.

Robert, Jason. 2004. *Embryology, Epigenesis, and Evolution*. Cambridge: Cambridge University Press.

Roberts, John. 2008. *The Law-Governed Universe*. Oxford: Oxford University Press.

Roberts, Lissa, ed. 1992. The Chemical Revolution: Contexts and Practices. *Eighteenth Century* 33 (3): 195–271.

Rorty, Richard. 1979. *Philosophy and the Mirror of Nature*. Princeton, NJ: Princeton University Press.

———. 1982. *Consequences of Pragmatism*. Minneapolis: University of Minnesota Press.

———. 1989. *Contingency, Irony and Solidarity*. Cambridge, MA: Harvard University Press.

———. 1991. *Objectivity, Relativism and Truth*. Cambridge: Cambridge University Press.

———. 1998. *Achieving Our Country*. Cambridge, MA: Harvard University Press.

———. 2007. *Philosophy as Cultural Politics*. Cambridge: Cambridge University Press.

Roth, Paul. 2003. Mistakes. *Synthese* 136: 389–409.

———. 2006. Naturalism without Fears. In *Handbook of the Philosophy of Science*, ed. S. Turner and M. Risjord. Vol. 15 of *Philosophy of Anthropology and Sociology*, 683–708. Dordrecht: Elsevier.

———. Forthcoming. What Would It Be to Be a Norm? In *Naturalism and Normativity in the Social Sciences*, ed. M. Risjord. New York: Routledge.

Rouse, Joseph. 1987. *Knowledge and Power*. Ithaca, NY: Cornell University Press.

———. 1996a. Beyond Epistemic Sovereignty. In *The Disunity of Science*, ed. P. Galison and D. Stump, 398–416. Stanford, CA: Stanford University Press.

———. 1996b. *Engaging Science*. Ithaca, NY: Cornell University Press.

———. 2002. *How Scientific Practices Matter*. Chicago: University of Chicago Press.

———. 2003. Power/Knowledge. In *The Cambridge Companion to Foucault*, 2nd ed., ed. G. Gutting, 95–122. Cambridge: Cambridge University Press.

———. 2004. Barad's Feminist Naturalism. *Hypatia* 19: 142–61.

———. 2006. Practice Theory. In *Handbook of the Philosophy of Science*, ed. S. Turner and M. Risjord. Vol. 15 of *Philosophy of Anthropology and Sociology*, 639–82. Dordrecht: Elsevier.

———. 2007. Social Practices and Normativity. *Philosophy of the Social Sciences* 37: 46–56.

———. 2009. Standpoint Theories Reconsidered. *Hypatia* 24: 200–209.

———. 2014. Temporal Externalism and the Normativity of Linguistic Practice. *Journal of the Philosophy of History* 8: 20–38.

———. Forthcoming. The Conceptual and Ethical Normativity of Intra-Active Phenomena. *Rhizomes* 28.

Rowlands, Mark. 2010. *The New Science of the Mind*. Cambridge, MA: MIT Press.

Rubenstein, Mary-Jane. 2014. *Worlds without End*. New York: Columbia University Press.

Saenger, Paul. 1997. *Space between Words*. Stanford, CA: Stanford University Press.

Salmon, Wesley. 1999. *Causality and Explanation*. Oxford: Oxford University Press.

Sapp, Jan. 1987. *Beyond the Gene*. Oxford: Oxford University Press.

Savage-Rumbaugh, Sue, Stuart Shanker, and Talbot Taylor. 1998. *Apes, Language and the Human Mind*. Oxford: Oxford University Press.

Schank, Roger. 1975a. The Primitive Acts of Conceptual Dependency. In *Proceedings of the 1975 Workshop on Theoretical Issues in Natural Language Processing*, ed. B. L. Nash-Webber and R. Schank, 34–37. Stroudsberg, PA: Association for Computational Linguistics.

———. 1975b. Using Knowledge to Understand. In *Proceedings of the 1975 Workshop on Theoretical Issues in Natural Language Processing*, ed. B. L. Nash-Webber and R. Schank, 117–21. Stroudsberg, PA: Association for Computational Linguistics.

Schear, Joseph, ed. 2013. *Mind, Reason, and Being-in-the-World*. New York: Routledge.

Scheffler, Israel. 1967. *Science and Subjectivity*. Indianapolis, IN: Hackett.

Scheffler, Samuel. 2013. *Death and the Afterlife*. Oxford: Oxford University Press.

Schrödinger, Erwin. 1983. The Present Situation in Quantum Mechanics. In *Quantum Theory and Measurement*, ed. J. Wheeler and J. H. Zurek, 152–67. Princeton, NJ: Princeton University Press.

Searle, John. 1969. *Speech Acts*. Cambridge: Cambridge University Press.

———. 1982. *Intentionality*. Cambridge: Cambridge University Press.

———. 1995. *The Construction of Social Reality*. New York: Free Press.

Sellars, Wilfrid. 1948. Concepts as Involving Laws and Inconceivable without Them. *Philosophy of Science* 15: 287–315.

———. 1957. Counterfactuals, Dispositions, and the Causal Modalities. In *Concepts, Theories and the Mind-Body Problem*, ed. H. Feigl, M. Scriven, and G. Maxwell. Vol. 2 of *Minnesota Studies in the Philosophy of Science*, 225–308. Minneapolis: University of Minnesota Press.

———. 1985. Toward a Theory of Predication. In *How Things Are*, ed. J. Bogen and J. McGuire, 285–322. Dordrecht: D. Reidel.

———. 1997. *Empiricism and the Philosophy of Mind*. Cambridge, MA: Harvard University Press.

———. 2007. *In the Space of Reasons*. Cambridge, MA: Harvard University Press.

Seyfarth, Robert, Dorothy Cheney, and Peter Marler. 1980. Monkey Responses to Three Different Alarm Calls: Evidence of Predator Classification and Semantic Communication. *Science* 210: 801–3.

Shapere, Dudley. 1984. *Reason and the Search for Knowledge*. Dordrecht: D. Reidel.

Shapin, Steven, and Simon Schaffer. 1985. *Leviathan and the Air Pump*. Princeton, NJ: Princeton University Press.

Shapiro, Lawrence. 2011. *Embodied Cognitive Science*. New York: Routledge.

Shieh, Sanford. Forthcoming. *Modality and Logic in Early Analytic Philosophy*.

Shulman, Seth. 2006. *Undermining Science*. Berkeley: University of California Press.
Simon, Herbert. 1981. *The Sciences of the Artificial*. 2nd ed. Cambridge, MA: MIT Press.
Smith, Crosbie, and M. Norton Wise. 1989. *Energy and Empire*. Cambridge: Cambridge University Press.
Smocovitis, Vassiliki. 1996. *Unifying Biology*. Princeton, NJ: Princeton University Press.
Sober, Elliott, and David Sloan Wilson. 1999. *Unto Others*. Cambridge, MA: Harvard University Press.
Solomon, Miriam. 2007. The Social Epistemology of NIH Consensus Conferences. In *Establishing Medical Reality: Methodological and Metaphysical Issues in Philosophy of Medicine*, ed. H. Kincaid and J. McKitrick, 167–78. Dordrecht: Springer.
———. 2011. Group Judgment and the Medical Consensus Conference. In *Philosophy of Medicine*, ed. Fred Gifford, 239–54. Amsterdam: Elsevier.
Springer, Elise. 2013. *Communicating Moral Concern*. Cambridge, MA: MIT Press.
Sterelny, Kim. 2003. *Thought in a Hostile World*. Oxford: Blackwell.
———. 2012. *The Evolved Apprentice*. Cambridge, MA: MIT Press.
Suárez, Mauricio, ed. 2009a. *Fictions in Science: Philosophical Essays on Modeling and Idealization*. New York: Routledge.
———. 2009b. Scientific Fictions as Rules of Inference. In *Fictions in Science: Philosophical Essays on Modeling and Idealization*, ed. M. Suárez, 158–78. New York: Routledge.
Sultan, Sonia. 2007. Development in Context: The Timely Emergence of Eco-Devo. *Trends in Ecology and Evolution* 22: 575–82.
———. Forthcoming. *Organism and Environment*. Oxford: Oxford University Press.
Suppe, Frederick, ed. 1977. *The Structure of Scientific Theories*. Urbana: University of Illinois Press.
Suppes, Patrick. 1967. What Is a Scientific Theory? In *Philosophy of Science Today*, ed. S. Morgenbesser, 55–67. New York: Basic Books.
Tanesini, Alessandra. 2014. Temporal Externalism: A Taxonomy, an Articulation, and a Defense. *Journal of the Philosophy of History* 8: 1–19.
Taylor, Charles. 1985. *Philosophy and the Human Sciences*. Cambridge: Cambridge University Press.
Teller, Paul. 2001. Twilight of the Perfect Model Model. *Erkenntnis* 55: 393–415.
Thompson, Charis. 2005. *Making Parents*. Cambridge, MA: MIT Press.
Thompson, Evan. 2007. *Mind in Life*. Cambridge, MA: Harvard University Press.
Tomasello, Michael. 1999. *The Cultural Origins of Human Cognition*. Cambridge, MA: Harvard University Press.
———. 2008. *Origins of Human Communication*. Cambridge, MA: MIT Press.
———. 2014. *A Natural History of Human Thinking*. Cambridge, MA: Harvard University Press.
Traweek, Sharon. 1988. *Beamtimes and Lifetimes*. Cambridge, MA: Harvard University Press.

———. 1992. Border Crossings. In *Science as Practice and Culture*, ed. A. Pickering, 429–66. Chicago: University of Chicago Press.

Turner, Stephen. 1994. *The Social Theory of Practices*. Chicago: University of Chicago Press.

———. 2010. *Explaining the Normative*. Cambridge: Polity.

———. 2014. *Understanding the Tacit*. New York: Routledge.

van Fraassen, Bas. 1980. *The Scientific Image*. Oxford: Oxford University Press.

———. 1989. *Laws and Symmetries*. Oxford: Oxford University Press.

Wagner, Günter. 2000. What Is the Promise of Developmental Evolution? Part I. Why Is Developmental Biology Necessary to Explain Evolutionary Innovations? *Journal of Experimental Zoology (Molecular and Developmental Evolution)* 288: 95–98.

———. 2001. What Is the Promise of Developmental Evolution? Part 2. A Causal Explanation of Evolutionary Innovations May Be Impossible. *Journal of Experimental Zoology (Molecular and Developmental Evolution)* 291: 305–9.

Wagner, Günter, and Hans Larsson. 2003. What Is the Promise of Developmental Evolution? Part 3. The Crucible of Developmental Evolution. *Journal of Experimental Zoology Part B (Molecular and Developmental Evolution)* 300B: 1–4.

Warwick, Andrew. 2003. *Masters of Theory*. Chicago: University of Chicago Press.

Weber, Marcel. 2007. Redesigning the Fruit Fly: The Molecularization of *Drosophila*. In *Science without Laws*, ed. A. Creager and E. Lunbeck, 23–45. Durham, NC: Duke University Press.

Weinberg, Steven. 1992. *Dreams of a Final Theory*. New York: Random House.

Westfall, Richard. 1980. *Never at Rest*. Cambridge: Cambridge University Press.

Wheeler, Samuel III. 2000. *Deconstruction as Analytic Philosophy*. Stanford. CA: Stanford University Press.

Wilson, Mark. 1982. Predicate Meets Property. *Philosophical Review* 93: 335–74.

———. 2006. *Wandering Significance*. Oxford: Oxford University Press.

Wilson, Robert. 2005. *Genes and the Agents of Life*. Cambridge: Cambridge University Press.

Winsberg, Eric. 2003. Simulated Experiments: Methodology for a Virtual World. *Philosophy of Science* 70: 105–25.

———. 2010. *Science in the Age of Computer Simulation*. Chicago: University of Chicago Press.

Winsberg, Eric, Mathias Frisch, Karen Merikangas Darling, Arthur Fine. 2000. Review of *The Dappled World* by Nancy Cartwright. *Journal of Philosophy* 97: 403–8.

Witt, Charlotte. 1989. *Substance and Essence in Aristotle*. Ithaca, NY: Cornell University Press.

Wittgenstein, Ludwig. 1953. *Philosophical Investigations*. Translated by E. Anscombe. Oxford: Basil Blackwell.

———. 1961. *Tractatus-Logico-Philosophicus*. Translated by D. Pears and B. McGuinness. London: Routledge and Kegan Paul.

Woodward, James. 2003. *Making Things Happen*. Oxford: Oxford University Press.

Woody, Andrea. 2004a. More Telltale Signs. *Philosophy of Science* 71: 780–93.
———. 2004b. Telltale Signs. *Foundations of Chemistry* 6: 13–43.
———. 2014. Chemistry's Periodic Law. In *Science after the Practice Turn in the Philosophy, History and Social Studies of Science*, ed. L. Soler, S. Zwart, V. Israel-Jost and M. Lynch, 123–50. New York: Routledge.
Woody, Andrea, and Clark Glymour. 2000. Missing Elements. In *Of Minds and Molecules*, ed. N. Bushan and S. Rosenfeld, 17–33. Oxford: Oxford University Press.
Wooldridge, Dean. 1963. *The Machinery of the Brain*. New York: McGraw-Hill.
Woolgar, Steve. 1982. Irony in the Social Study of Science. In *Science Observed*, ed. K. Knorr-Cetina and M. Mulkay, 239–66. London: Sage.
———. 1988. *Science: The Very Idea*. Chichester: Ellis Horwood.
Zammito, John. 2004. *A Nice Derangement of Epistemes*. Chicago: University of Chicago Press.

Acknowledgments

This book has taken more than a decade to write, and I have incurred many debts on the way. My philosophical work is conversational and collaborative, and what I have to say has been formed and reshaped in ongoing engagement with colleagues and friends. The book emerged in the course of extended conversations with Mark Okrent and the late John Haugeland. Neither bears responsibility for what is written here, but each did much to bring it about. John's premature death in 2010 was a great loss to me personally and to philosophy. My gratitude to both Mark and John has no bounds.

Many other conversational exchanges have also left discernible marks on what I have to say in ways that cannot be confined to formal acknowledgment in the footnotes, and so I hereby express my thanks to Hanne Andersen, Rachel Ankeny, Karen Barad, Bob Brandom, Nancy Cartwright, Hasok Chang, Steven Crowell, Bert Dreyfus, John Dupre, Gary Ebbs, Catherine Elgin, Paul Erickson, Arthur Fine, Ron Giere, Gillian Goslinga, Ian Hacking, Donna Haraway, Philipp Haueis, Steve Horst, Bryce Huebner, Henry Jackman, Rebecca Kukla, Mark Lance, Marc Lange, Lisa Lloyd, Michael Lynch, Denis McManus, Sandra Mitchell, Margaret Morrison, Alva Nöe, Jaroslav Peregrin, Laura Perini, Isabelle Peschard, Bjørn Ramberg, Mark Risjord, Paul Roth, Mary Jane Rubenstein, Sanford Shieh, Gil Skillman, Jan Slaby, Brian Cantwell Smith, Léna Soler, Elise Springer, Alessandra Tanesini, Paul Teller, Emiliano Trizio, Jennifer Tucker, Stephen Turner, Bas van Fraassen, Matt Walhout, Joan Wellman, Andrea Woody, Alison Wylie, and Jack Zammito.

Bill deVries, Mark Okrent, Steve Horst, Jan Slaby, Mark Lance, Philipp Haueis, and two anonymous readers for the Press read early versions of the manuscript in whole or in part, and their thoughtful comments saved me from some mistakes and many more misleading or otherwise infelicitous formulations. Steve Horst, Jan Slaby, and Gillian Goslinga also taught earlier versions of book chapters in their seminars, and so I have also benefitted from their students and their pedagogical reflections on where the book presented difficulties.

My colleagues in Wesleyan University's Philosophy Department and Science in Society Program deserve special mention for sustaining a stimulating and productive community of scholars and teachers as my intellectual home. Wesleyan is also one of a very small number of institutions of its size and collegial mission that maintain the research enterprise needed for PhD programs in the sciences. For three decades, the science faculty has generously welcomed this philosophical interloper into their weekly research presentations and departmental colloquia. I owe special gratitude to Annie Burke for the opportunity to audit her graduate seminar in developmental evolution and have benefitted from extensive conversations with her, Barry Chernoff, Fred Cohan, Martha Gilmore, Laura Grabel, Bill Herbst, Manju Hingorani, Bob Lane, Mike McAlear, Stew Novick, Suzanne O'Connell, Brian Stewart, Sonia Sultan, and Greg Voth.

Most of what I have to say in the book was tried out first in courses at Wesleyan University, whose students regularly provide astute and thoughtful responses to the material. I have long since learned that if I cannot make something clear to Wesleyan students, I do not really understand it. Special thanks for more extended engagement go to Brittany Allen, Alex Anthony, John Baierl, Adam Baim, Sam Bernhardt, Ali Burstein, Colleen Buyers, Lucas Carrico, Katherine Cohen, Stratton Coffman, Nina Cohodes, Rachel Connor, Caleb Corliss, Mike Dacey, Corey Dethier, Micah Dubreuil, Rebecca Ehrlich, Will Fraker, Nick Gerry-Bullard, Guy Geyer, Lauren Graber, Sam Grover, Beth Herz, Rebecca Hock, Ethan Hoffman, Emily Kaditz, Aaron Khandros, Bennett Kirschner, Gretchen Kischbauch, Josh Krugman, Liza Litvina, Lewis Lo, Yannig Luthra, Catherine MacLean, Samantha Maisel, Dana Matthiessen, Thomas McAteer, Chris McGinnis, Hannah Overton, Tom Peteet, Ed Quish, Keren Reichler, Efrain Ribeiro, Emily Rowan, Isabel Rouse, Nick Russell, Sophie Sadinsky, Evan Scarlett, Anna Talman, Adin Vaewsorn, Nora Vogel, Lily Walkover, Rachel Warren, Ben Weisgall, Alix Weisman, Ashley Williams, and Lily Zucker, along with apologies to others whom I inadvertently omitted.

I have benefitted from thoughtful comments and exchanges over the past decade with the audiences to whom I have had opportunities to present earlier versions of material from the book. Many thanks to the philosophy departments or science studies groups at the University of Bergen, Calvin College, the University of Chicago, Georgetown University, Harvey Mudd College, the University of Houston, the University of Kentucky, the Norwegian University of Science and Technology in Trondheim, the University of Oslo, Rice University, the University of South Florida, Texas A&M University, the University of Texas at Austin, along with the Center for the Humanities, the cognitive science group, and the Division of Natural Sciences and Mathematics at Wesleyan University. Thanks are similarly due to the participants in multiple sessions of the Society for the Philosophy of Science in Practice and the International Society for Phenomenological Studies and in the following conferences or workshops: "Rethinking Science after the Practice Turn" at Nancy-Université, France; "Naturalism and Normativity in the Social Sciences" at the University of Hradec Králové, Czech Republic; "Objectivity and Affectivity" at the Freie Universität, Berlin, Germany; "Tradition" at the Chinese Academy of the Social Sciences, Beijing, China; "Practical Realism" at the University of Tartu, Estonia; "Mind, Meaning and Understanding: The Philosophy of John Haugeland" at the University of Chicago; "The Experimental Side of Modeling" at San Francisco State University; "Reclaiming the World: The Future of Objectivity" at the University of Toronto; "After the Science Wars: Whither Science Studies?" at the Boston Colloquium for the Philosophy of Science; the Social Science Roundtable meeting at the University of California, Santa Cruz; "Standpoint Theory" at the Society for Analytical Feminism; "Naturalized Epistemology and Naturalized Philosophy of Science" at Soochow University, Taiwan; "Scientific Representation: Idealizations, Fictions and Reasons," Universidad Complutense de Madrid, Spain; "What Is Science Studies?" at the Franke Institute for the Humanities, University of Chicago; two sessions at the Pacific Division of the American Philosophical Association; the Loemker Conference on the Philosophy of Sociology and Anthropology at Emory University; and "Filosofia de las Prácticas Cientificas" at Instituto de Investigaciones Filosóficas, Universidad Nacional Autónoma de México.

Although none of the chapters of this book were previously published in this form, some material in the book has been reworked from published articles. I am grateful to Routledge and to Taylor and Francis for the reuse of material from the following articles:

Scientific Practices and the Scientific Image. In *Rethinking Science after the Practice Turn*, ed. Léna Soler, Sjoerd Swart, Michael Lynch, and Vincent Israel-Jost, 277–94. New York: Routledge, 2014.

What is the Phenomenon of Conceptual Understanding? In *Mind, Reason, and Being-in-the-World*, ed. Joseph Schear, 250–71. New York: Routledge, 2013.

Articulating the World: Experimental Systems and Conceptual Understanding. *International Studies in Philosophy of Science* 25 (2011): 243–54. (Taylor and Francis)

Laboratory Fictions. In *Fictions in Science: Philosophical Essays on Modeling and Idealization*, ed. Mauricio Suárez, 37–55. New York: Routledge, 2008.

It has been a great pleasure to work with Karen Merikangas Darling, the science studies acquisitions editor for the University of Chicago Press. Karen's support for the project has been unwavering and gratifying from our initial conversations, throughout the review process until the manuscript went into production.

My deepest and most enduring gratitude goes to my family. Throughout the years of writing, and well before that, Sally Grucan has been there with love, friendship, and unstinting support; sometimes the center does hold. I dedicate the book to our sons, Brian Grucan Rouse and Martin Grucan Rouse. It has been under way for half their lives, and they have brought great joy along with their continuing support, good humor, and forbearance. Writing and parenting proceed in parallel. An intensive labor of love culminates in the foci of that attention and effort going out into the world to stand on their own. Martin and Brian are making their own way with integrity, courage, and grace; what could be more gratifying than that?

Index